THEORIES OF VISION FROM AL-KINDI TO KEPLER

UNIVERSITY OF
CHICAGO HISTORY OF
SCIENCE AND
MEDICINE

ALLEN G. DEBUS, Editor

DAVID C.
LINDBERG

THEORIES OF VISION FROM AL-KINDI TO KEPLER

THE UNIVERSITY OF CHICAGO PRESS
CHICAGO AND LONDON

The University of Chicago Press
Chicago 60637
The University of Chicago Press, Ltd.
London

Paperback edition 1981
Printed in the United States of America
01 00 99 98 97 96 2 3 4 5

Library of Congress Cataloging in Publication Data:

Lindberg, David C
Theories of vision from al-Kindi to Kepler.

(University of Chicago history of science and medicine)
Bibliography: p.
Includes index.
1. Vision. 2. Optics, Physiological — History.
3. Optics — History. I. Title.
QP475.L66 612'.84 75-19504
ISBN 0-226-48235-9 (paper)

To my father,
Milton B. Lindberg,
and the memory of
my mother,
Beth M. Lindberg

CONTENTS

PREFACE

It is possible to view the science of optics simply as a narrow scientific specialty, occupying a modest place in the hierarchy of the sciences and responsible for a limited range of natural phenomena. But to many of its early practitioners it seemed much more than that. It appeared, rather, to be the most fundamental of the natural sciences, the key that would unlock nature's door and reveal her innermost secrets. Thus Roger Bacon could write that

> it is necessary for all things to be known through this science, since all actions of things occur according to the multiplication of species and powers from the agents of this world into material recipients; and the laws of these multiplications are known only through perspective. [*Opus tertium*]

And John Dee could add:

> Among these Artes, by good reason, Perspectiue ought to be had, ere of *Astronomicall Apparences*, perfect knowledge can be atteyned. And bycause of the prerogatiue of *Light*, beyng the first of *Gods Creatures:* and the eye the light of our body, and this Sense most mighty, and his organ most Artificiall and *Geometricall*. At *Perspectiue*, we will begyn therefore. . . . This art of *Perspectiue*, is of that excellency, and may be led, to the certifying, and executing of such thinges, as no man would easily beleve: without Actuall profe perceived. I speake nothing of *Naturall Philosophie*, which, without *Perspectiue*, can not be fully vnderstanded, nor perfectly atteined vnto. Nor, of *Astronomie*: which, without *Perspectiue*, can not well be grounded: Nor *Astrologie*, naturally Verified, and auouched. [John Dee's *Mathematicall Praeface*]

But even those who declined to regard it as the universal science judged optics to be a broad and important discipline. It posed philosophical questions of great interest, it had practical applications in such fields as astronomy, scenography, and ophthalmology, and it was therefore worthy of serious attention. Whatever the precise motivations, an optical tradition of considerable vigor developed in ancient Greece and flourished continuously through the Middle Ages to the seventeenth century and beyond. Its practitioners included many of the most distinguished natural philosophers of the Islamic and European worlds, and it penetrated a wide variety of neighboring disciplines. It is clear that we dare not ignore a tradition of such pervasive influence if we wish to gain a broad understanding of early European intellectual history.

Before 1600 the science of optics tended to coalesce around two interrelated, yet distinguishable, problems—the nature and propagation of light, and the process of visual perception. Either problem could serve as an effective starting point for an investigation of early optics, but the second is clearly the broader and more representative. The problem of vision not only embraces the anatomy and physiology of the visual system, the mathematical principles of perspective, and the psychology of visual perception, but it also requires us at least to touch upon the nature of light and the mathematics and physics of its propagation. It is thus a microcosm of the entire optical enterprise, and by submitting it to careful scrutiny we should come to understand the principal controversies, crucial developments, and major schools of thought within medieval and Renaissance optics.

But in all honesty I have not investigated the history of visual theory solely for its ability to instruct us about the science of optics in general. The problem of vision is interesting for its own sake. All early natural philosophers acknowledged that vision is man's most noble and dependable sense, and the struggle to understand its workings occupied large numbers of scholars for some two thousand years. Kepler's successful solution of the problem of vision early in the seventeenth century was a theoretical triumph comparable in significance to other, far more celebrated developments of the scientific revolution, and one of my principal aims in this book will be to reveal the exact nature and magnitude of Kepler's achievement by placing it firmly in historical perspective.

An investigation of this sort can easily get out of hand, and therefore on occasion I have had to impose fairly severe limits. In particular, I have struggled to resist the temptations of psychology and epistemology, which, if indulged, would have quickly engulfed and sunk the entire project. This limitation is not only expedient but also legitimate, since psychological and epistemological issues, though often raised within the context of visual theory, were never its central concerns. The early practitioners of visual theory considered their endeavor chiefly a matter of mathematics, physics, and physiology, and on these aspects of visual theory I too will concentrate.

Two methodological points must be made. First, the attempt to trace the development of an idea vertically over a long period of time runs the risk of becoming "Whig history." However, I do not believe we can permit the dangers to dissuade us from the journey; we must proceed but take appropriate precautions. In the present study there is great danger of viewing the period before Kepler from his vantage point—of selecting as the crucial or interesting features of earlier visual theories those elements that would survive and be incorporated into Kepler's theory of the retinal image. To the best of my ability I have attempted to avoid this pitfall; I have endeavored to face each theory on its own ground, to treat the matters that it treats, and to emphasize what it emphasizes. Indeed, a substantial portion of my effort has been devoted to discussing the shifting perspectives from which the problem of sight was considered. And yet I have not given up the goal of discovering connections and of tracing the influence of one theory upon another, for the ultimate purpose of understanding and appreciating Kepler's achievement.

Second, I must confess that I have investigated the visual theory of Arabic natural philosophers without benefit of a reading knowledge of Arabic. I am a Latinist, and I have had to rely on medieval Latin translations and modern translations into European languages. I will not be surprised if Arabists are indignant, but I would defend myself by pointing out that in no instance does my case rest on philological detail and that the argument of almost every text I have considered is sufficiently full and repetitious to make the author's intent perfectly clear, unless the translator was guilty of wholesale distortion of scarcely imaginable proportions. Moreover, in some cases (e.g., al-Kindi's *De aspectibus*) only the Latin text is extant, and these fall properly within the domain of the Latinist; in other cases I have been able to compare several translations and thus eliminate potential errors and ambiguities; and finally, in my study of Alhazen I have benefited from the generous counsel of A. I. Sabra, who knows the Arabic text intimately and has steered me away from a variety of pitfalls. If I were prudent, I might retreat to the claim that what I have written is a history of medieval and Renaissance visual theory *as seen from the West* (for virtually every text I have studied was available to Western scholars during the Middle Ages and the Renaissance); however, I am satisfied that I have adequately understood the intended meaning of the Arabic natural philosophers with whom I have dealt, and I am willing to present my interpretations for public discussion and criticism. If my judgments can be improved or corrected by those with direct access to the Arabic text, I will be more than pleased.

Two remarks concerning the typography and format of the book need to be made. First, in order to reduce publication costs, diacritical marks (except the ain and hamza) have been eliminated from transliterated Arabic names and words in the text. Ambiguity need not ensue, however, since all diacritical marks are included in the index. Second, it has been necessary (again in order to reduce costs) to place the notes at the end of the book. This is regrettable, since the notes contain considerable explanatory material and were meant to be read; I remain hopeful, however, that an occasional reader may find his way to that section of the book and benefit from its contents.

This book was begun in 1970-71 during a sabbatical year at the Institute for Advanced Study in Princeton; it was completed in 1975 during a semester of leave at the Institute for Research in the Humanities of the University of Wisconsin. To each of these institutions I owe a deep debt of gratitude. I must also express appreciation to the National Science Foundation for funding my research through a series of grants and to the Graduate School of the University of Wisconsin, which has contributed salary support and a research assistant. European libraries have also made a vital contribution to my work by supplying microfilms of manuscripts and printed books.

For permission to reproduce pages from manuscripts in their care, I wish to thank the Conservateur en Chef of the Bibliothèque de l'Institut de France (Paris), the directors of the Topkapi Palace Museum and The Sülemaniye Library (Istanbul), and the Astronomer Royal for Scotland (Edinburgh). Grateful acknowledgment is made to the following publishers for permission to quote from publications for which they hold copyright: Government Press, Cairo: *The Book of the Ten Treatises on the Eye Ascribed to Hunain ibn Is-haq (809-877 A.D.)*, edited and translated by Max Meyerhof (Cairo, 1928); Harcourt Brace Jovanovich, Inc., and Jonathan Cape Ltd.: *The Notebooks of Leonardo da Vinci,* translated by Edward MacCurdy (New York, 1939); Harvard University Press: *A Source Book in Medieval Science,* edited by Edward Grant (Cambridge, Mass., 1974); Phaidon Press Ltd.: *The Literary Works of Leonardo da Vinci,* edited and translated by Jean Paul Richter, 2d ed. (London, 1939). And for permission to reprint portions of previously published papers of my own, I thank the editors of the *Bulletin of the History of Medicine, Isis,* and Ohio State University Press.

Finally, I must express deepest gratitude to friends and colleagues from whom I have received generous advice and criticism. My greatest debt is to Stephen M. Straker, who read the manuscript chapter by chapter as I completed it and responded with long, perceptive epistles; we still have disagreements over the interpretation of Kepler, but my argument has been much improved because of his thoughtful commentary. Richard S. Westfall and Allen G. Debus also read the entire manuscript, which has benefited substantially from their attention. Others who read and commented on various chapters are M.-Th. d'Alverny, William B. Ashworth, Jr., Jean Bollack, Harold Cherniss, Marshall Clagett, William J. Courtenay, Samuel Y. Edgerton, Jr., David Hahm, Stephen C. McCluskey, A. I. Sabra, Emilie S. Smith, Nicholas H. Steneck, and Donald S. Strong. They have been responsible for introducing many improvements and preventing many errors. Over the past decade Bruce S. Eastwood and I have had many discussions about medieval optics, and I wish also to acknowledge his contributions to my understanding. To A. C. Crombie I owe a great debt of gratitude for his pioneering work in the history of medieval and Renaissance optics; indeed, it is he who first awakened my own interest in the subject. My wife, Greta, has patiently tolerated this vice of mine and has even taken my enthusiasms as her own. Christin and Erik, my children, have been cheerleaders since the beginning of the project, and that has vastly lightened the burden.

1 THE BACKGROUND: ANCIENT THEORIES OF VISION

Introduction

Optical phenomena have attracted the attention of mankind since the beginnings of human history. Speculations about the rainbow can be traced almost as far back as written records go. Mirrors and burning lenses dating from 1500 B.C. or before have survived in various parts of the world, and Egyptian ophthalmological recipes go back at least to Papyrus Ebers, copied before 1500 from considerably older sources. In China a systematic analysis of radiation, shadows, and the phenomena of reflection was already in existence by the fourth century B.C.[1]

In the Greek world, where in practical terms the Western optical tradition began, we find discussions of light and the visual process among the earliest extant philosophical fragments. We do not know all the motivations that underlie these earliest Greek discussions, but it seems clear that blindness and eye disease stimulated medical thought on the subject of vision and gave birth to the science of ophthalmology; that an interest in epistemology and psychology led philosophers to examine mankind's most noble and dependable sense, the sense of vision, and to analyze its functioning in physical terms;[2] and that the artist's concern for scenography and the astronomer's concern for accurate celestial observations induced mathematicians to formulate a mathematical theory of perspective.[3]

These three approaches to optical phenomena furnish a useful scheme for classifying Greek optical thought. Despite some overlapping, three broad traditions appear to contain the great bulk of Greek optics: a medical tradition, concerned primarily with the anatomy and physiology of the eye and the treatment of eye disease; a physical or philosophical tradition, devoted to questions of epistemology, psychology, and physical causation; and a mathematical tradition, directed principally toward a geometrical explanation of the perception of space. Later, as Greek civilization went into decline, these same three traditions were transmitted to Islam and Latin Christendom, where they provided the framework and the materials for the medieval science of optics. Therefore, if we wish to understand medieval and Renaissance theories of vision, we must begin by glancing at the Greek background.[4]

The Atomists

Among the first philosophers to propose a systematic doctrine of light and vision were the atomists. To be sure, within the atomic school there was considerable diversity of opinion, and it would be gross error to suppose that atomists from Leucippus to Lucretius spoke with a single voice. Nevertheless, there was a significant core of agreement, springing from the common premise that all sensation is caused by direct contact with the organ of sense and therefore that a material effluence must be conveyed from the visible object to the eye.[5] Aetius (second century A.D.?) reports that "Leucippus, Democritus and Epicurus say that perception and thought arise when images (*eidola*) enter from outside"; Alexander of Aphrodisias, writing in the first half of the third century, adds that Leucippus and Democritus "attributed sight to certain images, of the same shape as the object, which were continually streaming off from the objects of sight and impinging on the eye."[6]

There is no question about the corpuscular nature of these images or *eidola*. Theophrastus, Aristotle's successor at the Lyceum, attributes to Democritus (b. ca. 460 B.C.) the opinion that "the visual image does not arise directly in the pupil, but the air between the eye and the object of sight is contracted and stamped by the object seen and the seer; for from everything there is always a sort of effluence proceeding. So this air, which is solid and variously coloured, appears in the eye."[7] There are difficulties in the interpretation of this passage,[8] but it is at least clear that the image consists of material substance (consolidated air). Epicurus (ca. 341-270 B.C.) is even more explicit on the corpuscular character of the effluence:

> For particles are continually streaming off from the surface of bodies, though no diminution of the bodies is observed, because other particles take their place. And those given off for a long time retain the position and arrangement which their atoms had when they formed part of the solid bodies, although occasionally they are thrown into confusion. . . . We must also consider that it is by the entrance of something coming from external objects that we see their shapes and think of them. For external things would not stamp on us their own nature of colour and form through the medium of the air which is between them and us, or by means of rays of light or currents of any sort going from us to them, so well as by the entrance into our eyes or minds, to whichever their size is suitable, of films coming from the things themselves.[9]

Finally, Lucretius (d. ca. 55 B.C.) attempts to clarify the nature of the films (which he calls *simulacra*) coming from visible objects, through several comparisons: "amongst visible things many throw off bodies, sometimes loosely diffused abroad, as wood throws off smoke and fire heat; sometimes more close-knit and condensed, as often when cicalas [i.e., cicadas] drop their thin coats in summer, and when calves at birth throw off the caul from their outermost surface, and also when the slippery serpent casts off his vesture among the thorns."[10] If I seem to stress the corpuscular character of the effluences unnecessarily, it is because of a recent attempt to demonstrate that Democritus and Epicurus held to a noncorpuscular emanation.[11] I do not believe such a position can be taken seriously, for it is inconsistent not only with explicit statements of Democritus and Epicurus regarding the visual process, but also with their natural philosophies in general. In a word, without a material efflux, it isn't atomism.

Vision, then, is reduced to a species of touch.[12] Material replicas issue in all directions from visible bodies and enter the eye of an observer to produce visual sensation. If this intromission theory leaves many unanswered questions—How can *eidola* or *simulacra* pass through one another without interference? How can the image of a large object shrink enough to enter the pupil?—it nevertheless answers the principal question: namely, how the soul of the observer and the visible object make contact.

Before leaving the visual theory of the atomists, I must call attention to a conception of Democritus that was to have considerable subsequent influence.[13] Democritus attached great significance in the act of sight to the pupillary image, that is, the image of outside objects mirrored in the cornea. Aristotle reports:

> Democritus ... is right in his opinion that the eye is of water; not, however, when he goes on to explain seeing as mere mirroring. The mirroring that takes place in an eye is due to the fact that the eye is smooth, and it really has its seat not in the eye which *is seen*, but in that which *sees*. For the case is merely one of reflexion.... It is strange too, that it never occurred to him to ask why, if his theory be true, the eye alone sees, while none of the other things in which images are reflected do so.[14]

As von Fritz has pointed out, this theory undoubtedly originated in the observation that we can see in a person's eye a miniature reflected picture of his visual field.[15] If we are to understand what Democritus meant by it, we must construe reflection (*emphasis*) not according to our modern understanding, as a process of rebound or deflection, but simply as visibility or presence in something.[16] Thus for Democritus the mirror image in the eye signified nothing more than the real presence of the image on the cornea (or perhaps in the interior humors)[17] and hence the visibility of the object.

The Platonic Theory of Vision

The ancestry of Plato's theory of vision, according to which a stream of light or fire issues from the observer's eye and coalesces with sunlight, is not easily determined. The theory of a visual current coming from the eye has commonly been associated with the Pythagorean school, and in particular with Alcmaeon of Croton (early fifth century B.C.?).[18] Of Alcmaeon's theory of vision Theophrastus writes: "And the eye obviously has fire within, for when one is struck [this fire] flashes out."[19] However, Theophrastus goes on to say that according to Alcmaeon "vision is due to the gleaming—that is to say, the transparent—character of that which [in the eye] reflects the object,"[20] an apparent reference to the theory of pupillary images that was to be expressed again, later in the fifth century, by Democritus. If we take both statements at face value, we must suppose either (with Beare) that the fire goes forth to seize the object and then doubles back to form a mirror image, or (with Verdenius) that the impression received by the observer, represented by the mirror image, is returned to the object for certification.[21] In either case, visual fire is requisite to the process of visual perception. Democritus too must surely be seen as one of Plato's predecessors, at least if we can trust Theophrastus, who attributes to him the view that an image is produced in the air when the effluence from the object of sight meets an opposing effluence from the observer.[22]

Far more problematic is the case of Empedocles (ca. 493–ca. 433 B.C.). There is a much-debated fragment of a poem by Empedocles which can be interpreted as an expression of a theory of visual fire:

> As when a man, thinking to go out through the wintry night, makes ready a light, a flame of blazing fire, putting round it a lantern to keep away all manner of winds; it divides the blasts of the rushing winds, but the light, the finer substance, passes through and shines on the threshold with unyielding beams; so at that time [when Aphrodite created eyes] primeval fire, enclosed in membranes, gave birth to the round pupil in its delicate garments which are pierced through with wondrous channels. These keep out the water which surrounds the pupil, but let through the fire, the finer part.[23]

Aristotle viewed this passage as expressing the extramission or visual ray theory, remarking that "Empedocles seems sometimes to imagine that one sees because light issues from the eye."[24] However, as Aristotle also notes, Empedocles makes it quite clear elsewhere that visual perception involves the reception of effluences from the object in the pores of the eye;[25] the problem, then, is to reconcile the two conceptions. Modern interpreters have been unable to agree on the matter, and thought ranges from Cherniss, who argues that "the passage . . . is a simile concerning the structure of the eye without a word about vision or its cause,"[26] and Millerd, who sees no possibility of reconciliation,[27] to Verdenius, who assigns particular functions to both the incoming effluences and the outgoing ocular fire.[28] Perhaps the most appealing solution is that of Long, who argues that Empedocles' intraocular fire simply represents an extension of the principle that like perceives like: vision occurs only "when there is the right correspondence between internal fire . . . and external fire."[29] On this view, although an intraocular fire is required for visual perception, it is not a fire that issues forth from the eye. Nevertheless, all the proposed solutions contain a very large speculative element, and it is unlikely that there can ever be a final resolution of the matter.

Whatever its origins, the theory of intraocular fire reached its full development with Plato (ca. 427–347 B.C.). Plato's theory of vision was misunderstood as early as the third century B.C. by Theophrastus, who maintained that Plato conceived of two emanations, one from the eye and the other from the visible object, which meet and coalesce somewhere in the intervening space to produce visual sensation.[30] But this description ignores an absolutely essential element in Plato's theory, namely, daylight, which coalesces with the fire issuing from the eye. Plato presents his theory most fully in the *Timaeus*:

> Such fire as has the property, not of burning, but of yielding a gentle light, they contrived should become the proper body of each day. For the pure fire within us is akin to this, and they caused it to flow through the eyes, making the whole fabric of the eye-ball, and especially the central part (the pupil), smooth and close in texture, so as to let nothing pass that is of coarser stuff, but only fire of this description to filter through pure by itself. Accordingly, whenever there is daylight round about, the visual current issues forth, like to like, and coalesces with it [i.e., daylight] and is formed into a single homogeneous body in a direct line with the eyes, in whatever quarter the stream issuing from within strikes upon any object it encounters outside. So the whole, because of its homogeneity, is similarly affected and passes on the motions of anything it comes in contact with or that comes into contact with it, throughout the whole body, to the soul, and thus causes the sensation we call seeing.[31]

Visual fire emanates from the eye and coalesces with its like, daylight, to form "a single homogeneous body" stretching from the eye to the visible object: this body is the instrument of the visual power for reaching into the space before the eye. The stress in this passage is not on the emission of an effluence from both the eye and the object of vision, but on the formation of a body, through the coalescence of visual rays and daylight, which serves as a material intermediary between the visible object and the eye.

To be sure, there is also a motion or emanation coming from the visible object—revealed in the last sentence of the passage quoted above, where Plato says that the homogeneous body formed by coalescence of the visual fire with daylight "passes on the motions of anything it comes in contact with or that comes into contact with it" to the soul.[32] The emanation from the object is more clearly evident in Plato's *Theaetetus*, where we read that

> as soon . . . as an eye and something else whose structure is adjusted to the eye come within range and give birth to the whiteness together with its cognate perception . . . then it is that, as the vision from the eyes and the whiteness from the thing that joins in giving birth to the colour pass in the space between, the eye becomes filled with vision and now sees, and

> becomes not vision, but a seeing eye; while the other parent of the colour is saturated with whiteness and becomes, on its side, not whiteness, but a white thing.[33]

And in the *Meno* we read that "colour is an effluence of figures, commensurate with sight and sensible."[34] But what must be kept in mind is that vision results not from the coalescence of an emanation from the visible object with an emanation from the eye, but from an encounter between the emanation from the object and the "single homogeneous body" already formed by coalescence of the ocular emanation and daylight. Through this encounter, motions are transmitted to the soul, where they produce sensation.[35]

Finally, is there more to be said of these motions returned to the soul? In the *Timaeus* Plato explains that the motions are produced by particles of various sizes, which he apparently means to associate with the various colors:

> There remains yet a fourth kind of sensation which demands classification, since it embraces a great number of diversities. They are known by the general name of colour, a flame which streams off from bodies of every sort and has its particles so proportioned to the visual ray as to yield sensation. . . . The particles that come from other bodies and enter the visual ray when they encounter it, are sometimes smaller, sometimes larger than those of the visual ray itself; or they may be of the same size. Those of the same size are imperceptible—"transparent," as we call them. The larger, which contract the ray, and the smaller which dilate it, are analogous to what is cold or hot to the flesh. . . . These are black and white.[36]

Particles of different sizes produce different motions (dilation and contraction are Plato's examples), and different motions lead to the sensation of different colors.

Aristotle

We have no full and systematic discussion of vision before Aristotle (384–322 B.C.). From Plato and the Pre-Socratics we have only fragmentary quotations or brief statements in works devoted primarily to other matters. It is with Aristotle, and primarily in his two works on psychology, that we first observe the careful definitions and distinctions indispensable to an orderly discussion of vision.[37]

Aristotle firmly rejected earlier theories of light and vision. In the first place, light is not a corpuscular emanation: "it is neither fire, nor in general any body, nor an emanation from any body (for in that case too it would be a body of some kind)."[38] In the second place, vision is not produced by extramission—that is, by the issuance of a ray from the observer's eye:

> In general it is unreasonable to suppose that seeing occurs by something issuing from the eye; that the ray of vision reaches as far as the stars, or

> goes to a certain point and there coalesces with the object, as some think. It would be better to suppose that coalescence occurs in the very heart of the eye. But even this is foolish; what is the meaning of light coalescing with light? How can it occur? For the coalescence is not between any chance objects. And how could the light inside coalesce with that outside?[39]

And yet Aristotle by no means disagrees with the atomists and Plato (whose views he has been attacking) about the need for a physical intermediary between the visible object and the observing intellect. "For certainly," he wrote in *De sensu*, "it is not true that the beholder sees, and the object is seen, in virtue of some merely abstract relationship between them, such as that between equals. For if that were so, there would be no need [as there is] that either [the beholder or the thing beheld] should occupy some particular place."[40] Since he cannot accept a material effluence either from the visible object to the eye or the contrary, Aristotle directs his attention toward the medium between the observer and the visible object. He perceives the absolute necessity of this medium: neither an object separated from the eye by void space nor an object placed directly on the eye can be seen:

> The evidence for this is clear; for if one puts that which has colour right up to the eye, it will not be visible. Colour moves the transparent medium, *e.g.*, the air, and this, being continuous, acts upon the sense organ. Democritus is mistaken in thinking that if the intervening space were empty, even an ant in the sky would be clearly visible; for this is impossible. For vision occurs when the sensitive faculty is acted upon; as it cannot be acted upon by the actual colour which is seen, there only remains the medium to act on it, so that some medium must exist; in fact, if the intervening space were void, not merely would accurate vision be impossible, but nothing would be seen at all.[41]

The analysis of the medium leads Aristotle to precise definitions of transparency, light, and color. The medium of sight is the diaphanous or transparent, a nature or power found in all bodies, but especially in air, water, and certain solid substances. Aristotle defines the transparent as "that which is visible, only not absolutely and in itself, but owing to the colour of something else. This character is shared by air, water, and many solid objects; it is not *qua* water or air that water or air is transparent, but because the same nature belongs to these two as to the everlasting upper firmament."[42] Aristotle means that the transparent is visible insofar as it communicates to the observer the color of bodies on the other side of it; to be quite literal, it is not something that we see, but something through which we see.

Light (*phos*) is a state of the transparent, resulting from the presence of fire or some other luminous body. In particular, it is the actualization of the transparency, the achievement of that state in which transparency is no longer merely potential, but actual, so that bodies separated from the observer by the medium become visible.

> Of this substance [i.e., the transparent] light is the activity—the activity of what is transparent so far forth as it has in it the determinate power of becoming transparent; where this power is present, there is also the potentiality of the contrary, viz. darkness. Light is as it were the proper colour of what is transparent, and exists whenever the potentially transparent is excited to actuality by the influence of fire or something resembling "the uppermost body."[43]

As a state of the medium, rather than a substance, light requires no time for propagation, for the entire medium may be moved from potential to actual transparency at the same instant. Indeed, terms like "propagation" and "transmission," which imply progressive motion, are not appropriate for describing light; just as it is conceivable for water to be frozen simultaneously throughout, so the change of state resulting in light "may conceivably take place in a thing all at once, without half of it being changed before the other."[44] If Aristotle seems, at times, to claim that motion comes from the visible object to the eye,[45] this does not imply a temporal process, but simply identifies the source or efficient cause of the instantaneous change of state.

If light is that state of a transparent medium in which transparency is actualized, what is color? Color (*chroma*), according to Aristotle, is that which overlies the surface of visible objects and has the power to set in motion (i.e., to produce further qualitative change in) the actually transparent: "Every colour can produce movement in that which is actually transparent, and it is its very nature to do so. This is why it is not visible without light, but it is only in light that the colour of each individual thing is seen."[46] The color of a body "moves the [actually] transparent medium, *e.g.*, the air, and this, being continuous, acts upon the sense organ."[47] Thus what is visible, that is, the proper object of sight, is color. A transparent medium is first moved to actuality by the presence of a luminous body, such as fire; it is further moved or affected by the color of bodies in contact with it, and the change thus produced is communicated to the observer. There are, however, passages in which Aristotle associates light and color in a somewhat different way. That which is light in the interior of a transparent body is color at the surface: color, he concludes "will be the limit of the transparent in a defined body."[48]

Finally, in what does the act or process of visual sensation consist? Aristotle's answer, in the meagerness of its physiological content, is typical of ancient theories of vision (excepting Galen's) and, to a lesser extent, of medieval and Renaissance theories. The eyes are composed primarily of water so as to participate in the transparent and thus be receptive of light and color. There is thus a continuous medium, the transparent, from the visible object all the way to the interior of the eye. The color of the visible object moves the medium, and the medium, "being continuous, acts upon the sense organ."[49] And this sense organ, that is, the sense organ of the soul, "does not reside in the surface of the eye, but must evidently be within; consequently the part within the eye must be transparent and receptive of light."[50] Sight thus occurs when the watery substance of the eye, which Aristotle explicitly identifies as the seeing part of the eye,[51] is moved by the visible object and assumes its qualities; in Cherniss's phrase, the sense organ "becomes the sensible object."[52]

It should be evident, even from this brief discussion, that a comparison of the theories of Plato and Aristotle cannot be reduced to a simple conflict over the extramission and intromission of rays. It is true, of course, that Aristotle objects to Plato's theory of streams issuing in opposite directions from the visible object and the observer's eye[53] and argues that light is an incorporeal state of the medium that involves no temporal process and therefore has nothing at all to do with emanation or change of place. But there is another issue of equal or greater importance on which Plato and Aristotle are in considerably closer agreement. For Plato, the coalescence of visual rays and daylight produces an effective optical medium between the observer and the observed—a "sympathetic chain," as Cornford puts it,[54] linking the visible object and the soul of the observer. Now Aristotle, although denying the emission and coalescence of rays, adopts Plato's emphasis on the creation of an optical medium; that it is a new state of the old medium rather than an altogether new medium does not alter the basic fact that this medium serves as the essential connecting link between the visible object and the observer. Aristotle, like Plato, solves the problem of vision by arguing that the eye and external media become parts of a homogeneous chain capable of transmitting motions (in the broadest sense) to the intellect of the observer.

The Stoics and Galen

The theory of vision of the Stoics is closely related to their conception of the pneuma, an all-pervasive active agent composed of a mixture of air and fire.[55] An optical pneuma, it was supposed, flows from the seat of consciousness (the *hegemonikon*) to the eye and excites the air adjacent to the eye, putting it in a state of tension or stress. Through this stressed air, when it is illuminated by the sun, contact is made with the visible object. Indeed, there are indications that in the Stoic view the simultaneous action of the optical pneuma and sunlight on air transforms it into an instrument of the soul and makes it percipient; it becomes an extension of ourselves. As Cicero put it, "the air itself sees together with us."[56]

Now it is evident from this brief account that the Stoics, like Plato and Aristotle, focused their attention on the medium between the observer and the visible object. Alexander of Aphrodisias makes this quite clear in his description of the Stoic theory: "Some people explain vision by the stress of air. The air adjoining the pupil is excited by vision and formed into a cone which is stamped on its base by an impression of the object of vision, and thus perception is created similar to the touch of a stick."[57] Thus visual perception does not occur by the actual emanation of pieces of the visible object to the eye, but by qualitative changes produced by the object in a medium suitably prepared to receive them; this, it appears to me, is the common teaching of Plato, Aristotle, and the Stoics.[58]

The basic features of the Stoic theory of vision were adopted and elaborated by Galen (ca. 129–ca. 199 A.D.). In *De placitis Hippocratis et Platonis*, Galen described what he regarded as the two principal alternatives among theories of vision: "A body that is seen does one of two things: either it sends something from itself to us and thereby gives an indication of its peculiar character, or if it does not itself send something, it waits for some sensory power to come to it from us."[59] The first alternative was unacceptable for several reasons: it would be impossible to determine the size of an object from "some portion or power or image or quality" coming from it to us, for the image of a very large mountain would have to shrink drastically in order to enter the pupil; moreover, it is absurd to suppose that the image of a mountain could reach a multitude of observers simultaneously.[60] Since the mountain will not come to the observer, the observer must go to the mountain; he must, that is, send forth his sensory power to perceive it. This occurs, Galen argues, not through extension of visual pneuma from the eye to the object, but by means of the intervening air, which "becomes for us the kind of instrument that the nerve in the body is at all times."[61]

Galen discusses at length the transformation of the air into a visual instrument. Visual or optic pneuma is sent from the brain through the optic nerves, which are hollow,[62] to the eye. Pneuma emerges from the eyes and "is united at the first impact with the surrounding air and alters the air to its own peculiar nature but does not itself extend out very far."[63] This is an instantaneous transformation similar to the effect of sunlight on air: "For sunlight, touching the upper limit of the air, transmits its power to the whole; and the vision that is carried through the optic nerves has a substance of the nature of pneuma, and when it strikes the surrounding air it produces by its first impact an alteration that is transmitted to the furthest distance."[64] As a result, the air itself becomes percipient; it "becomes for the eye the same kind of instrument for the proper discrimination of its sense-objects, as the nerve is for the brain."[65] Galen makes his view that the air itself becomes endowed with sensory power particularly clear when he expresses disapproval of the Stoic analogy of the walking stick:

> The nerve itself is a part of the brain, like a branch or offshoot of a tree, and the member to which the part is attached receives the power from the part into the whole of itself and thus becomes capable of discerning the things that touch it. Something similar happens also in the case of the air that surrounds us. When it has been illuminated by the sun, it is already an instrument of vision of the same description as the pneuma arriving from

> the brain; but until it is illuminated it does not turn into a sympathetic instrument in accordance with the change effected by the outflow of the pneuma into it. The Stoics, then, must not say that we see by means of the surrounding air as with a walking-stick.[66]

The trouble with the walking-stick analogy is that walking sticks merely transmit pressures back to the hand, whereas sight "reaches out through the intervening air to the colored body."[67]

Galen not only elaborated and corrected the Stoic theory of vision; he also introduced, for the first time, a wealth of anatomical and physiological detail into the theory of vision.[68] I have already referred to his theory of the optic nerves, which carry pneuma to the eyes. Galen argued also that the crystalline lens is the chief organ of sight in the eye: "the crystalline humor itself is the principal instrument of vision, a fact clearly proved by what physicians call cataracts, which lie between the crystalline humor and the cornea and interfere with vision until they are couched."[69] Finally, Galen discussed in detail the structure and function of the remaining ocular organs—the retina, cornea, iris, uvea, vitreous and aqueous humors, and eyelids.[70]

The Mathematicians: Euclid, Hero, and Ptolemy

The theories of vision of the atomists, of Plato and his predecessors, of the Stoics and Galen, and of Aristotle are almost entirely devoid of mathematics. It is true, of course, that in the *Physics* Aristotle subordinates "optics" to mathematics, pointing out that "optics investigates mathematical lines, but *qua* physical, not *qua* mathematical."[71] However, it is clear that Aristotle here employs the term "optics" in a narrow sense to denote perspective and catoptrics and hence to exclude investigation of the nature of light and the act of visual perception; in short, optics is the mathematical portion of the larger study of light and vision, a portion to which Aristotle himself devoted almost no attention.[72] An awareness of the difference between the mathematical and nonmathematical approaches to visual phenomena is evident also in prefatory remarks attached by Galen to his brief discussion of the geometry of visual perception:

> I have explained nearly everything pertaining to the eyes with the exception of one point which I had intended to omit lest many of my readers be annoyed with the obscurity of the explanations and the length of the treatment. For since it necessarily involves the theory of geometry and most people pretending to some education not only are ignorant of this but also avoid those who do understand it and are annoyed with them, I thought it better to omit the matter altogether. But afterward I dreamed that I was being censured because I was unjust to the most godlike of the instruments and was behaving impiously toward the Creator in leaving unexplained a

great work of his providence for animals, and so I felt impelled to take up again what I had omitted and add it to the end of this book.[73]

Galen then proceeds to a brief summary of what is evidently Euclidean optics, but without satisfactorily integrating it into his pneumatic theory.[74]

The first full-fledged exposition of a mathematical theory of vision is found in the *Optica* of Euclid (fl. 300 B.C.).[75] Indeed, Euclid's approach to vision was so strictly mathematical as to exclude all but the most incidental references to those aspects of the visual process not reducible to geometry—the ontology of visual radiation and the physiology and psychology of vision. Lejeune comments that Euclid's *Optica*

systematically ignores every physical and psychological aspect of the problem of vision. It restricts itself to that which can be expressed geometrically.... Its model is the treatise on pure geometry, and its method that of the *Elements*: a few postulates all fully necessary, from which follow deductively and with full mathematical rigor a series of theorems of a traditional form.[76]

Euclid thus represents the opposite side of the coin from the atomists, Plato, Aristotle, and Galen.

The postulates on which Euclid bases the geometrical theorems of the *Optica* are seven:

Let it be assumed

1. That the rectilinear rays proceeding from the eye diverge indefinitely;
2. That the figure contained by a set of visual rays is a cone of which the vertex is at the eye[77] and the base at the surface of the objects seen;
3. That those things are seen upon which visual rays fall and those things are not seen upon which visual rays do not fall;
4. That things seen under a larger angle appear larger, those under a smaller angle appear smaller, and those under equal angles appear equal;
5. That things seen by higher visual rays appear higher, and things seen by lower visual rays appear lower;
6. That, similarly, things seen by rays further to the right appear further to the right, and things seen by rays further to the left appear further to the left;
7. That things seen under more angles are seen more clearly.[78]

The first three postulates define the visual process and cast it in a geometrical mold: rays proceed in straight lines from the eye, the collection of such rays constituting a cone; to be seen, an object must intercept a visual ray. The rectilinearity of the rays, which Euclid assumes in the first postulate, permits the development of a theory of vision along geometrical lines.[79] This simple rule governing the propagation of light having been given (as well as the law of reflection, introduced in proposition 19), it is possible to employ the straight lines of a geometrical diagram to represent visual rays and thus transform optical problems into geometrical problems. The process of geometrization is completed in postulates 4-6: the apparent size of a visible object is determined by the angular separation between the visual rays that encounter its extremes, and the position of a visible object in space is determined by the location, within the visual cone, of the rays by which it is perceived. Finally, in the seventh postulate Euclid provides an explanation for variations in the clarity of visual perception: objects seen under more angles—that is, encountered by more visual rays and hence seen under more of the angles formed between adjacent visual rays—are seen with greater clarity.

With few exceptions, the fifty-eight propositions of the *Optica* are based upon these seven postulates. The majority of them treat problems of perspective—that is, the appearance of an object as a function of its spatial relationship to the observer. In only one proposition is there any suggestion of the possibility of depth perception (proposition 57), and, as many commentators have observed, this is probably a later interpolation;[80] Euclid's own view seems rather to be expressed in proposition 5, where it is admitted that if equally large objects are observed from different distances, the closer one will appear larger. Euclid thus ignores, for the most part, the physical problems associated with the nature of visual rays and their encounter with visible objects, the physiology of sight, and the psychological factors influencing the perception and localization of visible objects. His is a geometrical theory of vision.

However, the claim that Euclid ignored those aspects of vision not reducible to geometry is true only to a first (or perhaps second) approximation. For, as even Euclid's ancient commentators recognized, the postulates and several propositions of the *Optica* have inescapable implications vis-à-vis the ontology of visual rays and thus spill over into the physical realm. In the first place, it is apparent from such expressions as "proceeding from the eye" and "those things . . . upon which visual rays fall" that vision is the result of rays issuing from the observer's eye; there is no warrant, as far as I can see, for construing these as awkward metaphors, intended (but failing) to convey purely geometrical truths.[81] The eye is thus the active participant in the visual process, reaching out to apprehend its object; and on this account many have identified Euclid as a member of the Platonic tradition. Second, within the cone of visual rays there are sentient and insentient regions. In the first proposition it is asserted that an object is not visible in its entirety at one time because of spaces between the visual rays, and in the second that any given object can be removed to a distance from which it will no longer be visible because it falls between adjacent visual rays. It is clear that rays having such properties cannot be mere constructions intended to represent the geometry of sight; they must be the physical agents of sight. A third and final nongeometrical claim can be extracted from the seventh postulate and the second proposition, where Euclid offers a physical explanation of what we would now regard as a psychophysiological phenomenon. The clarity of perception, he asserts, depends on the number of angles under which an object is seen—or, to put it in the clearer terms of the second proposition, on the number of visual rays intercepted by the object.

Euclid devotes the remainder of the *Optica* to problems of visual perspective, avoiding further discussion of physical issues. He thus leaves it to others to resolve the serious problems raised by his conception of discrete rays emanating from the eye. Euclid also fails to say what role in vision he would assign to external light, though in several propositions he refers to the existence of such light, and in proposition 18 he even has a solar ray terminating at the observer's eye. It may be that he would have agreed with Ptolemy and Damianos, who argued that luminous and visual rays are identical in nature and that luminous rays, as well as visual rays, have a share in the visual process,[82] but his silence prohibits us from reaching any firm conclusion. Apparently Euclid intended to formulate a theory of vision restricted to geometry, and the only nongeometrical claims it contained are those that slipped in along with the conception of visual rays. It was left to Euclid's followers and commentators to enlarge the theory and invest it with additional physical content.[83]

The avowed purpose of the *Catoptrica* of Hero of Alexandria (fl. 62 A.D.) was geometrical—to treat the theory and applications of mirrors. At the beginning of the *Catoptrica* Hero wrote:

> The science of vision is divided into three parts: optics, dioptrics, and catoptrics. Now optics [i.e., the process of vision] has been adequately treated by our predecessors and particularly by Aristotle, and dioptrics we have ourselves treated elsewhere as fully as seemed necessary. But catoptrics, too, is clearly a science worthy of study and at the same time produces spectacles which excite wonder in the observer. . . . The study of catoptrics, however, is useful not merely in affording diverting spectacles but also for necessary purposes.[84]

It is noteworthy that Hero regards the geometrical portion of optics not as an alternative to Aristotle's physical analysis of the visual process but as complementary. Aristotle treated the natures and causes; Hero will provide the mathematics, and he will do so in an area hardly touched by Aristotle—the science of mirrors.[85] Nevertheless, Hero, like Euclid, could not avoid making statements with physical implications. He revealed his acceptance of the visual ray theory when he wrote of the doubts of some "as to why rays proceeding from our eyes are reflected by mirrors and why the reflections are at equal angles," and when he argued that the visual rays move so rapidly that "when, after our eyes have been closed, we open them and look up at the sky, no interval of time is required for the visual rays to reach the sky."[86] He revealed his belief in the material nature of visual rays in his account of the cause of reflection, pointing out that visual rays incident on unpolished mirrors enter the porosities of the bodies and are not reflected. However, if

> these mirrors are polished by rubbing until the porosities are filled by a fine substance; then the rays incident upon the compact surface are reflected. For just as a stone violently hurled against a compact body, such as a board or wall, rebounds, whereas a stone hurled against a soft body, such as wool or the like, does not,... so the rays that are emitted by us with great velocity ... also rebound when they impinge on a body of compact surface.[87]

The greatest optician of antiquity was undoubtedly Claudius Ptolemy (fl. 127-148 A.D.), who not only extended Euclid's mathematical analysis of vision, but also enlarged it to include additional physical, physiological, and psychological elements.[88] However, those who would interpret Ptolemy's theory of vision must labor under severe handicaps: we lack the entire first book of Ptolemy's *Optica*, which treated the more physical aspects of vision, and we possess the remaining four books only in a difficult and occasionally incoherent Latin translation by Admiral Eugene of Sicily from an Arabic translation of the Greek original.[89] It is primarily through the heroic efforts of Albert Lejeune in reconstructing the broad outline of Ptolemy's first book that we have the pleasure of understanding the principal features of Ptolemy's theory of light and vision.[90]

Ptolemy attributed sight to the action of a visual flux issuing conically from the observer's eye; in this regard, as well as in his mathematical approach to optics, he must be counted among the followers of Euclid. However, Ptolemy gave the visual radiation a physical interpretation going far beyond anything in Euclid (if we accept Lejeune's reconstruction of book 1 of the *Optica*),[91] and it is apparent that we must also view him as a follower of Plato and the Stoics. By piecing together remarks from books 2-5 of the *Optica* and a brief resume of book 1 appearing at the beginning of book 2, Lejeune has been able to show that Ptolemy assigned the visual flux to the same genus as external light—the luminous—and hence attributed physical reality to it. This interpretation is confirmed by a later commentator, Simeon Seth, who wrote: "Ptolemy says in his *Optica* that the visual pneuma is something of ether, belonging to the quintessence."[92] If one recalls Aristotle's association of light with the quintessence,[93] the bearing of Seth's comment on the relationship between light and the visual flux becomes clear. In addition, Lejeune has argued that Ptolemy assigned the mechanical properties of *virtus* and *motus* to the visual radiation, although he confesses to some doubt about the precise interpretation that should be attached to these terms.[94] Finally, Lejeune argues that Ptolemy conceived the visual radiation (as well as external light) as a transfer of energy, concluding: "It seems that he conceives all radiation to be endowed with a certain quantity of energy, which its emitter communicates to it and which is diminished by the impact that accompanies reflection and by the resistance to penetration offered more or less by all transparent media. Applied to a single point of the object, the heterogeneous energies add their effects."[95]

There is no feature of Ptolemy's theory of vision thus far described that could not be construed as a natural extension of Euclid's teachings. However, there is another point, a very crucial one, on which Ptolemy and Euclid disagree totally. Whereas Euclid had posited discrete rays, separated by spaces that increase with distance from the eye, Ptolemy maintained that the visual rays form a continuous bundle or cone; if this were not so, he argued, objects could not be seen in their entirety at a glance. But this is to deny that discrete, numerable visual rays (whether one-dimensional entities or thin three-dimensional pencils) exist at all. Rays may represent the geometry of sight, but they have nothing to do with the physical reality. This position is evident in Ptolemy's comment that "it is necessary to recognize that the nature of the visual ray . . . is necessarily continuous rather than discrete."[96] This continuous visual energy emerging from the eye has the power to perceive the objects that it encounters with a clarity dependent on the strength of the radiation.[97]

Any physical theory of sight must consider the participation of the visible object and radiation from the sun (or some other external luminous body) in the visual process. The color of the object, in Ptolemy's theory of vision, performed the same function as it had for Aristotle and soon would for Galen. Color, according to Ptolemy, is an inherent property of bodies, a quality, which produces a modification (*passio*) in the visual cone. Color is the proper object of vision, and it is through patterns of color and their effect on the visual radiation that other sensible characteristics of bodies (shape, for example) are perceived.[98] However, as in the theories of Plato, Aristotle, and Galen, color cannot affect the visual cone (or the transparent medium in the case of Aristotle) without the presence and cooperation of external light.[99] In the extant books of the *Optica*, Ptolemy does not inform us of the mode of action of external light or how it makes vision possible; he simply asserts that "color is not seen unless light cooperates."[100]

If the physical and physiological aspects of Ptolemy's theory of vision went beyond what has already been expressed, they are lost with the first book of the *Optica*. Vision, so far as the extant books reveal, is simply the result of the interaction of visual radiation and color, assisted by external illumination. Many important questions are thus left unanswered: we are not informed of the character of the modification or *passio* introduced into the visual radiation by the color of the viewed object; we do not know whether Ptolemy regarded the visual cone as sentient, in the manner of the Stoics, or simply as a medium for transmitting impressions back to the observer; nor do we have any description of ocular anatomy or any physiological theory. However,

Ptolemy did make two additional geometrical points that were destined to be of great importance in subsequent ages. First, he recognized that there are variations in sensitivity within the visual cone, commenting that "that which is over the axis should be seen more clearly than that which is observed on the sides through the lateral rays."[101] It is thus to Ptolemy that we owe the observation that visual perception is less acute as the visual flux diverges from the central ray or axis of the cone.[102] Second, Ptolemy fixed the precise location of the apex of the visual cone. Archimedes already had observed that the eye sees from a sector, rather than a point, on the surface of the eye,[103] and it follows from this observation (for an extramissionist) that the apex of the visual cone is within the eye. But where? Ptolemy's line of argument on this question is lost, but his conclusion is evident from the extant portions of the *Optica*. He asserts that the apex of the visual cone is at the center of curvature of the cornea and also at the center of rotation of the ocular globe. In all probability, this location was determined by the desirability of having the visual radiation unrefracted as it emerged from the eye and perhaps also of having an invariable separation between the apexes of the two visual cones (in binocular vision) despite variability in the point of fixation.[104]

2 AL-KINDI'S CRITIQUE OF EUCLID'S THEORY OF VISION

The Place of Optics in al-Kindi's Thought

The translation of Greek philosophical, mathematical, and medical texts into Arabic reached its culmination in the ninth century A.D.—whereupon Muslim scholars began in earnest the task of assimilating and criticizing their new heritage. The first great philosopher of the Islamic world, and surely the first to undertake serious optical studies, was Abu Yusuf Ya'qub ibn Ishaq al-Kindi. Al-Kindi was probably born late in the eighth century in the city of al-Kufa, where his father was governor. It seems that as a young man he moved to al-Basra and later to Baghdad. In Baghdad he pursued a scholarly career under the patronage of three caliphs (813–47): al-Ma'mun, al-Mu'tasim, and al-Wathiq. He fell out of favor during the reign of al-Mutawakkil (847–61) and died about 866.[1]

Al-Kindi was a leader in the endeavor to communicate Greek learning to Islam. Not only did he encourage and patronize the translating activity of the ninth century, but he also attempted to integrate Greek philosophy with Mu'tazilite theology and thus, in Walzer's phrase, "to naturalise Greek philosophy in the Islamic world."[2] Al-Kindi's deep respect for ancient thought is revealed in the preface to one of his works on metaphysics:

> It is fitting then [for us] to acknowledge the utmost gratitude to all those who have contributed even a little to truth not to speak of all those who have contributed much. If they had not lived, it would have been impossible for us, despite all our zeal, during the whole of our lifetime, to assemble these principles of truth which form the basis of the final inferences of our research. The assembling of all these elements has been effected century by century, in past ages down to our own time. . . . It is fitting then for us not to be ashamed to acknowledge truth and to assimilate it from whatever source it comes to us, even if it is brought to us by former generations and foreign peoples.[3]

His own task, as he conceived it, was to communicate, complete, and correct this body of ancient learning. In the same preface, he wrote,

> It is fitting then [for us] to remain faithful to the principle which we have followed in all our works, which is first to record in complete quotations all that the Ancients have said on the subject, secondly to complete what the Ancients have not fully expressed, and this according to the usage of our Arabic language, the customs of our age and our own ability.[4]

It was in this spirit that al-Kindi wrote some 260 works on all branches of learning.[5] Among them was a work on optics, entitled *De aspectibus* in its Latin translation, which was to exert a strong and continuing influence on Islamic and Western optics throughout the Middle Ages.[6] In the preface to *De aspectibus*, al-Kindi reveals that one of his motives for writing on optics was the desire to correct and communicate to Islamic society the full legacy of ancient learning: "Since we wish to complete the theoretical sciences, to express what the ancients have given us of them, and to augment what they have begun, . . . it is necessary for us to speak . . . concerning differences of appearance [i.e., optics or perspective] according to the measure of our ability."[7]

However, optics fit into al-Kindi's philosophical program in yet another way. That is, optics was not simply a branch of Greek learning, on a par with the others, to be communicated to Islam in its turn, but had a place of special importance in al-Kindi's philosophy of nature. The necessity of seeing al-Kindi's work in optics against the background of the natural philosophy expressed in his *De radiis stellarum* (or *De radiis stellatis* or *stellicis*) has been emphasized only recently by G. F. Vescovini, who has also provided a very useful analysis of this treatise.[8] The basic theme of *De radiis stellarum* is the universal activity of nature, exercised through the radiation of power or force. "It is manifest," al-Kindi asserts, "that everything in this world, whether it be substance or accident, produces rays in its own manner like a star. . . . Everything that has actual existence in the world of the elements emits rays in every direction, which fill the whole world."[9] This radiation binds the world into a vast network in which everything acts upon everything else to produce natural effects. Stars act upon the terrestrial world; magnets, fire, sound, and colors act on objects in their vicinity. Even words conceived by the mind can radiate power and thus produce effects outside the mind.[10] This is a natural philosophy that was destined to influence Robert Grosseteste and Roger Bacon and to reappear in their doctrine of the multiplication of species.[11]

Optics, then, is of special significance because it treats the most fundamental of all natural phenomena, the radiation of power. The laws of radiation are the laws of nature, and optics is consequently a prerequisite to other studies. This view, I believe, lies behind a passage from al-Kindi's *De temporum mutationibus sive de imbribus* to which Vescovini and Lemay have called attention: "Man is not instructed in philosophy until he can divine superior impressions, and he cannot ascend to that knowledge until he has mastered the quadrivial sciences, which are the introduction to philosophy—and they are mathematical."[12]

Al-Kindi's Defense of the Extramission Theory

Historians have long recognized that al-Kindi's *De aspectibus* is based upon, or more exactly is a response to, the *Optica* of Euclid. It is not, however, merely a recension of Euclid's *Optica*, as several have suggested. Rather, it is a thorough and determined critique of Euclid's theory of vision, an attempt to remove several serious lacunae and to correct Euclid on a number of basic points.[13] Let us first consider the lacunae.

Euclid had made several fundamental assumptions in the *Optica* without attempting to justify them. In the first postulate he had asserted that the rays issuing from the eye are rectilinear.[14] Al-Kindi, who regarded rectilinear propagation as demonstrable, devoted the first six propositions of *De aspectibus* to removing this lacuna in Euclid's work. Strangely, however, he attempted to demonstrate the rectilinearity not of visual rays, but of luminous rays—and not, as we shall see, because he intended to deny the existence of visual rays or give priority in the process of vision to luminous rays. Al-Kindi's adoption of such a mode of argument can be explained by noting that he was here following the preface to Theon's recension of Euclid's *Optica*;[15] for a logical justification, however, one must surmise that implicit to his argument is belief in the identity of luminous and visual rays, or at least of their modes of propagation.[16]

In propositions 1-3 of *De aspectibus*, al-Kindi attempts to demonstrate the rectilinear propagation of luminous rays by considering the shadows cast by opaque bodies exposed to luminous bodies: a body equal in size to the luminous body casts a cylindrical shadow, whereas bodies smaller and larger than the luminous body cast converging and diverging shadows. The shadows thus conform to straight lines drawn tangent to the illuminating and illuminated bodies, and from this the rectilinear propagation of light follows directly.[17] A closely related demonstration appears in proposition 4, where al-Kindi points out that the straight line bisecting any of the shadows passes through the centers of both the opaque body and the luminous body. Propositions 5 and 6 provide two more demonstrations based on the same principles. If a candle is placed higher than an opaque body so that it casts a shadow on a horizontal surface, the length of the shadow (*GB* in fig. 1) is in the same proportion to the height of the obstacle (*AB*) as the horizontal distance between the candle and the end of the shadow (*GE*) to the height of the candle (*DE*); this would not be so if line *GAD*, representing a ray emanating from the candle and grazing the top of the opaque obstacle, were not straight. Finally, al-Kindi argues that if a candle *ABG* (fig. 2) is placed opposite aperture *UZ*, behind which is screen *HT*, a straight line drawn from *K* (the outer edge of the illumination on the screen) will graze the edge of the aperture at *U* and encounter the very periphery of the candle at *B*—thus proving, once more, that light is propagated in straight lines.

A second undemonstrated assumption made by Euclid takes us to the very heart of the visual process: the first and third postulates of Euclid's *Optica* claim that the rays by which objects are perceived issue not from the objects,

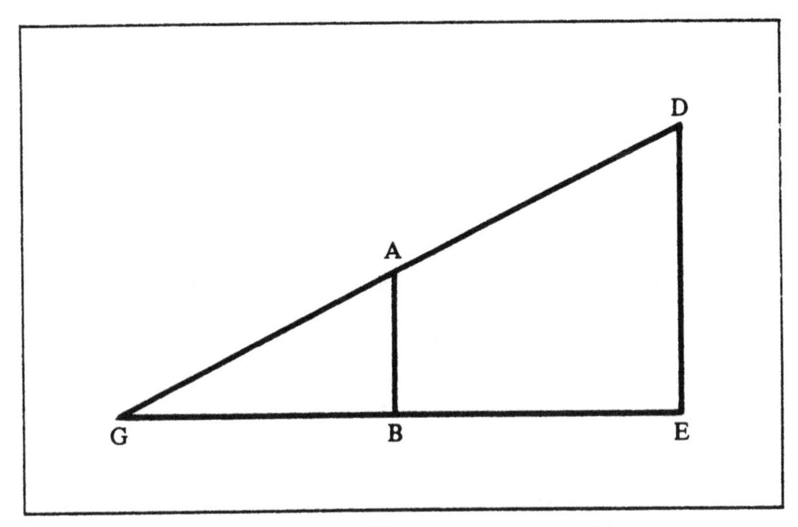

Fig. 1

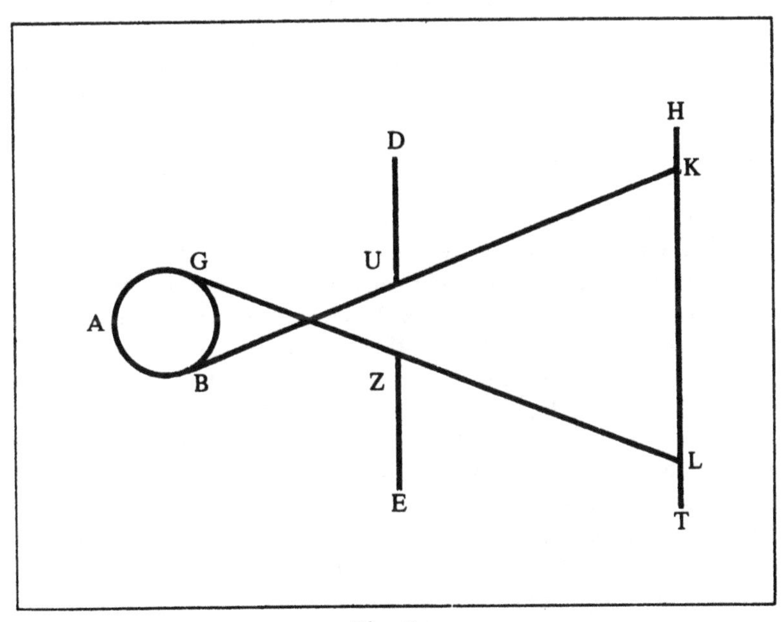

Fig. 2

but from the observer's eye.[18] In defense of this claim, al-Kindi develops an elaborate series of arguments in propositions 7-10. He begins with a brief summary of the various alternative theories formulated in antiquity:

Therefore I say that it is impossible that the eye should perceive its

> sensibles except [1] by their forms travelling to the eye, as many of the ancients have judged, and being impressed in it, or [2] by power proceeding from the eye to sensible things, by which it perceives them, or [3] by these two things occurring simultaneously, or [4] by their forms being stamped and impressed in the air and the air stamping and impressing them in the eye, which [forms] the eye comprehends by its power of perceiving that which air, when light mediates, impresses in it.[19]

The first of these alternatives is clearly the intromission theory of the atomists; the second is the extramission or visual ray theory of Euclid and Ptolemy; the third is the combined intromission-extramission theory of Plato; and the fourth is the mediumistic theory of Aristotle. Al-Kindi's strategy will be to disprove the first, third, and fourth theories and thus show that only the Euclidean theory is consistent with the known facts of visual perception.

All the theories except the Euclidean include at least a component of intromissionism, and al-Kindi hopes to refute them by making a basic attack on the intromission idea. He repeats Aristotle's argument about weak-sighted people who see their own image before them because "the power proceeding from sight, when it cannot penetrate the air because of weakness, is made to return by the air to the body of the observer."[20] He also argues, following Theon of Alexandria, that the structure of a sense organ implies the mode of its functioning. The ears, for example, are hollow in order to collect the air that produces sound. But God made the eye spherical and mobile. It was therefore designed not to collect impressions, but, through its mobility, to shift itself about and select the object to which it will send its ray.[21]

Another argument is that only the extramission theory can explain the selectivity of sight and why acuity depends on position within the visual field. When we read a book, we must strain to locate a particular letter, and we perceive it only after an interval; it is thus evident that we perceive objects in the visual field in temporal sequence rather than all at once. Moreover, objects at the side of the visual field or far from the observer are poorly perceived. Now if sight occurred through an impression made in the eye by the form of the visible object (i.e., by any of the intromission theories), we would see everything within the visual field simultaneously and with equal acuity; for once forms have entered the eye it does not matter how far they have traveled or from what direction they have come. Furthermore, if objects as far off as celestial bodies are visible, surely objects at the distance of a palm or cubit (such as the letters of a book) must impress their forms on the eye all the more clearly and should not have to be sought by the eye. One must conclude, therefore, that the intromission theory is false; rather, a visual power issues from the eye, weakening as it diverges from direct opposition to the center of the eye, and selects its objects successively.[22]

But al-Kindi's key argument against the intromission theory is a mathematical one. If sight occurred through intromission of the forms of sensible things, he argues, a circle situated edgewise before the eye would impress its form in the eye and consequently would be perceived in its full circularity. But this does not occur. "On the contrary, when circles and observer are in the same plane, the circles are by no means seen [except as straight lines]. Therefore it remains that a power proceeds from the observer to the visible objects, by which they are perceived."[23] This power proceeds from the eye in straight lines and falls only on the edges of the circles, perceiving them as straight lines.

Now this is an obscure and difficult argument, but it will pay us well to pause and consider its meaning and import. If we are to grasp the argument at all, we must first understand what al-Kindi means by "form." What he does *not* mean is a composite impression produced by a large number of individual rays, as in the modern conception of an optical image. Rather, it seems that forms are coherent images or likenesses, not susceptible of analysis into individual rays, which (according to any of the intromission theories, al-Kindi believes) would be impressed in the observer's eye. They bear a strong resemblance to the *eidola* or thin films of the Epicurean theory of vision, except that they represent the entire object rather than merely the surface facing the observer.[24] There is no justification, of course, for al-Kindi's attribution of the same conception of forms to Plato and Aristotle, who made no attempt to treat the perception of shape.[25] But whether fair to his predecessors or not, al-Kindi's conception of vision by intromission (i.e., of all theories except the Euclidean) is clear: if a circle placed edgewise before the eye were seen by the entrance of its form into the eye, this would not be because radiation from each point on the near edge of the circle entered the eye to produce an image (as in the modern view); rather, the form of the circle would enter the eye as a unit, and there its spatial orientation would have nothing to do with its perception, for within the eye the laws of perspective no longer apply.[26] Indeed, this appears to be the essential point of al-Kindi's argument: if the perception of an object is to depend on its spatial orientation, if visual theory is to be submitted to mathematical analysis, one must hold to the theory of visual rays. In short, al-Kindi sees no means by which the intromission theory, which for him is the theory of coherent forms, can be made compatible with the laws of perspective.

This may seem to be a fatuous argument. After all, is it not self-evident that light radiates independently from each point on the surface of a visible object, and that on this conception of radiation we can build an intromission theory in which perception depends on the spatial orientation of the perceived object? But this self-evidence is purely hindsight, for we have learned from Alhazen and Kepler how to construct an intromission theory on the punctiform analysis of the visible object. What would surely have been self-evident to al-Kindi and his ancient predecessors is that a coherent visual impression can result only from a coherent process of radiation, and therefore that the image must depart from the visible object as a unit.[27] Al-Kindi has thus taken the intromission theory as he conceived it (or, more specifically, the only intromissionist account of the perception of shape yet presented)—the theory of coherent forms—and demonstrated that it is not viable from a mathematical standpoint.

The Nature of Visual Radiation

Thus far al-Kindi has been in full agreement with Euclid, attempting only to supply demonstrations that Euclid had omitted. Both have agreed that rectilinear rays emanate from the eye in the form of a cone. However, on the nature of the visual cone al-Kindi finds the Euclidean theory untenable. Whereas Euclid conceived of the cone as a composite of discrete rays, separated by spaces,[28] al-Kindi regards any such conception as absurd. The visual cone, as he shows in a series of arguments, must be conceived as a continuous body of radiation.[29]

The first phase of al-Kindi's attack on the conception of discontinuous visual radiation is based on the assumption that Euclid's visual rays are geometrical lines, having length but no width:

> Certain of the ancients have judged that many rays issue from the observer along straight lines, between which there are intervals. From which opinion follows an absurdity, namely that the definition of a line, according to the one who advanced this opinion and also according to others who are subtle in learning, is a magnitude having one dimension, namely length without width, whereas a ray is the impression of luminous bodies in dark bodies, denoted by the name "light" because of the alteration of accidents produced in the bodies receiving the impression. Therefore a ray is both the impression and that in which the impression is. However, the impressing body has three dimensions—length, width, and depth.[30]

Thus on account of their nature as impressions made by three-dimensional bodies, rays must themselves be three-dimensional; it is therefore self-contradictory to speak of one-dimensional rays issuing from the eye.[31]

This is a confusing and somewhat ambiguous argument because of its reference to both luminous and visual radiation, but it is followed immediately by another argument restricted entirely to visual rays. If that which proceeds from sight to perceive objects is an infinity of lines having no width (separated by an infinity of intervals), al-Kindi argues, then these lines terminate in points having no part. Since the rays are in contact with the visible object only in a point, they are capable of perceiving only a point. "But a point is not perceived, since it possesses neither length nor width nor depth; and what lacks length and width and depth is not perceived by sight.[32] Therefore these lines [i.e., the visual rays] perceive what is not perceived, which is . . . a very horrendous absurdity."[33] It must therefore be argued to the contrary: since visual rays perceive points (which, because they are perceived, must in reality be small areas), and since rays perceive only that on which they fall, they must have width as well as length.

Al-Kindi has thus far denied that visual rays are devoid of width. In a further argument, deriving from the nature of the eye and the visual power, he demonstrates that visual rays not only are three-dimensional, but consititute a single continuous radiant cone:

> If the parts of the instrument of sight [the eye] are continuous, i.e., of one substance, then the visual[34] power is in the whole instrument. What then is it that forms the cone into lines, since the instrument impressing it is a single continuous thing, in which there are no intervals, and the [visual] power would not be in certain places and not others? If part of the instrument impressed a ray in the air and part did not, then the power, having two parts, is diverse; for a power of one part produces one effect. But if the power of the instrument is one and its whole consists of one substance [which al-Kindi obviously believes to be the case], then it produces one impression in that in which it makes an impression, and not two, one of which would be the line of the ray and the other not.[35]

However, the cause of discontinuous radiation might be in the medium rather than the source. If this were so, "then the air would consist of lines of two different substances, some of which admit light and some of which do not."[36] But that air should "consist of various lines, fixed, unmoved, and not overflowing [into one another], and that the extremity of each of those lines should be fixed over a point, which would be sought by the center of the instrument of sight of each observer until it touches it, . . . is very unseemly, and all who hear of it would laugh."[37]

The continuity of visual radiation can be established not only from the nature of the eye and the medium, but also from the phenomena of sight. Let us suppose, al-Kindi suggests, that Euclid did not mean to assert that rays are geometrical lines, that what he meant was only that three-dimensional pencils (separated by intervals) move *along* straight lines or with the rectitude of straight lines. Even this latter view falls into absurdity, for it predicts that we will observe only those portions of the visual field on which visual rays fall; between them will be blank spaces where no ray is present. In short, we will receive a spotted impression of the visual field.[38] But what we actually observe is something quite different: we see clearly only that which is directly opposite the center of the eye, and as objects increasingly diverge from opposition, they are seen with decreasing clarity. Therefore the opinion that the visual cone is composed of discrete pencils of radiation violates the teachings of sense and "is worthy of much derision."[39] It is not Euclid himself, however, who is worthy of derision, but rather certain opinions attributed to him. For there is still another possible interpretation of his position that will altogether exonerate him from the charge of error.[40] Euclid must have meant, al-Kindi argues, that although radiation itself is continuous, its overall shape can be defined by an infinity of discrete geometrical lines, because "the boundaries [i.e., the lateral surface] of the cone-shaped figure impressed in the air by the visual power proceed with the rectitude of straight lines separated by intervals."[41]

For Euclid's followers, however, al-Kindi has only harsh words. One of them, in the attempt to save Euclid from the absurdity of discrete, one-dimensional rays, has led him into equal or worse absurdity. This interpreter, whom al-Kindi does not identify,[42] maintains that there is a visual cone consisting of rays separated by intervals, but concedes (in order to avoid the absurdity of discrete rays) that only the central ray of the visual cone, the axis, perceives objects. Al-Kindi's response is that if visual rays (other than the axis) are neither perceived nor able to perceive, there is no evidence for their existence; and it is no less absurd to postulate a visual cone for which there is no evidence than to suppose that perception occurs through discrete, one-dimensional rays. Thus if one wishes to defend the existence of a visual cone, he is obliged to maintain also that perception takes place through all of its rays.

Euclid's followers are guilty of even more. The same people who maintain that only the axis of the visual cone perceives objects also claim that objects nearer the axis are more clearly perceived than objects far from the axis.[43] "But if places near the center of the circle [i.e., the base of the visual cone] are perceived, then the ray that perceives visible things is not a single line [i.e., the axis]."[44] Therefore these followers of Euclid have contradicted themselves, and the absurdity of their teaching is manifest.

Variations in Sensitivity within the Visual Cone

Al-Kindi has treated the geometry of the visual cone and has concluded that the cone is a continuous beam of radiation, sensitive throughout. But al-Kindi's polemic against Euclid's followers has led him to acknowledge variations in the sensitivity of different regions of the cone, which must now be explained. Why should objects near the axis of the visual cone be perceived more clearly than objects on the periphery? Not for the reasons offered by those followers of Euclid who "have regarded themselves as acute teachers in a learned discipline before they have been students."[45] They have argued that the axis of the visual cone is the shortest of all rays,[46] and therefore that it perceives most strongly, the assumption being that strength of perception varies inversely with the length of the ray.[47] The trouble with this explanation, al-Kindi argues, is that it ignores several obvious phenomena. An object at *E*, on the edge of the visual cone (fig. 3), is closer to the eye than another object on the axis at *D*, and yet the latter object is seen more clearly than the

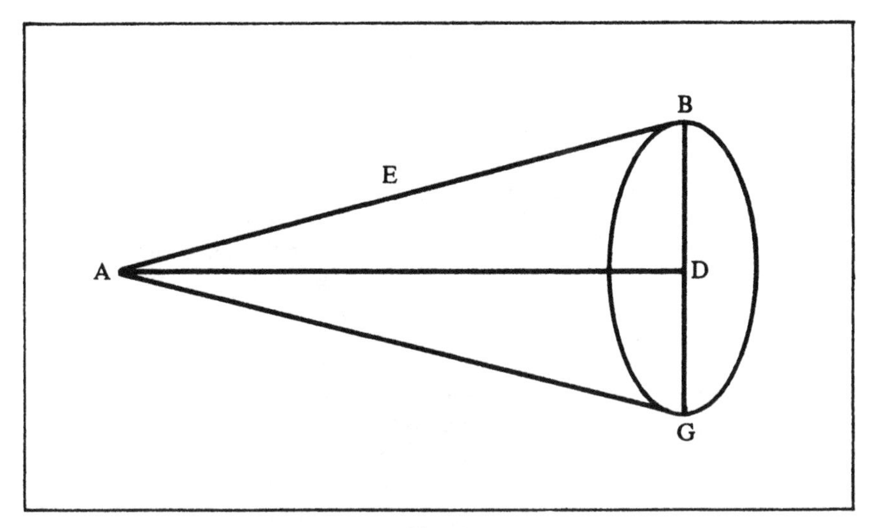

Fig. 3

former. Indeed, a nearby object at the side of the visual field is not seen as clearly as an object in the orb of the fixed stars situated on the axis of the visual cone. Therefore the factor that determines clarity of vision cannot be the length of the ray.

The true cause is discerned through the analogous behavior of light and color. Color is not perceived unless illuminated by light, and the stronger the light the clearer the perception of color. Similarly, the stronger the visual ray, the clearer its perception of color. But what does it mean for a visual ray to be stronger? Where does its strength lie? The operation of sight is to produce a conversion or transformation in the surrounding medium—though on the nature of the transformation al-Kindi does not at this point elaborate.[48] A strong ray, then, is simply that which produces a perfect or complete transformation, and a weak ray that which produces an imperfect or incomplete transformation. The axial ray produces the most perfect transformation of the medium and therefore perceives its object most clearly; other rays produce transformations that decrease in perfection with the distance of the ray from the central axis. Al-Kindi's explanation may raise as many questions as it answers—Why are peripheral rays less capable of transforming the medium than central rays?—but he has at least associated clarity of perception with the object's position in the visual field rather than with its distance from the eye.

However, two propositions later (in proposition 14) al-Kindi takes up the problem again, this time offering a geometrical explanation of variations in the strength of visual radiation and hence in the clarity of visual perception. Here again al-Kindi uses the analogy of external illumination. Just as two candles illuminate the same place better than one, so also places "illuminated" by more visual radiation are more clearly seen.

> For example, let the instrument of vision, namely the eye, be circle *ABG* [fig. 4], the center of which is *D*. And let the part [of the eye] to which belongs the power of comprehending the visible object, namely the aforesaid exterior gibbosity of the eye [i.e., the cornea], be arc *ABG*. Let us draw *ZE* tangent to *B*, which divides arc *AG* into two halves, and line *HT* tangent to point *A*, which is one of the two ends of the external gibbosity of the eye, and line *IK* tangent to point *G*, which is the other end of the aforesaid gibbosity of the eye. And let the observed body be arc *HEITZK*. Let there be a point[49] directly opposite the center of the eye, point *L*; and when a line is drawn from *L* to *B*, let it be perpendicular to line *EZ*.
>
> Now every part of arc *HT* is illuminated by part *A* [of the eye]; therefore part *L* is illuminated by part *A*. Also, every part of arc *IK* is illuminated by part *G* [of the eye]; therefore *L* is also illuminated by part *G*. Also, every part of arc *EZ* is illuminated by part *B*; therefore *L* is also illuminated by part *B*. Therefore part *L* is illuminated by three parts [of the eye], *A*,*B*, and *G*.
>
> But arc *EI* is common simultaneously to two arcs, *HT* and *EZ*; therefore it is simultaneously illuminated by only two parts, *A* and *B*. With the same arrangement it is demonstrated that arc *ZT* is illuminated only by the two parts *G* and *B*. Arc *HE*, however, is part of arc *HT* alone; therefore it is illuminated only by part *A*. And according to this [same] arrangement it is shown that arc *ZK* is illuminated only by part *G*.
>
> Thus point *L* is illuminated by three parts, namely *A*, *B*, and *G*. And every point in [either of] the two arcs *EI* and *TZ* is illuminated by two parts, as we have said. Therefore *L* is illuminated more strongly than any point in the two arcs *EI* and *ZT*, and the illumination of the two arcs *EI* and *ZT* is stronger than the illumination of the two arcs *HE* and *ZK*. . . . Therefore it has been made clear that the center is illuminated more strongly, and that which is closer to it is more [strongly] illuminated than that which is further from it; for more light falls on it because it is illuminated by more luminous parts.[50]

Thus al-Kindi traces the apparent strength of axial rays ultimately to the fact that the central region of the visual field receives more visual radiation (from various points on the surface of the eye) than any other region; it is not true, apparently, that individual points of the eye emit rays in various directions that actually vary in strength.

We must note several of the consequences of this argument. In the first place, al-Kindi indicates that radiation is emitted in all directions from every point on the surface of the cornea. This reveals that the part of the eye active in vision is the outer surface; and, indeed, al-Kindi says quite plainly in the passage quoted above that "the part [of the eye] to which belongs the power of comprehending the visible object" is "the aforesaid exterior gibbosity of the

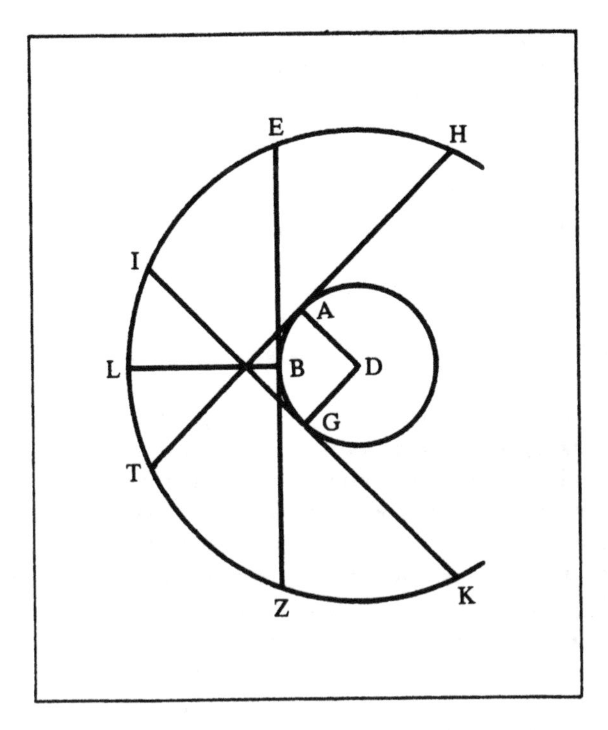

Fig. 4

eye." Al-Kindi thus disagrees with Euclid, Damianos, and Ptolemy, who located the source of visual rays within the eye.[51]

Second, al-Kindi's argument calls into question the very idea of rays as entities useful for describing the physical process of radiation. I am not, of course, referring to the one-dimensional rays of Euclidean optics or even to three-dimensional, but discrete, pencils of radiation, both of which al-Kindi has already explicitly rejected, but rather to the idea that within the cone of continuous radiation there is a unique, physical correspondence between areas of the visual field perceived and portions of the visual cone or the eye that perceive it—for example, that an object on the axis of the visual cone is perceived by radiation issuing from the center of the eye, whereas objects at the periphery of the visual field are perceived by radiation issuing from the periphery of the eye.[52] This entire idea, I say, is called into question, for al-Kindi teaches (in this proposition) that every part of the visual field is "illuminated" by radiation from every part of the eye that has straight-line access to it. It follows that visual radiation does not, after all, proceed along the visual rays of Euclidean or Ptolemaic optics; the single visual cone of Euclidean and Ptolemaic optics is no longer held to represent the physical mode of radiation, but only to symbolize the perception of space geometrically. Or to state the same point somewhat differently, if we want the visual cone to represent the physical process of radiation as conceived by al-Kindi, there will be not one cone, but many, emanating from every point on the surface of the observer's eye.[53]

I do not know how far along this line of argument al-Kindi would have been willing to follow. Since it contradicts what appears to be his clear teaching in other places, I suspect that the answer is, not very far. At this point in his argument, al-Kindi was attempting only to explain how radiation could vary in strength over the visual field, and there is no reason why the uncomfortable consequences of his explanation must have come to his attention.

One more point must be made regarding al-Kindi's explanation of variations in the clarity of perception over the visual field. Al-Kindi bases his claim that rays issue in all directions from every point of the surface of the eye on the analogous behavior of external light. In fact, the proposition just before the one we have been discussing is devoted wholly to the radiation of external light. In this proposition (the thirteenth), al-Kindi reveals an explicit understanding of the principle that luminous rays issue in all directions from every point on the surface of a luminous body. He remarks in conclusion, "Therefore we have illustrated how each part of the luminous body illuminates that which is opposite it, namely that to which a straight line can be drawn."[54] Now such a conception (I call it the principle of "punctiform analysis") had perhaps been implicit to optical theory (at least within certain schools of thought) since its inception, but so far as I have been able to discover al-Kindi was the first to state it explicitly.[55] This is one of the most fundamental principles of optics, and we must not permit its present familiarity and self-evidence to obscure its importance. In Alhazen's hands it was to become an alternative to the coherent forms of ancient intromission theories, the very foundation of a new theory of vision;[56] nevertheless, for al-Kindi its relationship to vision remained simply that of an analogy, helpful in elucidating the emission of visual rays. Ironically, al-Kindi has thus formulated the basis of the new conceptual scheme that would eventually supplant his own extramission theory.

Al-Kindi as a Galenist

Although *De aspectibus* is principally a geometrical treatise, on several occasions al-Kindi has found it necessary to comment briefly on the physical nature of visual rays. In his attack on the idea of discontinuous radiation, al-Kindi defines a ray as "the impression of luminous bodies in dark bodies, denoted by the name 'light' because of the alteration of accidents produced in the bodies receiving the impression," and argues that "if part of the instrument [i.e., the eye] impressed a ray in the air and part did not, then the power, having two parts, is diverse."[57] In his analysis of variations in the sensitivity of the visual cone, he argues that the effect of sight is "that it converts that which is opposite."[58] Finally, in discussing reflection al-Kindi refers to the transformation of the medium provoked by sight as a "resolution."[59] It is apparent from these remarks that in al-Kindi's view the ray is not a substantial entity issuing from the eye—no corporeal thing is actually transported from the eye to the visible object—but a transformation of the ambient air produced by the visual power of the eye.

What is important about this position is that it takes al-Kindi beyond the Euclidean tradition and identifies him with the Galenic or Stoic theory of vision.[60] Hunain ibn Ishaq, al-Kindi's younger contemporary and one of the earliest Arabic authors to expound the Galenic theory, claims that "when [the visual spirit] meets the air in the moment in which it goes forth from the pupil, it transforms it immediately it encounters it, and that which arises from the change runs through it [the air] for a very long distance. . . . So the change in the air caused by the [action of the] visual spirit penetrates the whole air."[61] It is precisely this position that al-Kindi adopts. I do not wish to argue, however, that al-Kindi is a full-fledged Galenist in his theory of vision. Far from it! Al-Kindi does not suggest that the air itself becomes percipient, as Galen and his followers maintained.[62] Al-Kindi retains the basic geometrical framework of Euclid and Ptolemy and avoids the anatomical and physiological orientation of the Galenists. Moreover, it is clear from his favorable references to Euclid that al-Kindi regards himself as a loyal representative of the Euclidean theory of vision. He has simply given the Euclidean theory a Galenic interpretation at a point where Euclid himself had been entirely silent, namely, on the physical nature of visual rays. This is in no sense a violation of the Euclidean theory, but only an extension of it along Galenic lines.

Al-Kindi's Influence and the Fate of the Extramission Theory

Al-Kindi's *De aspectibus* became a popular textbook in Islam[63] and influenced the course of optics for centuries to come. The strength of its influence is revealed by the continuing popularity of the extramission theory of vision among Islamic scholars. Among those who explicitly defended the extramission theory were al-Farabi (d. 950), 'Ubaid Allah ibn Jibril ibn Bakhtyashu' (d. 1058), Ibn Hazm (d. 1064), Nasir al-Din al-Tusi (d. 1274), al-Qarafi (d. after 1285), and Ahmad ibn Abi Ya'qubi (fourteenth century).[64] Another, Salah al-Din ibn Yusuf (fl. 1296), wrote a treatise entitled *The Light of the Eyes*, in which he developed the extramission theory of vision at considerable length; however, he then appended Avicenna's attack on the extramission theory and took no firm stand on the controversy himself.[65]

In the West, the extramission theory had considerably less success. Although it was known and discussed, as numerous manuscript copies of

Euclid's *Optica* and al-Kindi's *De aspectibus* attest, it was seldom adopted as the true theory of vision.[66] But al-Kindi's influence cannot be measured solely in terms of converts to a pure form of the extramission theory of vision. There were many natural philosophers, particularly in the West, who looked to al-Kindi for support in their defense of a combined intromission-extramission theory.[67] Grosseteste, an early defender of such a combined theory, was in all likelihood familiar with al-Kindi's *De aspectibus*[68] and probably had al-Kindi in mind when he wrote: "However, mathematicians and physicists [by contrast with natural philosophers], whose concern is with those things that are above nature, . . . maintain that vision is produced by extramission."[69] Later in the thirteenth century Roger Bacon and John Pecham also appealed to the authority of al-Kindi to support their contention that rays issue from, as well as enter, the observer's eye.[70]

3 GALENISTS AND ARISTOTELIANS IN ISLAM

Introduction

Euclid's mathematical extramission theory of vision—revised and extended by Ptolemy and al-Kindi—faced formidable rivals in Islam. The predominant school of thought on the subject of vision was undoubtedly the Galenic, committed to another form of the extramission theory; this school was led by Hunain ibn Ishaq and 'Ali ibn 'Isa and prevailed especially in medical circles. The popularity of the Galenic theory is explained not simply by the fact that it had been taught by the "Prince of Physicians" and therefore found a natural audience among medical men, but also by the fact that it included anatomical and physiological detail not present in any other ancient theory of vision. If one's purpose was to treat diseases of the eye, what mattered was not so much the geometrical features of the perception of space as the anatomical features of the eye.

But there was another important scientific current in Islamic thought (besides the medical and the mathematical)—that of natural philosophy, rooted principally in the Aristotelian tradition. Thus, by the beginning of the tenth century the Aristotelian theory of vision (an intromission theory, by contrast with the Euclidean and Galenic theories) had reared its head, and in the eleventh and twelfth centuries it found particularly forceful and articulate representatives in Avicenna and Averroes; indeed, in Avicenna's works we find a blistering attack directed at the extramission theory in both its Euclidean and its Galenic forms. Nevertheless, even Aristotelians recognized the necessity of saying something about the anatomy of the eye and the physiology of sight, topics that Aristotle had largely ignored, and consequently we find Avicenna and Averroes incorporating significant Galenic elements into their optical thought.

Hunain ibn Ishaq and the Reception of Galenism

One of the branches of medicine pursued most vigorously in Islam was ophthalmology, undoubtedly because of the frequency of eye disease in the Near East. The first Arabic works on ophthalmology appeared in the ninth century, and several dozen more appeared in the next five hundred years.[1] The earliest known writer on ophthalmology in Islam (though himself a Christian) was the court physician at Baghdad, Yuhanna ibn Masawaih (d.

857, known in the West as Johannes Mesue).[2] Far more important for the subsequent history of Islamic ophthalmology, however, was Yuhanna's pupil, Hunain ibn Ishaq.

Hunain (d. 877, known in the West as Johannitius) was the son of a Nestorian Christian. He studied medicine under Yuhanna ibn Masawaih, traveled widely, and eventually settled in Baghdad as one of the outstanding translators of Greek scientific and philosophical works into Syriac and Arabic.[3] He was a contemporary of al-Kindi, and since both lived in Baghdad, employed by the same patrons, it is scarcely conceivable that they were not acquainted. Hunain wrote, as well as translated, scientific works, and his own compositions include two on ophthalmology: the *Ten Treatises on the Eye* (or *On the Structures of the Eye, Its Diseases, and Their Treatment*) and the *Book of the Questions on the Eye*.[4] The importance of these two books is not that they contain anatomical or physiological novelties or new departures in visual theory, but that they represent the penetration of Galenic theories into Islam.[5] Moreover, since Hunain's *Ten Treatises* was translated into Latin very early and Galen's *De placitis Hippocratis et Platonis* and book 10 of *De usu partium* (in which the eye and vision are treated) quite late, it was chiefly through Hunain that medieval ophthalmologists in the West obtained their Galen.[6]

In the first of the *Ten Treatises*, Hunain presents a wholly traditional account of ocular anatomy and physiology, drawn primarily from book 10 of Galen's *De usu partium*. The central feature of the eye is the icelike or crystalline humor (our crystalline lens), which is uncolored, transparent, luminous, and round.[7] Its roundness is not complete, however, for there is a certain flattening, which "enables it to receive impressions of more perceptible objects than would be the case if it were perfectly round; for a flattened body meets more of the objects which are in its path than does a perfectly spherical body."[8] The crystalline humor occupies the central position in the eye not only so that it may be served by the other ocular humors and tunics, but also as a measure of its rank as seat of the visual power.[9]

Behind the crystalline humor is the glasslike or vitreous humor. Its principal function is to nourish the crystalline humor by mediating between it and the blood vessels of the retina. Hunain's explanation of the necessity of such mediation is an interesting example of ancient and medieval physiological reasoning:

> Nutrition is effected in this wise, *viz.* that the member receives an addition of substance resembling its own nature. This accretion, however, can only resemble the nature of the member if the latter transmutes it according to its own nature. A substance is most quickly transmuted into the thing which resembles its own nature most closely. Since the lens without doubt requires nutriment and since, as we mentioned already, this humour is white, transparent and luminous, it is impossible for it to receive its nutrition

direct from the blood. It requires an intermediary between its nature and that of the blood; and such is the glass-like humour [the vitreous] as it is nearer to the white colour and transparency than [is] the blood. Therefore the vitreous is adjacent to the lens without any partition, and it [the lens] is half submerged in it [the vitreous].[10]

Behind the vitreous humor are three tunics: the retina, the choroid or secundina, and the sclera. The retina, or netlike tunic, arises from the optic nerve and encloses the vitreous humor. Through its veins and arteries it supplies nourishment to the vitreous humor and ultimately to the crystalline humor, and through its nerve (i.e., the optic nerve) it conducts the visual spirit to the crystalline humor.[11] As for the remaining membranes, the choroid, which arises from the inner coat of the optic nerve, covers and nourishes the retina; the sclera, which arises from the outer coat of the optic nerve, is hard and serves to protect the eye from injury.[12]

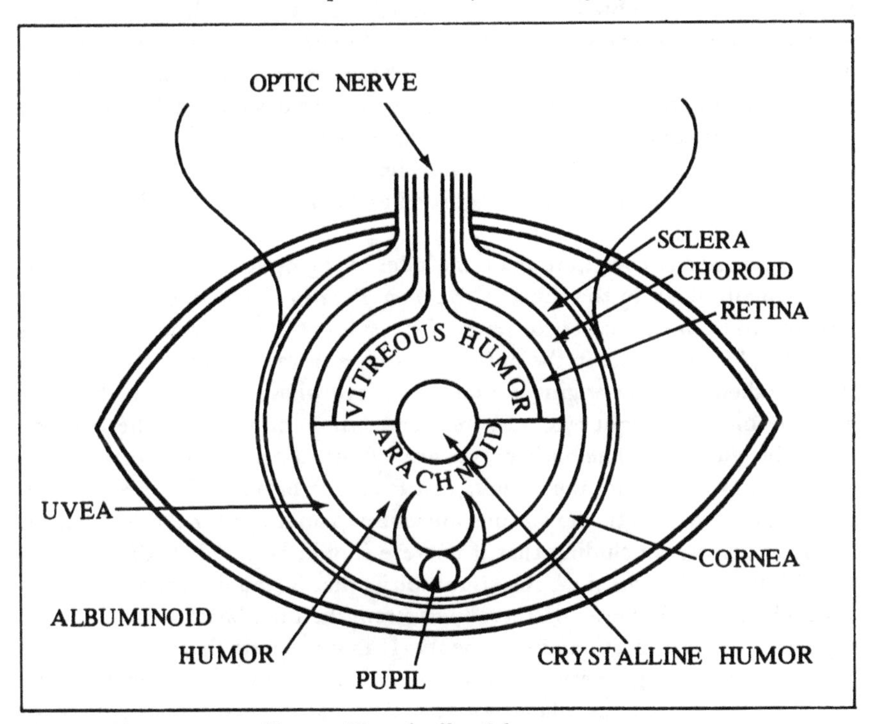

Fig. 5. The eye according to Hunain ibn Ishaq

There are also three tunics and one humor in front of the crystalline. The albuminoid (or aqueous) humor, so named because it resembles the white of an egg, resides in the aperture of the iris (the pupil) and serves to separate the crystalline humor from the cornea and to nourish and moisten the crystalline and the iris. Outside the albuminoid humor is the uvea or grapelike tunic, which grows out of the choroid. It nourishes the cornea, separates the crystalline humor from the cornea so that the latter will not injure the former through friction, and has an aperture in front to permit the visual spirit to issue forth. Outside the uvea is the cornea, growing out of the sclera; it is transparent to allow the passage of visual spirit or light, and it is hard in order to protect the eye from outside objects. Finally, the outermost tunic is the conjunctiva, which arises from the membrane that covers the skull.[13] Hunain's conception of the eye is represented very roughly in figure 5, copied from a manuscript of the *Ten Treatises* dated to A.D. 1196.[14]

In a passage that was to give rise to innumerable debates among Islamic and Western ophthalmologists, Hunain refers to diversity of opinion on the number of ocular tunics.[15] Six tunics have thus far been described, but some people identify a seventh in the arachnoid membrane, a thin husk resembling an onion skin or a cobweb, which separates the crystalline humor from the aqueous humor.[16] There are also those, Hunain points out, who would reduce the number of tunics to five, four, three, or even two. They do this by denying that certain of the membranes deserve to be designated "tunics" or by maintaining that certain pairs of membranes regarded as distinct tunics are in fact the anterior and posterior hemispheres of a single tunic. The sclera and cornea, for example, can be regarded as a single tunic, and likewise the uvea and choroid.

The optic nerves constitute the remainder of the optical apparatus, and to them Hunain turns in the third of the *Ten Treatises*. The optic nerves take their origin from the "posterior parts of the sides of the anterior ventricles of the brain," briefly unite in the optic chiasma, and revert to the eyes directly opposite their cerebral origin, so that "the nerve whose origin is on the right side goes on to the right eye, and the nerve whose origin is on the left side enters the left eye."[17] The optic nerves are hollow, the only such nerves in the body,[18] so as to conduct the visual spirit from the brain to the eye. Hunain discusses at some length the various ancient explanations for the junction of the optic nerves, concluding that if one eye should be blinded, the junction permits its visual spirits (i.e., the visual spirits that it would normally receive) to pass into the other eye and thereby to strengthen the latter's vision. As an experimental confirmation of this theory, Hunain reports that if an observer "shuts one eye, his sight with the other one becomes clearer and sharper. The reason of this is that the whole power which was divided between both of them . . . now enters into this one eye alone."[19] Further confirmation is found in the fact that if one eye is closed, the pupil of the other is immediately enlarged, supposedly because the uvea is distended by an increased quantity of visual spirit.[20]

A second purpose of the junction of the optic nerves is to produce a single impression in binocular vision. An object will be seen singly only if the rays proceeding from the two eyes have a common origin, from which they are conducted in straight lines through the optic nerve and the pupil to the object, and of course they must then fall on the latter identically. The common point of origin is the junction of the optic nerves, and "if the glances proceed from this single origin and then come out into the pupils and regard the perceived object, they reach it in its place and see it as one."[21] However, if either of the eyes is displaced from its position, the visible object appears double.[22]

Having provided a full description of the anatomy and physiology of the eyes and optic nerves, Hunain turns, in the latter half of the third treatise, to vision itself and to the defense of the Galenic theory. He begins with a summary of the ancient alternatives:

> We say: the object of vision can be seen only in one of the following three ways: [*i*] by sending out something from itself to us by which it indicates its presence so that we know what it is; [*ii*] by not sending anything out but remaining steady and unchanged in its place; then the faculty of perception goes out from us to it, and we recognize what it is through this medium; [*iii*] by there being another thing . . . intermediate between us and it; it is this which gives us information about it, so that we learn what it is. And we shall now see which of these three [theories] is the right one.[23]

The first of these alternatives appears to encompass both the atomic theory of *simulacra* or *eidola*, in which substantial images are held to emanate from the visible object and to penetrate the eye, and the Aristotelian theory of vision, in which the emanation consists of a form or quality. Both, Hunain argues, are untenable because they imply that the form or outline of a very large body, say a mountain, must enter the eyes not only of one observer, but of as many as ten thousand observers at the same time. Such an event Hunain considers exceedingly improbable:

> All people acknowledge and agree that we see only by the hole which is in the pupil. Now, if this hole had to wait until something coming from the seen object reached it, or a power emanating from it, or a form, an outline or a quality, as some people maintain, we should not know, in looking at an object, either its extent or its volume. . . . Its entering into the eyes is something which reason does not comprehend and of which nobody has ever heard, for according to this hypothesis a complete form or outline of the viewed object would necessarily reach and enter into the eye of the beholder at the same moment. Supposing then that a great many people looked at it, say, for example, ten thousand persons, it would have to return to the eye of every one of them, and its form and outline would have to enter completely into them. But this is very far from probable and must therefore be ranked among the untenable hypotheses. This being so, there is no possibility that something proceeding from the seen object reaches and enters the pupil.[24]

The second theory, which is apparently supposed to be that of Euclid and Ptolemy, is no better and receives a terse dismissal from Hunain: "It is not possible that the visual spirit extends over all this space [between the eye and a distant visible object] until it spreads round the seen body and encircles it entirely."[25]

The third theory, which upon further elaboration proves to be that of Galen and the Stoics, is the true one. There is no flow of material substance either from the visible object to the eye or in the opposite direction. Rather, the air that already fills the space between the object and the observer becomes the instrument of the eye, thus mediating between the eye and its object. Hunain's solution to the age-old problem of contact between the observer and the visible object is thus, in broad outline, the same as Aristotle's and Galen's:[26] there is no flow of substance from one to the other; rather, contact is established through the transparent medium that already intervenes. But two conditions must be met before air can perform this function. First, the air must be transformed by sunlight.[27] Second, and here Hunain takes leave of Aristotle to follow Galen, the air must be transformed by an encounter with the visual spirit issuing from the eye. The visual spirit "meets the air in the moment in which it goes forth from the pupil" and transforms it, "and that which arises from the change runs through it [the air] for a very long distance. . . . So the change in the air caused by the [action of the] visual spirit penetrates the whole air."[28] For Aristotle, the air is influenced only by external light and color and remains simply a transparent medium capable of conveying forms to the eye; for Hunain and Galen it is so transformed by visual spirit as to become an organ of sight.[29]

It must be emphasized that in Hunain's theory it is not visual spirit that is propagated from the eye to the visible object, but a transformation of the medium provoked by the visual spirit. The visual spirit itself apparently emerges from the eye,[30] whereupon it transforms the air; but it extends no farther.[31] However, if the visual spirit is not transported substantially from the eye to the visible object, is there any reason to suppose that the substance of the visual spirit flows from the brain to the eye in the first place? Why not suppose that the substance of the visual spirit remains in the brain, while its power alone flows through the optic nerve to the eye? Hunain maintains that "as for the lucid spirit [i.e., visual spirit] which goes to the eyes, it is not only its power but its actual substance which reaches them by way of the two canals piercing the optic nerves, the quantity of it reaching them being [just] the quantity necessary to fulfill the function of vision."[32] The proof of this claim is in the observation that when one eye is closed, the pupil of the other is dilated; this results from distension of the uvea by visual spirit, and "it is not possible that this enlargement should have any other cause than this."[33]

Hunain remains silent, for the most part, on the precise nature of the transformation of air produced by the visual spirit. His most revealing statement is the following: "When it [the lucid or visual spirit] has entered the eye and comes out of it, so that it meets the surrounding air, it strikes it as it were with the shock of a collision, transforms it and renders it similar to itself [i.e., to the visual spirit]."[34] As a result of this transformation, the air becomes "for the time being a homogeneous and coherent organ of vision,"[35] through which "vision extends itself . . . until it reaches the coloured body."[36] Hunain illustrates this conception with the analogy (already old in the ninth century) of the walking stick:

> If a person is walking in the dark and holds a stick in his hand and stretches it out full length before him, and the stick encounters an object which prevents it from advancing further, he knows immediately by analogy that the object preventing the stick from advancing is a solid body which resists anything that comes up against it. . . . It is the same with vision.[37]

The air, transformed by sunlight and visual spirit, is the instrument of the eye in the same sense as the cane is the instrument of the person's hand for extending its reach into the space before it. But this analogy has connotations that I do not believe Hunain intended, for it may seem to imply that the air is a passive (and inanimate) medium through which pressures or impressions are returned to the eye. In short, it is likely to have, for us, the connotations of Cartesian mechanism, whereas in fact Hunain intended nothing of the kind.[38] In Hunain's view, the air itself becomes sensitive in precisely the same sense as the eye is sensitive.[39] Just as the power of vision proceeds through the optic nerve to the eye, transported by the visual spirit, so it is transmitted through the transformed air "until it reaches the coloured body."[40] Hunain insists that "the relation of the brain to the nerve proceeding from it is exactly the same as that of the eye to the air surrounding the human body."[41] The visual power is thus conveyed to the object rather than the object (or its impression) being conveyed to the power of vision.

Having been conducted to the visible object, how does the visual power perceive its various qualities? This is not a question with which ancient and medieval psychologies ordinarily concerned themselves. The problem of establishing contact between the visual power and the perceived body seemed of the utmost importance, but how contact led to sensation was not generally regarded as a problem.[42] The analogy of perception by contact in the sense of touch seemed to establish to nearly everybody's satisfaction that contact was tantamount to sensation, and it was not apparent that further explanation was required.[43] Hunain points out that the visual power perceives the size, shape, and color of the bodies that it encounters simply because it has that capacity.[44] Nevertheless, with regard to the perception of color Hunain adds a small amount of additional detail. He claims that color produces another transmutation in the air (now the instrument of vision), making three such transmutations in all: "In the same way in which the air is entirely transformed by the sunlight until it becomes luminous . . . it is also instantaneously transformed by the colours, and in the same way that it is transformed by the colours, it is also transformed, with the utmost rapidity, by the luminous spirit."[45] As an example of this process, Hunain notes that if a man lies beneath the branches of a tree, his clothes take on the color of the tree.[46]

The visual process is not completed when the air (now the instrument of the visual power) has perceived the size, shape, and color of an object, for perceptions must be "returned" to the eye and ultimately to the seat of consciousness in the brain. Unfortunately, Hunain says little about this aspect of the visual process, and we must attempt to reconstruct his view from fragmentary evidence; since he is clearly attempting to follow Galen in the matter, we must also probe Galen's works for clues. Visual perceptions or impressions are first returned to the crystalline humor by a process which one must assume is optical: since the medium is air, a process of radiation seems to be the only possible mode of transmission. That the crystalline humor receives the impressions is evident from Hunain's remark that its flatness "enables it to receive the impressions of more perceptible objects than would be the case if it were perfectly round."[47] Similar remarks were made by Galen, who argued that visible objects produce alterations in the crystalline humor.[48]

If the transmission of visual impressions or perceptions from the visible object to the crystalline humor is optical, what is the nature of the remainder of the transmission, from the crystalline humor to the brain?[49] Is it also optical—a process of radiation through the crystalline humor, into the transparent vitreous humor, and finally through the visual spirit occupying the hollow optic nerves—or is it what we would today call "neurophysiological," taking place through the tissues (rather than the transparent humors) that lead from the crystalline humor to the brain?[50] Neither Galen nor Hunain ever addressed himself unambiguously to this question, but I believe it can be argued with some probability that they regarded the post-crystalline transmission as neurophysiological. This is suggested as a strong possibility by the admission of both Galen and Hunain that the crystalline humor is physically attached at its equator to the retina, which in turn is but an expansion of the optic nerve; indeed, Galen comes very close to settling the matter when he writes that "the purpose for which it [the retina] was sent down has now been accomplished, and it is inserted into the crystalline body, the changes in which it can faithfully report to the encephalon."[51] To this may be added that when Galen and Hunain discuss the utility of the vitreous humor they link its transparency to its ability to nourish the crystalline humor but say not a word about the transmission of visual impressions through it; moreover, both admit that the vitreous humor falls short of the perfect transparency of the crystalline humor.[52]

Further support for this neurophysiological interpretation of the post-crystalline transmission comes from Galen's and Hunain's claim that the crystalline humor is the chief organ of sight. Hunain (echoing Galen) argues:

"And further proof that the [power of] vision is in this [crystalline] humor, and not in any other part of the eye, lies in the circumstance that ... vision ceases when cataract intervenes between it [the lens] and the perceptible object, and that ... vision returns when the cataract is removed from it by couching [operation]."[53] Now if the transmission of visual impressions from the visible object to the brain were regarded as optical in its entirety, the crystalline humor would seem to be just another transparent medium through which visual spirit and visual impressions must pass, with no function except optical transmission and therefore with no special sensitivity. If, on the other hand, we assume that the crystalline humor represents the end of the optical phase of transmission and the beginning of the nonoptical or neurophysiological phase, the claim that the crystalline humor is the principal seat of visual power makes perfect sense. Finally, interpretation of the post-crystalline transmission as a neurophysiological process seems more consistent with Galen's claim that the principal function of the retina is "to perceive alterations of the crystalline humor"[54]—for what could this mean on the theory that the post-crystalline transmission is simply a matter of optical radiation through the vitreous humor and the visual spirit filling the optic nerves?[55]

Hunain's influence on both Islamic and Western ophthalmology was exceedingly strong. In Islam, the *Ten Treatises* was cited by the Persians Abu Bakr Muhammad ibn Zakariya al-Razi (d. 923/24, known in the West as Rhazes), Abu Ruh Muhammad ibn Mansur al-Jurjani (fl. 1087, also known as Zarrin-Dast), and 'Ali ibn Ibrahim ibn Bakhtyashu' (fl. second half of the eleventh century); by the Syrian Khalifa al-Halabi (fl. 1256); by the Andalusian Abu Ja'far Ahmad ibn Muhammad al-Ghafiqi (d. 1165); and by many others.[56] 'Ali ibn 'Isa (fl. first half of the eleventh century) and Alcoati (fl. 1159) claimed Hunain among their principal sources.[57] In the West, Hunain was no less influential. His *Ten Treatises* was translated into Latin near the end of the eleventh century by Constantinus Africanus and circulated widely under the name of Constantinus or that of Galen.[58] It was cited by Benvenutus Grassus, Bartholomaeus Anglicus, Vincent of Beauvais, and Roger Bacon, and it is apparent that directly or indirectly it influenced almost every member of the Western optical and ophthalmological tradition before the seventeenth century.[59]

Hunain, as I have pointed out above, made no significant departures from Galen on the subject of vision, but transmitted an essentially pure version of the Galenic theory. Nor, as far as I have been able to determine, did the Galenic theory undergo significant development between Hunain and the sixteenth century. Valiant efforts by Julius Hirschberg, Pierre Pansier, and Max Meyerhof to trace the development of Islamic ophthalmology have not uncovered any fundamental novelty with regard to the theory of vision. It might be debated how many tunics the eye comprised, but it was clearly recognized even by the combatants that this was not a disagreement about the real structure of the eye, but a debate about what constituted a tunic.[60] There might also be discussion about the distance traversed by visual spirit after it emerged from the eye (did it proceed all the way to the visible object or only a short distance beyond the eye?), but this did not affect the basic claim that visual spirit transformed the air into the instrument of the eye. There is nothing mysterious, of course, about this lack of development of visual theory within the ophthalmological tradition; ophthalmologists are chiefly concerned about diseases of the eye, and theoretical questions about the nature of the visual process were quite properly seen as peripheral (if not irrelevant) to this central concern.

Early Opposition to the Extramission Theory of Vision

The Euclidean and Galenic theories of vision, though rivals, were in agreement on one of the most fundamental issues: each maintained that the visual power issues from the eye and is sent to the visible object in order to perceive it. After the reception and assimilation of Aristotelian psychology, which understood sense organs to be passively involved in the process of perception, it was inevitable that these two major forms of the extramission theory should come under attack.

We receive our first hint of a refutation of extramission theories of vision from Abu Bakr Muhammad ibn Zakariya al-Razi (d. 923/24, known in the West as Rhazes). According to the historian of Islamic medicine Ibn Abi Usaibi'a (first half of the thirteenth century), al-Razi wrote a number of works on optics and ophthalmology, including *On the Nature and Method of Seeing*, *On the Reason Why the Pupil Contracts in Light and Dilates in Darkness*, *On the Form of the Eye*, and *On the Conditions of Vision*.[61] Unfortunately, none of these works survives, but Ibn Abi Usaibi'a makes the tantalizing remark that in the first of them al-Razi shows that vision does not occur by the emanation of rays from the eye. We catch a glimpse of the same position in an extant work of al-Razi, the *Kitab al-Mansuri* or *Liber ad Almansorem*, where it is stated that the pupil contracts or dilates according to the amount of light (presumably external light) required by the crystalline humor;[62] this is in marked contrast to the position of Hunain, that the pupil is dilated by the pressure of the emerging (or about to emerge) visual spirit.

We catch sight of another early alternative to the Euclidean and Galenic theories in the work of al-Razi's younger contemporary Abu Nasr al-Farabi (d. 950, known in the West as Alpharabius), the first great Islamic expositor of Aristotle.[63] In his *Catalogue of the Sciences*, where he attempted to classify the various sciences, al-Farabi retains the Euclidean theory of vision, for he remarks:

> Everything which is observed or seen is seen only through a ray which pierces the air and every transparent body [going] from a point where it

> touches our sight up to a point where it falls on the object seen. . . . Direct rays are those which upon leaving the sight are extended in the straight line of sight until they arrive at the end [the object] and so are cut off.[64]

However, in his *The Model State* we see traces of an alternative view, which appears to be predominantly Aristotelian in its inspiration. Al-Farabi writes that

> vision is a potency and a disposition in matter. Before seeing, it is only potential vision; and colors, before being seen, are visible only potentially. In the nature of the visual power that is in the eye, there is no aptitude to become vision in actuality; nor in the nature of colors an aptitude to be visible in actuality. The sun imparts to the eye a light that illuminates it, and to colors a light that illuminates them. Through the light that it receives from the sun, vision becomes seeing in actuality.[65]

Al-Farabi goes on to say that it is through the action of this external light, rather than the action of vision, that the light itself, as well as the sun from which it issues and all potentially visible things, becomes observed in actuality. He compares the relationship of the sun to vision with that of the agent intellect to the material or passive intellect, noting that it is the agent intellect that makes an impression on the material intellect and thereby raises it to a state of actuality;[66] he thereby makes it quite clear that in his opinion the eye and the power of sight participate passively in the visual process.[67]

Avicenna's Defense of the Aristotelian Theory of Vision[68]

The attack on the extramission theory of vision reached full flower in the work of Avicenna (Abu 'Ali al-Husain ibn 'Abdullah ibn Sina, 980–1037). probably the most influential natural philosopher in all of Islamic history.[69] Born in the eastern regions of Islam, near Bukhara, Avicenna pursued an adventurous career as advisor, administrator, and physician for a variety of princes. Entirely self-taught after the age of about fourteen, he succeeded in mastering the whole corpus of Greek learning (he claims to have finished Ptolemy's *Almagest* before reaching age sixteen) and then proceeded to turn out his own extraordinary literary production amid administrative duties, while in prison, and during several longer periods of relative tranquility. He dealt with the theory of vision in a variety of still extant works, including the *Kitab al-Shifa* (*The Book of Healing*, known in the West as *Sufficientia*), *Kitab al-Najat* (*The Book of Deliverance*), *Maqala fi 'l-Nafs* (*Epistle* or *Compendium on the Soul*), *Danishnama* (*Book of Knowledge*), and *Kitab al-Qanun fi 'l-Tibb* (*Liber canonis* or *Canon of Medicine*).[70]

Before considering the content of these works, a word is necessary about their composition and hence their interrelationships. It appears to have been established that the *Compendium on the Soul* was Avicenna's first work.[71] The *Shifa* and *Liber canonis* were works of Avicenna's maturity and clearly represent his most complete thought on medical and natural subjects.[72] Finally, the *Najat* and *Danishnama* were abridgments of longer works (the *Najat* an abridgment of the *Shifa*) prepared toward the end of his scholarly career; their importance is that they represent Avicenna's own view of what was essential to the longer works.

Even the briefest perusal of Avicenna's writings on vision reveals his Aristotelian sympathies. However, he devoted his major effort not to defending the Aristotelian theory, but rather to refuting the alternatives—the extramission theory in its various forms. The most general statement of the theory to be refuted is simply that sight occurs through the emission of a power or ray from the eye. The theory might be defended in this general, simplified form by noting that an object in direct contact with the eye cannot be observed and therefore that separation between the eye and its object is an indispensable condition of sight. The same can be said of any substantial emanation from the object (because it too is an object), and the accidents of the object (e.g., its color and shape) cannot be transported without matter.[73] Therefore only one alternative remains: "the sensitive power must extend to the place of the perceived object to encounter it."[74]

But how can the sensitive power be extended to the visible object? Here we must qualify the general statement that sight occurs through the emission of a ray by defining the nature of the ray; we thereby elaborate the extramission theory into its particular versions—the Euclidean, Galenic, and so forth. Avicenna attempts this process of elaboration in each of the works where he treats vision, but unfortunately with little consistency, and no two works contain identical schemata; even the *Najat* and the *Shifa*, the one conceived as an abridgment of the other, are not organized along identical lines.[75] However, if we look beyond the organization of Avicenna's arguments to their content, it becomes apparent that all the theories to which Avicenna addresses himself fall loosely into two main classes, covered by the rubrics "Euclidean" and "Galenic."[76] I propose, then, to recount Avicenna's refutation of these two versions of the extramission theory.[77]

The term "Euclidean," which I have used to denote the first version of the extramission theory, is to be broadly construed to mean that "something issues from the eye, meets the object of sight, takes its form from without—and that this constitutes the act of seeing."[78] Moreover, it is essential to this theory, in Avicenna's view, that the ray issuing from the eye be a material substance, because "the sensitive power cannot be transported except by the mediation of body."[79] But in order to disprove the Euclidean theory, Avicenna finds that he must further divide it into four subcategories, each separately refutable: (1) The radial corporeal substance emanating from the eye constitutes a single homogeneous conical body, which is in contact with the entire visible object and also with the observer's eye. (2) That which issues from the eye is a continuous substance, which makes contact with the entire visible object but loses contact with the observer's eye. (3) The substance issuing from the eye consists of separate rays or parts, not in mutual contact. These rays thus touch only certain portions of the visible object. (4) The radial corporeal substance does not make contact with the visible object at all.[80]

Avicenna rejects the first version of the Euclidean theory as absurd, because it suggests that from something as small as the eye can emerge a continuous substance large enough to fill a hemisphere of the world: "there will have emerged from the eye, despite its smallness, a conical body of immense size, which will have compressed the air and repulsed all the heavenly bodies, or [else] it will have traversed an empty space."[81] Moreover, this process must be repeated each time the eyes are opened; that is, either the radial substance must repeatedly issue forth and return, or else a new ray must be sent forth each time the eyes are opened. A further difficulty of this version of the Euclidean theory is that the extension of a continuous radial body as far as the fixed stars would require that the intervening air (and even the celestial spheres) be swept out of the way, unless, of course, this entire space were void; but to Avicenna neither possibility is conceivable.[82]

Avicenna concludes his assault on the first version of the Euclidean theory by arguing that if the visible object were perceived by a conical body issuing from the eye, then the remoteness of the visible object should not affect its apparent size and shape, and the size of the angle intercepted by the visible object at the eye would be irrelevant.[83] In his opinion, the extramission theory of Euclid entails that the power of perception is located in the base of the visual cone and therefore perceives the magnitude of the visible object by contact;[84] consequently, if the extramission theory were true, the actual magnitude of a remote object should be directly perceived, and the laws of perspective would not apply. This argument, which strikes at the very foundations of the mathematical theory of vision, would surely have surprised Euclid, al-Kindi, Alhazen, and other practitioners of the mathematical approach, for their position was precisely the opposite, namely, that only an extramission theory (through its visual cone) makes the perception of magnitude intelligible.[85] This radical difference in interpretation can be understood only by recognizing that the real issue between Avicenna and the mathematicians was over the criteria by which to evaluate a theory of vision. Their position was that only the extramission theory (with its visual cone) can make sense of visual perception *mathematically*. His position was that the mathematics of the visual cone is made irrelevant and inapplicable by the fact that the visual power is not, according to the version of the extramission theory under consideration, fixed in the eye, where it can perceive the angle between rays touching the extremes of the object, but in the base of the visual cone, where it is able to acquire immediate knowledge of the object's true size; thus a *physical* understanding of the extramission theory reveals its inability to explain such facts of perception as the diminution of objects with distance from the observer.

The second and fourth versions of the Euclidean theory are both vulnerable to the same objection and can therefore be treated together. If that which issues from the observer's eye should lose contact with the eye or fail to make contact with the visible object, what purpose could it serve, since the whole point of the extramission theory is to establish contact between the observer and the visible object?[86] The only salvation for either version of the theory is to suppose that contact is established through a medium (the air) that transmits impressions or forms from the object to the far end of the ray or from the near end of the ray to the eye. But if air has this property, "why does it not return [the form] to the pupil and avoid the labor of spirit issuing into the air?"[87] The second and fourth versions of the Euclidean theory are thus unacceptable because they are redundant.

The remaining version of the Euclidean theory, the third, which maintains that the substance issuing from the eye consists of discrete rays in contact with both the eye and the visible object, is the closest of Avicenna's four versions to Euclid's own teaching. This version of the theory escapes the troublesome problem of a continuous body half as large as the universe emanating from the eye, since the rays can be regarded as diffused to whatever extent is required to explain their increased volume. However, there is also a most serious difficulty: rays perceive only what they encounter (for encounter between the visual power and the object is the essence of perception), which means that the observer "will perceive the spots where that ray falls to the exclusion of the spots where it does not fall, so that he will only partially perceive the body, sensing some points here and there but missing the major part."[88] The difficulty can be escaped only by supposing that the rays make use of the medium that occupies the spaces between rays, endowing it with the power of perception by changing it into their own nature and "becoming [with it] as one thing."[89] However, this maneuver has serious difficulties of its own:

> What would we say concerning the heaven when we observe it? Can we say that the heaven is changed into the nature of the issuing ray so as to be sentient with it, as though they are one thing, so that it encounters Saturn and sees the whole of it, and also Jupiter and the other great stars? The falsity of this is obvious.[90]

If one argues that the medium is not united with the ray (to become sentient), but only endowed with the power to return forms to the ray, then of what use are visual rays in the first place? Why not suppose that the medium is transformed directly from the surface of the eye and returns forms to the eye without the mediation of visual rays? Moreover, the air intermediate between two rays would necessarily return the same form to each of them, so that the same object would be perceived twice.[91]

A second argument against the theory of discrete visual rays (this third version of the Euclidean theory) is derived from the nature of matter, both celestial and terrestrial. It is known that void is nonexistent and therefore that the celestial spheres are without vacuities or pores. It is therefore impossible for them to be penetrated by rays, and if vision were to occur through the passage of material rays from the observer to the object there would be no way of seeing the celestial bodies beyond them. A similar problem arises in vision through a body of water. On the theory that discrete rays are emitted from the eye, how is it possible to see the earth in a continuous fashion beneath a body of water, since there is no void?[92] If rays force their way into the water, creating passages where previously there were none, why doesn't the bulk of water increase because of the addition of the radial corporeal substance? Moreover,

> if there were void there [in the water], how great would the magnitude of its vacuities have to be in view of the fact that water is a heavy body, which [naturally] descends into vacuities and fills them? It is evident that the whole water would have to be vacuous, or its greater part, or [at least] half, so that this issuing substance could penetrate to the whole of that which is beneath it.[93]

Avicenna thus concludes the first and most extensive phase of his refutation of the extramission theory. By demonstrating the absurdity of each of the four possible versions of the Euclidean theory, he claims to prove that vision cannot occur through the emission of a substance that proceeds from the observer's eye to the object of sight.

In the second phase of his refutation, Avicenna disputes the other major form of the extramission theory of vision—the Galenic theory, according to which the ray issuing from the eye does not itself perceive the visible object, but uses the intervening air (and any other transparent media present) as its instrument.[94] Avicenna's fullest refutation of this theory appears in the *Shifa*, where he points out that the air may become the instrument of the eye in either of two senses: it can either become a true optical medium, capable of transmitting visual impressions to the eye, or be converted into a visual organ, percipient in itself.[95] Before pursuing these alternatives, however, Avicenna presents what he calls his general judgment or universal proposition: that in either case the air cannot acquire a new disposition or state whereby it comes to possess a certain quality or property in itself. This is impossible because a state (even a temporary one, which exists only so long as its efficient cause[96] continues to act) would exist in relation to all observers, whereas vision is clearly an individual phenomenon. For example, if the effect of the visual power were a new state of the medium, it would follow

> that weak-sighted people would see better when they congregate . . . and that a man of weak sight would see better when he is near another man whose vision is stronger. . . . However, we observe that a weak-sighted man

> is not aided at all in his sight by congregating with those who see better or by joining many people of weak sight. Therefore it is evident that this [opinion] is false.[97]

To return to our two alternatives, then, the effect of the visual power must be to convert the air into a medium for transmitting visual impressions (though only in relation to one observer) or else to convert the air into a sentient organ of sight.[98] The latter alternative is impossible, Avicenna points out in the *Najat*, because if the air itself were sentient, then any disturbance of the air would necessarily distort vision;[99] furthermore, the laws of perspective would not hold because the sentient power would be in direct contact with the visible object.[100] In the *Shifa* he argues that it would be absurd to maintain

> that air is altered so as to possess sensibility, so that it would perceive the fixed stars and return to sight what it has perceived. Besides, air does not touch everything that is seen, for we see the fixed stars, which air does not touch; and it would be absurd to say that the heavens, which are between [us and the fixed stars] are affected by our sight and become its instrument.[101]

Nor can it be reasonably maintained, as an alternative version of the theory, that light is a body dispersed throughout the air and the heavens, which is united to our eyes and becomes their instrument; for this would entail that the heavens contain pores through which the light passes,[102] which in turn would entail that only parts of the stars are visible. Finally, air and light are not conjoined to only one observer; why then would they return their perceptions to one observer and not another?[103]

The other alternative, that the visual power perfects the air as a medium, cannot stand either. What kind of affection could air receive from sight so as to become capable of transmitting impressions (and, what is more, transmitting them only to the observer whose visual power produced the transformation in the air)? Surely air cannot receive "the power of life," since it remains a simple element.[104] Nor is it reasonable to suppose that sight renders the air transparent in actuality, since the sun is much more efficacious in that regard. Sight cannot result from heating or cooling of the air, for if it did, the contrary effect (as a result of the presence of other hot or cold bodies) would terminate vision.[105] Perhaps, then, sight produces some quality that lacks a name; but "how, then, could proponents of this view know of it, and how did they apprehend it?"[106] Moreover, as Avicenna points out, he has already refuted (in his "general judgment") the possibility of changes that constitute a state or disposition of the medium. The conclusion then seems inescapable: if the air is actually transparent, if colors exist in actuality, and if the eye is healthy, nothing more is required for the occurrence of sight. As Avicenna expresses this conclusion in the *Danishnama*, "since the air itself is in contact with the eye, it goes without saying that it transmits [the image] to the eye, and there is no need for a ray to issue [from the eye]."[107] In the final analysis, the Galenic theory must be rejected because it is redundant.

The true theory of vision, in Avicenna's opinion, is the Aristotelian. In the *Shifa* he outlines its essential elements:

> Just as other sensibles [than color and light] are not perceived because something extends from the senses to them and encounters them or is joined to them or sends a messenger to them, so vision does not occur because a ray issues forth in some way to encounter the visible object, but because the form of the thing seen comes to sight, transmitted by a transparent medium.[108]

In the *Najat* Avicenna writes:

> But true philosophers hold the view that when an actually transparent body, i.e. a body which has no colour, intervenes between the eye and the object of sight, the exterior form of the coloured body on which light is falling is transmitted to the pupil of the eye and so the eye perceives it. This transmission is similar to the transmission of colours by means of light being refracted from a coloured thing and giving its colour to another body. The resemblance is not complete, however, for the former is more like an image in a mirror.[109]

The theory outlined in these passages is identical in its principal features to that of Aristotle, who wrote in his *De anima*: "Colour moves the transparent medium, e.g., the air, and this, being continuous, acts upon the sense organ.... For vision occurs when the sensitive faculty is acted upon; as it cannot be acted upon by the actual colour which is seen, there only remains the medium to act on it."[110] If Avicenna's account of his own theory seems unduly economical, the reason is that Aristotle had been equally terse.

But Avicenna goes beyond the Aristotelian theory in several respects. In the passage quoted above from the *Najat*, Avicenna compares sight to image-formation in a mirror. He enlarges on this point in the *Danishnama*, asserting that

> the eye is like a mirror, and the visible object is like the thing reflected in the mirror by the mediation of air or another transparent body; and when light falls on the visible object, it projects the image of the object onto the eye.... If a mirror should possess a soul, it would see the image that is formed on it.[111]

When the mirror image is formed in the eye, it is received by the crystalline humor, which transmits it ultimately to the seat of visual power in the hollow nerve.[112]

The chief advantage of this theory is that it permits one to understand how remoteness of the visible object affects the perception of its size. Avicenna gives the following geometrical example in the *Danishnama*. An object situated at *HD* (fig. 6) will cast an image at *AB* on the surface of the eye (represented by the circle); if the same object is moved to position *KZ*, the image will be restricted to *TY*, smaller than *AB*. "And everything formed on a smaller arc is also seen smaller; therefore the image at *ZK* is smaller."[113] Avicenna concludes his discussion of the theory of vision by mirroring with the following remark: "It is strange that the people who defend the theory of rays [emanating from the eye] also speak of the angle [formed in the eye by the visible object]; for the angle is of use when [one judges that] the image comes toward the eye, but not when [one supposes that] sight advances toward the image."[114] Thus Avicenna maintains that only the intromission theory of Aristotle is consistent with the geometrical facts embodied in the conception of the visual cone or pyramid, and he thereby purports to demonstrate the validity of the Aristotelian theory of vision. It is remarkable to see him attempting to turn the tables on proponents of the extramission theory by claiming that only his theory can account for the geometrical features of vision that had been associated since antiquity with their theory of visual rays.[115]

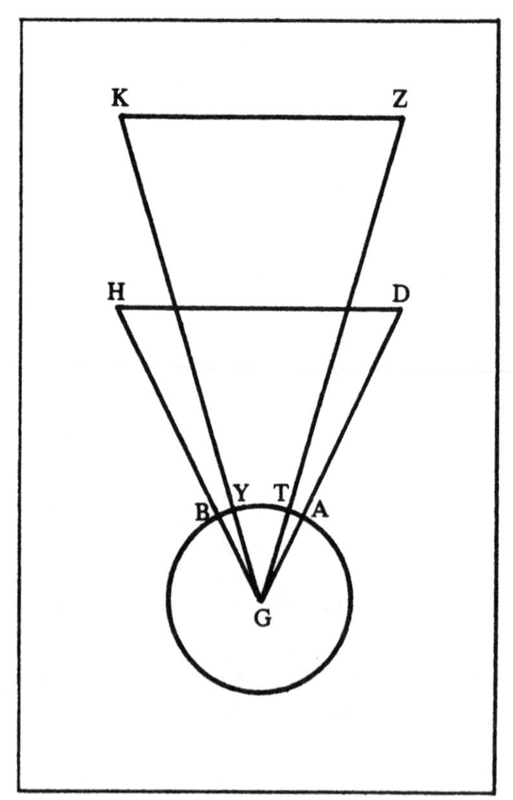

Fig. 6

In describing Avicenna's theory of vision, I have not yet mentioned the most influential of his works (in the West, at least), the *Liber canonis*.[116] The reason for this omission is that the optical portions of the *Canon* are devoted almost entirely to ophthalmology and deal only incidentally with theoretical problems of vision. However, in the tradition of Hunain, the section on the eye opens with a chapter on the anatomy of the eye, which begins as follows:

> The power of sight and the matter of the visual spirit proceed through the two hollow nerves about which anatomy has taught you and reaches the eye. When the nerves and their coats reach the orbits [of the eyes], the extremity of each is dilated and enlarged so that it encloses the humors of the eye, the central of which is the glacial [or crystalline] humor, which is as clear as a hailstone. The glacial humor is round, but its roundness is diminished in front by compression; and it is compressed so that the image formed on it might be of more suitable size and so that small objects that are observed might have a larger place on which to form [their images].[117]

This passage is followed by a traditional description of the other tunics and humors of the eye. What is important about these passages from the *Liber canonis* is that nothing in them suggests a departure from the Galenic theory of vision typically found in the medical tradition.[118] Indeed, the entire quotation might have come directly from Hunain's *Ten Treatises* or 'Ali ibn 'Isa's *Notebook for Oculists*. Visual spirit proceeds through the hollow nerves to the eye, bearing the power of sight. Even the emphasis on the glacial or crystalline humor as the organ on which images are formed has its counterpart in Hunain's statement that the "flattened form [of the crystalline humor] enables it to receive the impressions of more perceptible objects than would be the case if it were perfectly round" and 'Ali ibn 'Isa's claim that the visual light returns to the eye "and is stamped on the glacial humor."[119] This is not to suggest that anything in the *Liber canonis* violates the Aristotelian teaching of the *Shifa* or the short compendia, but only to point out that no traces of Avicenna's polemic against the extramission theory are evident in the *Liber canonis*.

In conclusion, we must inquire how successfully Avicenna has combatted the extramission theory of vision. The "Euclidean theory," in Euclid's hands, was largely a mathematical theory of vision, effective as long as physical issues were set aside.[120] Although Ptolemy and al-Kindi invested the theory with additional physical content, it remained primarily a mathematical theory, whose success was to be judged by mathematical criteria.[121] By contrast, Avicenna ignores the mathematics of the Euclidean theory and makes a series of devastating physical points. He argues that if the rays remain in contact with the observer's eye, they must either form a continuous body and fill all of the space up to the fixed stars (an obvious physical impossibility) or separate and thus achieve a spotted impression of the visible object (an apparent observational falsehood); if they do not remain in contact with the observer's eye, they are useless in explaining perception. The theory of a material substance emanating from the eye is thus inconsistent both with the medieval (as well as modern) conception of material substance and with the universally accepted facts of visual perception. To this argument there is no reply.[122] Although he has not discredited the mathematical utility of the Euclidean theory of visual rays, Avicenna has adequately established that it does not represent the physical nature of the visual process.

Avicenna presents an equally convincing refutation of the Galenic theory. The claim that air becomes a sentient instrument of vision is untenable because it cannot explain how one manages to see clearly when the wind is blowing, or how the same air can serve simultaneously as the visual instrument for a whole crowd of observers (To which one will the air return its perceptions?), or what kind of change occurs to make the air an instrument of vision. Nor is there escape in claiming that the air is not percipient, but only a medium for the transmission of visual impressions. This maneuver not only leaves unanswered the question about what kind of change could occur in the medium to give it the power of transmitting impressions to the observer (and only to the one observer whose visual power provoked the transformation of the medium), but makes the whole Galenic theory redundant, for it is perfectly obvious (or was to an Aristotelian, at any rate) that the medium is capable of returning impressions to sight without the issuance of visual rays. With a few strokes, Avicenna has deftly destroyed both the Euclidean and Galenic versions of the extramission theory as viable physical explanations of the process of sight. The Aristotelian theory was the only remaining alternative.

Averroes and the Aristotelian Theory of Vision

The Aristotelian theory of vision, restored to a position of prominence (if not leadership) by Avicenna, was further cultivated by Averroes (Abu-l-Walid Muhammad ibn Rushd, 1126–98).[123] Averroes, the most prolific and influential of Aristotle's medieval commentators, wrote commentaries on all the Aristotelian works available to him. Twenty-six of these are extant, including an epitome of the *Parva naturalia* (containing *De sensu et sensato*) and an epitome, middle commentary (i.e., of intermediate length), and long commentary on *De anima*.[124] Since the epitomes were not commentaries in the strict sense but summaries of Aristotle's teaching on a given subject, amplified, clarified, and even corrected when necessary, they offered Averroes considerable freedom to express his own views. Consequently, I will concentrate on the *Epitome of the Parva naturalia* (which contains a concise summary of Averroes's theory of vision and has the additional advantage of having had a Western influence during the Middle Ages) as well as Averroes's medical encyclopedia, the *Kitab al-Kulliyat* (or *Colliget*).[125]

In view of the common stereotypes regarding Averroes's slavish adherence to the letter of the Aristotelian text, we must note that the *Epitome of the Parva naturalia* amply reveals Averroes's independence. Most frequently his departures consist of relatively minor extensions of the Aristotelian theory, but on occasion they represent unmistakable corrections. The first kind of innovation is exemplified in Averroes's reference to visual pneuma, never mentioned by Aristotle, to explain the transmission of visual power from the brain to the eye; the second kind is seen in Averroes's claim that the common sense, which Aristotle had located in the heart, resides "behind the retina."[126] In both cases, the innovation represents Galenic influence, and it is essential to recognize that Averroes (like Avicenna) frequently bends the Aristotelian theory in a Galenic direction.

It remains, however, that Averroes's theory of vision is principally Aristotelian. Averroes summarizes his view in the *Epitome of the Parva naturalia* as follows:

> We maintain that the air, by means of light, receives the forms of objects first and then conveys them to the external coat of the eye, and the external coat conveys them to the remaining coats, until the movement reaches the innermost coat behind which the common sense is located, and the latter perceives the form of the object.[127]

Against the alternative views, he repeats not only several of Aristotle's arguments, but also a number of others that had been developed in the intervening fifteen centuries. The Platonic theory that the forms of sensible objects already exist in the soul and need only be called forth by the memory when stimulated by external objects is untenable because it would then be possible for the soul to recall the forms even without the external object; moreover, this theory makes the sensory organs useless, and nature does not create in vain.[128] The atomists maintain that forms are imprinted on the soul through the influx of corpuscles, but "the absurdity of their view can be demonstrated by the fact that the soul can receive the forms of contraries at the same time, whereas if they were bodies, this would be impossible."[129] Furthermore, it is evident from our ability to see bodies much larger than the eye (so large that corpuscles emanating from all parts of them could not conceivably fit into the eye simultaneously) that the colors of bodies are conveyed to the eye not materially, but spiritually.[130]

Averroes refutes the extramission theory of Euclid, Ptolemy, Galen, and al-Kindi by arguing that this theory implies the ability to see in the dark and that if the soul issues forth to perceive objects it should perceive far and near bodies with equal acuity, contrary to fact.[131] He also argues that what issues forth must be either body or light. If it is body, time would be required for the corpuscles to reach a remote object, and it is absurd in any case to suppose that body can issue from the eye to fill a hemisphere of the world; in addition, this body would have to serve as a substrate for sight, but the only animal body in which soul or its faculties can inhere is natural heat, which would be cooled (and thus extinguished) if it emerged from the eye.[132] If that which issues from the eye is light, it could not carry the soul with it, because only substance can serve as the substrate for the soul; therefore the soul must remain behind in the eye, which means that impressions must be sent back to it through a material medium (which cannot be light, since light is not material), and that leaves us with what amounts to the Aristotelian theory.[133]

Many other passages in this and other works reveal Averroes's basic Aristotelian loyalties, but of more interest are his comments on aspects of the visual process not treated by Aristotle. The most important of these is the reception of forms in the eye and their subsequent transmission to the seat of consciousness in the brain. It has been maintained by Fukala and Koelbing that Averroes identified the retina, rather than the crystalline humor, as the basic photosensitive organ of the eye and that Platter and Kepler (late in the sixteenth and early in the seventeenth century) are to be credited only with rediscovering this fact,[134] and there are indeed passages in Averroes's works that might seem to support such a view. In the *Epitome of the Parva naturalia*, Averroes writes:

> The innermost of the coats of the eye [i.e., the retina] must necessarily receive the light from the humors of the eye, just as the humors receive the light from the air. However, inasmuch as the perceptive faculty resides in the region of this coat of the eye, in the part which is connected with the cranium and not in the part facing the air, these coats, that is to say, the curtains of the eye, therefore protect the faculty of the sense by virtue of the fact that they are situated in the middle between the faculty and the air.[135]

Later in the same treatise, Averroes comments that "this water [i.e., the vitreous humor], which is conjoined to the innermost curtain [i.e., the retina], is the final part of the eye, and through it the common sense receives the forms coming to it from outside."[136] Finally, in the *Colliget* he writes:

> And you know that the sense of sight receives the forms of things in this manner. First air, when light mediates, receives the forms of things and transmits them to the anterior tunic [i.e., the cornea], which conveys them to the other tunics until this motion reaches the final tunic [the retina], behind which is situated the common sense, which apprehends the forms.[137]

Each of these passages attributes to the retina an important role in the reception of the forms of visible objects, and by themselves they might lead one to suppose that it is the principal role.

To understand Averroes's true position, we must take into account several other passages, where Averroes attaches special importance to the crystalline humor in transmitting forms from the visible objects to the common sense. In the *Epitome of the Parva naturalia*, he writes

> In the middle of these coats [i.e., in the center of the eye] lies the crystalline coat [the crystalline humor], which is like a mirror, partaking equally of the nature of air and of the nature of water. This coat, therefore, receives the forms from the air, since it is like a mirror, and it conveys them to the water [the vitreous humor].[138]

Similarly in the *Colliget*, Averroes argues that

> in the middle of these tunics is the glacial coat [the crystalline humor], which is like a mirror, a mean between the natures of air and water, and therefore it receives the form of the air, because it is like a mirror, and it conveys the forms to the water [the vitreous humor] because its nature participates in those two natures. . . . And thus the common sense receives the form.[139]

In yet another passage Averroes makes it quite clear that the crystalline humor, rather than the retina, is the *primary* photosensitive organ: "And it appears that the proper instrument of that sense [sight] is the round humor, called the crystalline or glacial humor, or else the fine spider's web that is placed over it."[140] The "spider's web (*reticula aranea*)" is a hypothetical membrane covering the anterior capsule of the crystalline;[141] thus both alternatives amount to very nearly the same thing, and it remains that the crystalline humor or its covering is the principal instrument for receiving the forms of outside objects.[142] The case is clinched when we note that the functions Averroes assigns to the retina in the chapter of the *Colliget* on the parts of the eye and their functions do not include the reception or perception of visible forms:

> But the first function of the retina is to transport the visual spirit through its nerves . . . and through the two nerves [the optic nerves] that go [from the brain] to the eyes. The retina also nourishes the crystalline humor in the manner of dew[143] and provides natural heat through its arteries.[144]

It is true that the *Colliget* is more explicit on retinal function than is the *Epitome*, and I have been forced to quote from the former in order to establish the finer details of what I take to be the genuine Averroistic theory; but since the two works contain perfectly compatible descriptions of vision, there is no reason whatsoever to doubt that, although differing in completeness, they were intended to express a single theory.

Thus far I have argued solely on the basis of evidence internal to Averroes's writings, but the case can be made still stronger by reference to the tradition that lies behind them. In book 10 of *De usu partium* Galen, after arguing explicitly that "the crystalline humor . . . is the principal instrument of vision," discusses the retina, noting that "its principal and greatest usefulness, that for the sake of which it was brought down from above, is to perceive the alterations of the crystalline humor and in addition to convey and transmit nutriment to the vitreous humor."[145] Now this passage says as much (and, indeed, the same things) about retinal sensitivity as any passage that can be found in the writings of Averroes, and it is apparent that if we wish to maintain that Averroes regarded the retina as the primary sensitive organ, then we must admit that so too did Galen.

What then did Galen mean by his claims concerning retinal sensitivity? According to Galen, not only is visual power propagated outward to the visible object through the optic nerve, eye, and other transparent media, but perceptions or impressions are in some way returned to the brain along the same route; there is a chain of media, conducting visual power to the observed object and visual impressions back to the brain.[146] Now in Galen (and consequently in many of his followers), there was considerable ambiguity about the functions of particular organs involved in this visual transmission. If there is a continuous chain of instruments between the observer's brain and the visible object, is it possible for particular instruments (the crystalline humor or the retina) to have unique functions? Although the matter was never fully clarified, it is clear that on certain occasions Galen and his followers did assign a particular function, and sometimes a particularly important function, to one or another of the organs involved in the transmission process. It is in this context, then, that one must understand Galen's claim that the "principal and greatest usefulness" of the retina "is to perceive the alterations of the crystalline humor." But in weighing this claim, we must recall that Galen also mentions the sensitivity of the crystalline humor and, indeed, maintains that the crystalline is the *principal* sensitive organ: "I have . . . said that the crystalline humor itself is the principal instrument of vision, a fact clearly proved by what physicians call cataracts, which lie between the crystalline humor and the cornea and interfere with vision until they are couched."[147]

Thus Galen and his medieval followers could speak of retinal sensitivity—that is, they could assign the retina a significant function in transmitting visual impressions from the visible object to the brain—without regarding the retina as the principal instrument of vision.[148] It is this—no more and no less—that Averroes intended when he stated in the *Colliget* and the *Epitome* that the retina receives light from the humors of the eye. The retina, lying at the rear of the eye, necessarily participates in transmitting visible forms to the common sense, which resides behind it; it may even mark the termination of the optical phase of this transmission.[149] But this does not mean that the retina is the principal sensitive organ of the eye—a function that Averroes, no less than Galen and Hunain, reserved for the crystalline humor.

Conclusion

The principal significance of the revival of the Galenic and Aristotelian theories of vision in Islam is that it represents the assertion of the anatomical and physiological concerns of the physicians and the physical and psychological concerns of the natural philosophers against the mathematical preoccupations of the Euclideans. The trichotomy is far from complete, of course; interests inevitably overlapped, and it was not uncommon for the physician, the natural philosopher, and the mathematician to be wrapped up in the same person. Moreover, the Euclidean theory of vision (especially in its Ptolemaic and al-Kindian forms) was not devoid of physical content, nor were the Galenic and Aristotelian theories devoid of mathematical content. Furthermore, as I have argued, both Euclideans and Aristotelians incorporated Galenic elements into their optical thought. Nevertheless, this threefold classification defines the principal battle lines among medieval theories of vision. I would not minimize the importance of other issues, such as the intromission-extramission controversy, but the dispute among physicians, natural philosophers, and mathematicians was the more fundamental in that it brought into question the very aims of optical theory.

The nonmathematical orientation of the Galenic and Aristotelian theories of vision should be emphasized. In them we find no geometrical analysis of any significance, no science of catoptrics or dioptrics, hardly more than a mention of the rectilinear propagation of light.[150] In place of the geometrical achievements of Euclid and Ptolemy, we find a discussion of ocular anatomy and physiology and an analysis of physically possible modes of radiation. We are here able to perceive the chasm separating various approaches to the problem of sight. The Euclidean theory on the one hand and the Galenic and Aristotelian theories on the other are simply incommensurable, because they have fundamentally different aims. It was the achievement of Avicenna's contemporary, Alhazen, to demonstrate how the anatomical, the physical, and the mathematical could all be integrated into a single theory of vision.

4 ALHAZEN AND THE NEW INTROMISSION THEORY OF VISION

THE INTROMISSION THEORY BEFORE ALHAZEN

Alhazen was undoubtedly the most significant figure in the history of optics between antiquity and the seventeenth century. But if we are to appreciate his achievement, we must first grasp the state of the intromission theory before his work.

One version of this theory had been formulated by the atomists, who argued that collections of atoms, likened by Lucretius to the skin of a snake or cicada, issue in all directions from all objects and enter the eyes of observers to produce visual sensation.[1] The essential feature of this theory is that the atoms streaming in various directions from a particular object form coherent units—films or *simulacra*—which communicate the shape and color of the object to the soul of an observer; encountering the *simulacrum* of an object is, as far as the soul is concerned, equivalent to encountering the object itself.

The other version of the intromission theory was that of Aristotle, who maintained that no corpuscular emanation from bodies can be responsible for sight, but that colored bodies produce qualitative changes in a transparent medium (actualized by the presence of a luminous body), and that these qualitative changes are instantaneously transmitted to the observer, where they give rise to visual sensation.[2] Whereas in the atomistic theory mechanical contact is established between the observer and the observed, Aristotle proposes that "the sentient object . . . is potentially such as the object of sense is actually" and that "at the end of the process [of being acted upon by the object] it has become like that object, and shares its quality."[3]

But both versions of the intromission theory are plainly inadequate. Both succeed in identifying a link between object and observer, but neither satisfactorily explains how that link can account for the facts of visual perception. One of the major difficulties in the atomistic theory is its requirement that the material image of an object larger than the eye (which might even be as large as a mountain) shrink sufficiently to enter the observer's eye; even if the physics of this shrinkage could be satisfactorily worked out, there is no way on mechanistic principles to explain how the amount of shrinkage is perfectly correlated with the remoteness of the observer so that the laws of perspective apply. The Aristotelian theory sidesteps the problem of shrinkage, but substitutes problems of equal severity. For example, how, on the Aristotelian theory, can individual parts of the visual field be distinguished? If colored bodies produce qualitative changes in all parts of a transparent medium to which they have rectilinear access, how do we see one object here and another object over there? What is the nature or source of the directional capabilities of sight? Aristotle's theory of vision provides no answer to these questions. Either the watery part of the eye, which Aristotle identifies as its sensitive organ,[4] must be affected by qualitative changes produced by all objects in the visual field, or else the various objects (or parts of objects) in the visual field must compete for the right to introduce their particular qualitative change into the eye; in either case, there is no mechanism that can explain how we perceive the infinite variety of the visual field or how we localize its various features. In short, if sense "becomes the sensible object," which of the objects in the visual field shall it become?[5]

It should be apparent, then, that the intromission theory, in either its atomistic or its Aristotelian form, cannot accomplish all that is required of a theory of vision. Hunain ibn Ishaq expressed the dissatisfaction of many when he wrote of the intromission theory:

> All people acknowledge and agree that we see only by the hole which is in the pupil. Now, if this hole had to wait until something coming from the seen object reached it, or a power emanating from it, or a form, an outline or a quality, as some people maintain, we should not know, in looking at an object, either its extent or its volume, whether it were, for example, a very high mountain, or the like. . . . Its entering into the eyes is something which reason does not comprehend and of which nobody has ever heard, for according to this hypothesis a complete form or outline of the viewed object would necessarily reach and enter into the eye of the beholder at the same moment. Supposing then that a great many people looked at it, say, for example, ten thousand persons, it would have to return to the eye of every one of them, and its form and outline would have to enter completely into them. But this is very far from probable and must therefore be ranked among the untenable hypotheses.[6]

Hunain was perfectly correct: the intromission theory of sight was indeed untenable. If Avicenna had demonstrated the absurdity of the alternatives—the extramission theories of Euclid and Galen—that did not establish the viability of the intromission theory, but only suggested that no satisfactory theory had yet been devised. One may object, however, that dissatisfaction with the intromission theory was based on an insufficient understanding of its possibilities, which were simply not revealed by the atomistic and Aristotelian versions of it. Obviously any theory that treats the visible object as a unit, from which issues a material film or a qualitative alteration of the medium, rather than as a collection of individual points or small areas, each of which sends forth its own ray, cannot hope to succeed. Quite true; but the point is that before Alhazen, the intromission theory *was* the theory of coherent images or forms.[7] Alhazen was the first to utilize the analysis of the visible object into point sources, each of which sends forth its ray, as the basis of an intromission theory of vision.[8] If such a step seems trivial today, that is because we are Alhazen's intellectual progeny.

However, Alhazen's achievement was not simply to submit the visible object to punctiform analysis, but also (as I have pointed out above and hope to demonstrate below) to integrate into a single (and highly successful) theory the mathematical, anatomical, and physical approaches to sight.[9]

Alhazen's Life and Works

Abu 'Ali al-Hasan ibn al-Hasan ibn al-Haytham (known in medieval Europe as Alhazen or Alhacen) was born in Basra about 965 A.D.[10] The little we know of his life comes from the biobibliographical sketches of Ibn al-Qifti and Ibn Abi Usaibi'a, who report that Alhazen was summoned to Egypt by the Fatimid Khalif, al-Hakim (996–1021), who had heard of Alhazen's great learning and of his boast that he knew how to regulate the flow of the Nile River.[11] Although his scheme for regulating the Nile proved unworkable, Alhazen remained in Egypt for the rest of his life, patronized by al-Hakim (and, for a time, feigning madness in order to be free of his patron). He died in Cairo in 1039 or shortly after.

Alhazen was a prolific writer on all aspects of science and natural philosophy. More than two hundred works are atrributed to him by Ibn Abi Usaibi'a, including ninety of which Alhazen himself acknowledged authorship. The latter group, whose authenticity is beyond question, includes commentaries on Euclid's *Elements* and Ptolemy's *Almagest*, an analysis of the optical works of Euclid and Ptolemy, a resume of the *Conics* of Apollonius of Perga, and analyses of Aristotle's *Physics, De anima,* and *Meteorologica*. Other books attributed to Alhazen treat physics, mathematics, astronomy, cosmology, meteorology, optics, medicine, metaphysics, and theology.[12]

Alhazen's extant works on the subject of optics are the following: (1) *Kitab al-Manazir* (*Book of Optics*), translated into Latin as *De aspectibus* or *Perspectiva,*[13] (2) *On the Paraboloidal Burning Mirror*, translated into Latin as *De speculis comburentibus,*[14] (3) *On the Spherical Burning Mirror,* (4) *On the Burning Sphere*, (5) *On Light,*[15] (6) *On the Rainbow and Halo*, (7) *On the Nature of Shadows*, (8) *On the Form of the Eclipse*, which deals with the theory of radiation through apertures, (9) *On the Light of the Moon*, and (10) *On the Light of the Stars*. Finally, a section on optics appears in Alhazen's (11) *Doubts concerning Ptolemy*. Another work frequently attributed to Alhazen, *On the Twilight* (translated into Latin as *De crepusculis*) was actually written by Abu 'Abd Allah Muhammad ibn Mu'adh.[16] Among those works of Alhazen that are, so far as is known, no longer extant, we find the following that bear on optics or the visual process:[17] (12) an analysis of the optical knowledge of Euclid and Ptolemy, (13) *On the Nature of Sight and the Manner in Which*

Sight Occurs, (14) *On Optics according to the Method of Ptolemy,* (15) an analysis of Aristotle's *De anima,* (16) an analysis of Aristotle's *Meteorologica,* and (17) *On the Perfection of the Art of Medicine*, containing chapters on diseases of the eye, on the usefulness of the parts of the body (probably based on Galen's *Usefulness of Parts*), and on the opinions of Hippocrates and Plato (probably based on Galen's *De placitis Hippocratis et Platonis*).[18] What must impress one about this list of works is Alhazen's remarkable scope. He wrote on every aspect of the preceding optical tradition—on two of Aristotle's works, on two of Galen's, and on the works of Euclid and Ptolemy, on geometrical optics, meteorology, psychology, and ophthalmology. It is a pity we lack so many of these works, but even a glance at the titles makes it apparent that Alhazen, perhaps more than any man before him, had the resources to develop a theory of vision along new lines—a theory that would incorporate elements from all the optical traditions of the past.

Besides the works thus far mentioned, what others did Alhazen have at his disposal? It is apparent from the titles of his own writings that he was widely read in Greek mathematics and mathematical science, for he wrote commentaries on Euclid's *Elements*, Apollonius of Perga's *Conics*, Archimedes' *On the Sphere and Cylinder*, and Ptolemy's *Almagest*. As for optical sources, the wide circulation of al-Kindi's *De aspectibus* and Hunain's *Ten Treatises* makes it probable that Alhazen was familiar with them. Moreover, the former contains theoretical contributions that were essential to Alhazen's new intromission theory, and I regard it as most unlikely that Alhazen arrived at these independently. It is ironic that Avicenna and Alhazen, two of the greatest Islamic writers on optics, should make their contributions contemporaneously. Few of their works can be dated precisely enough to allow us to determine the possible pattern of influence, though it appears that Avicenna wrote (or at least began) his principal work, the *Shifa,* some fifteen to twenty years before Alhazen wrote his great synthetic work, the *Kitab al-Manazir.*[19] However, we know nothing about the immediate circulation of Avicenna's works, and I can find no parallels between the works of Avicenna and Alhazen sufficiently striking to persuade me that either of them influenced the other.

Alhazen against the Extramission Theory

Alhazen's refutation of the extramission theory of vision is certainly not as extensive as Avicenna's.[20] Nevertheless, it had great persuasive force, as its reception in the West was to reveal,[21] and it merits our close attention. The Latin version of Alhazen's *Kitab al-Manazir*, the *De aspectibus* or *Perspectiva*, begins with an account of the pain and injury experienced by a person who observes bright lights:[22]

> We find that when the eye looks into exceedingly bright lights, it suffers greatly because of them and is injured; for when an observer looks at the

> body of the sun, he cannot behold it well, since his eye experiences pain because of its light. Similarly, when he looks into a polished mirror, above which rises the light of the sun, and his eye is in the place to which the light is reflected by the mirror, he will again experience pain because of the reflected light reaching his eye from the mirror, and he will not be able to open his eye to observe that light.[23]

Since it is the nature of injury to be inflicted from without, it is apparent that in the visual process the eye is the recipient of an external action.

The phenomenon of the afterimage supports the same position. Alhazen argues that

> when an observer looks at a pure white body, over which rises the light of the sun, and it remains in his view [for a time], after which he transfers his gaze from it to a dark place that is weakly illuminated, we find that he will scarcely be able to perceive visible things there with a true perception, and it will be as though there is an obstacle between sight and those things; then gradually the obstacle will be removed, and sight will revert to its [proper] disposition. Again, when an observer looks at a bright fire . . . and allows it to linger in his vision for a long time, if he then transfers his gaze to a dark place that is weakly illuminated, he will still see the same thing [i.e., the brightness of the fire].[24]

It makes no difference what source of light is employed—the sun, fire, or indirect daylight—nor does it matter whether the gaze is shifted to a dark place or the eye is simply closed. Indeed, the effect is the same if one looks at bright colors instead of light: if an observer looks at a garden thick with herbs and then shifts his gaze to a dark place, he finds in that dark place "a form colored with the green of those herbs."[25] Or if he shifts his sight from a brightly colored object to a white object in the shade, he finds the bright colors and the whiteness intermixed. Thus in all cases, looking at bright light or color leaves an impression in sight, which lingers for a time before fading away. "All these things indicate," Alhazen concludes, "that light produces some effect in the eye."[26]

But this is not yet a disproof of the extramission theory of vision—nor did Alhazen conceive it to be.[27] After all, a Galenist like Hunain had allowed for the return of visual impressions to the eye,[28] and it is abundantly clear that his version of the extramission theory was capable of assimilating the phenomena of the afterimage and the pain experienced in looking at bright lights. What Alhazen has demonstrated is simply that light and color affect the eye,[29] without establishing whether that effect is exercised through radiation passing from the object to the eye or through visual power emerging from the eye and reaching forth to seize the light and color of the object (and thereafter returning its impressions to the eye). He has demonstrated, as he was to put it elsewhere in *De aspectibus*, "that it is a property of light to act on the eye and that it is the nature of the eye to be affected by light."[30] This, as we shall see, is only the first element in Alhazen's refutation of the extramission theory of vision.

The second element is the claim, defended most fully in a portion of the Arabic text not translated into Latin but frequently echoed in other passages, that light and color (or the forms of light and color) issue in all directions from every point of every self-luminous or illuminated object.[31] Moreover, it is the nature of transparent bodies, such as air and the tunics of the eye, to admit and transmit light and color.[32] Now if the eye is placed opposite a luminous or illuminated object, with only transparent substances intervening, the light and color of the object (or their forms) will reach and penetrate the eye. Since it has already been demonstrated that it is the nature of light and color to act on the eye and of the eye to be affected by light and color, it is "proper that vision of the color of the visible object and of the light that is in it should occur only through the mixed form of light and color coming to it [i.e., to the eye] from the surface of the object."[33] The only reasonable account of vision is thus an intromission theory.

But could one not argue that this process of intromission is only the second half of an extramission-intromission sequence? In short, why not adopt the Galenic theory that rays issue from the eye to transform the medium between the eye and the visible object and that this medium then returns visual impressions to the eye?[34] Alhazen addresses himself to this question with considerable precision and skill. "Let it be supposed," he urges, "that rays issue from the eye and pass through the transparent medium to the object of sight and that perception occurs by means of those rays."[35] Either these rays take something from the object and return it to the eye, or they do not. If they do not, "then the eye does not perceive. But the eye does perceive the object of sight, and [we have supposed that] it perceives only by the mediation of rays. Therefore those rays that perceive the visible object [must] transmit something to the eye, by means of which the eye perceives the object."[36] Crucial to this argument is the premise, implicit here but defended earlier, that in the process of vision the eye itself must ultimately be the recipient of an action; that is, that even on the visual ray theory the act of perception occurs in the eye and mind of the observer rather than where the ray meets the object.[37] It follows that "the eye does not perceive the light and color in the visible object unless something comes to the eye from the light and color in the object," and if, for the sake of argument, we assume the visual ray theory, it is perfectly apparent that "this [something] is delivered by the [visual] rays."[38]

> Therefore, according to all possibilities, sight does not occur unless something of the visible object comes from the object, whether or not rays issue from the eye. It has already been shown that sight is achieved only if the body intermediate between the eye and the visible object is transparent, and it is not achieved if the medium between them is opaque. . . . Since, as we have said and as has been demonstrated, the forms of the light and color

> in the visible object reach the eye (if they are opposite it), that which comes from the visible object to the eye (through which the eye perceives the light and color in the visible object whatever the situation [with respect to visual rays]) is merely that form, whether or not rays issue [from the eye]. Furthermore, it has been shown that the forms of light and color are always generated in air and in all transparent bodies and are always extended to the opposite regions, whether or not the eye is present. Therefore the egress of rays [from the eye] is superfluous and useless.[39]

There is no purpose to be served by the transmission of the visual power to the object of vision, because the forms of light and color of the visible object are transmitted to the observer's eye without its help. Alhazen, like his contemporary Avicenna, thus rejects the Galenic theory of vision because of its redundancy.[40]

The arguments presented thus far might seem to rule out the Euclidean theory of vision as well as the Galenic, but Alhazen nevertheless formulates a distinct refutation of the former. He argues briefly, but effectively, that

> if sight is due to something issuing from the eye to the visible object, that thing is either corporeal or incorporeal. If it is corporeal, it follows that when we look at the sky and see the stars in it, corporeal substance issues from our eye in that time and fills the whole space between heaven and earth, and [yet] the eye is in no way destroyed; and this is [obviously] false. Therefore sight does not occur through the passage of corporeal substance from the eye to the visible object. But if that which issues from the eye is incorporeal, it will not perceive the object, since there is no perception except in corporeal things. Therefore nothing issues from the eye to the visible object to perceive that object.[41]

Corporeal rays cannot account for vision because if material substance were to issue from the eye to fill the whole space between the eye and the heavens, the eye would be dissolved in the process.[42] Incorporeal rays cannot account for vision because perception is a corporeal process.[43] Alhazen appears to have formulated this argument with extraordinary care, and it is essential that we notice its precise content. In concluding that "sight does not occur through the passage of corporeal substance from the eye to the visible object" and that "nothing issues from the eye to the visible object *to perceive that object*," he apparently does not claim to have demonstrated the nonexistence of visual rays, but only their inability to explain perception. Although he clearly disbelieved in the existence of visual rays, the caution displayed at this point was to provide Western writers with room to maneuver; thus Roger Bacon and John Pecham would argue that although visual rays are not sufficient to account for vision, they nevertheless exist and perform useful functions. This permitted them to reconcile (or to feel that they had reconciled) the teaching of the extramissionists and the intromissionists.[44]

In due course, however, Alhazen took the inevitable final step of denying the very existence of visual rays. His argument was simply that if visual rays cannot explain anything, there is no reason to suppose they exist:

> Now it is evident that sight occurs through the eye; and since this is so, and [by hypothesis] the eye perceives the visible object only when something issues from the eye to the visible object, and since [by the previous arguments] that which issues forth does not perceive the object, therefore that which issues from the eye to the visible object does not return anything to the eye, by which the eye perceives the object. And [therefore] that which issues from the eye is not sensible but conjectural, and nothing ought to be believed except for a [sufficient] reason.[45]

People have devised the theory of visual rays because they have recognized that perception occurs only by contact and that space intervenes between the eye and its object. Therefore they have conjectured that something must issue from the eye and proceed to the visible object, "so that the thing issuing forth perceives the object in place of the eye, or rather receives something from the visible object and returns it to the eye."[46] Since that which perceives the object must be corporeal, and no corporeal substance can issue from the eye to perceive the object (by arguments presented above in refutation of the Euclidean theory), there has seemed to be no alternative except to suppose that an immaterial something issues from the eye and receives from the visible object something, which it then returns to the eye. However,

> since it has been demonstrated that air and transparent bodies receive the form of the visible object and transmit it to the eye and to every facing body, that which they conjecture to return something from the visible object to the eye is nothing but air and the [other] transparent bodies between the eye and the object of vision. . . . And since air and transparent bodies do this without requiring that something issue from the eye, and, moreover, since the air and transparent bodies are extended between the eye and the visible object without defect, it is useless to suppose that something else returns something from the object of vision to the eye. Therefore it is useless to say that [visual] rays exist.[47]

Visual rays must be denied as physically existing entities because they serve no theoretical purpose.

Alhazen appears to have made his position regarding the theory of visual rays unambiguously clear, and yet confusion continues to surround the question. Vasco Ronchi has argued that

> Alhazen, after having resolutely affirmed that vision does not occur through rays emitted from the eye . . . , comes later to a sort of compromise and affirms that vision seems to occur simultaneously through rays received and emitted. He explains that the existence of rays coming from the object toward the eye is insufficient and that it is necessary for the eye to be directed toward the object so as to be able to receive the rays.[48]

However, this is by no means Alhazen's position. The origin of the confusion is a caption inserted by Friedrich Risner into his edition (Basel, 1572) of Alhazen's *De aspectibus*. Although the original Arabic version and its Latin translation were divided only into books and chapters, Risner divided each chapter into subsections or propositions and gave a title to each.[49] Section 24 of book 1 he entitled: "Vision seems to occur through συναύγειαν, that is, rays simultaneously received and emitted."[50] Since συναύγεια (simultaneous radiation) was a term employed in antiquity to describe Plato's theory of sight,[51] the implication that Alhazen has compromised his denial of the visual ray theory is obvious. Morever, the text immediately following Risner's caption begins as follows:

> It has been asserted on account of this [i.e., the argument of the previous section] that both schools of thought [extramission and intromission] speak the truth and that both beliefs are correct and consistent; but one does not suffice without the other, and there can be no sight except through that which is maintained by both schools of thought.[52]

But to suppose that Alhazen has here granted a certain validity to the visual ray theory as a description of physical reality is to miss the point—as Risner apparently missed it. Alhazen's position is that mathematicians, who are concerned with a mathematical description of the phenomena rather than with the real nature of things, may properly use visual rays to represent the geometrical properties of sight. Indeed, these rays or lines are indispensable if one is to understand the visual process, for through them one is able to visualize "the nature of the arrangement according to which the eye is affected by the form [of light and color]."[53] However, Alhazen continues, all mathematicians who postulate visual rays "use nothing in their demonstrations except imaginary lines, and they call them radial lines; . . . and the belief of those who consider radial lines to be imaginary is true, and the belief of those who suppose that something [really] issues from the eye is false."[54] Thus visual rays (or radial lines) are mere geometrical constructions, useful in demonstrating the properties of sight. They can serve as a mathematical hypothesis, but they have no physical reality.[55] And yet, if these rays are imaginary, why imagine them to issue from the eye rather than from the visible object? Alhazen's position undoubtedly represents a concession, first, to traditional geometrical optics as practiced by Euclid, Ptolemy, and al-Kindi, which had been pursued within the framework of the extramission theory, and, second, to the natural intelligibility of a center of perspective from which rays emanate to perceive visible things.

Alhazen's refutation of the extramission theory was certainly less comprehensive—if not less convincing—than Avicenna's. Whereas Avicenna carefully elaborated the Euclidean and Galenic theories of vision into all of their possible versions and systematically refuted each, Alhazen was content to wage a more diffuse and generalized campaign. But perhaps the battle was not to be won at this level of engagement anyway. Rarely is a theory dislodged by a negative attack alone—by a mere demonstration of its weakness or futility—for most theories are pliable enough to be bent and remodeled by their partisans to meet almost any criticism. It is a historiographical commonplace that what is required is an alternative theory that will displace the old by virtue of greater theoretical power and fecundity. In the struggle against the extramission theory of vision, it was Alhazen, more fully than Avicenna, who developed an alternative theory, convincingly and in great detail. Alhazen's most effective refutation of the extramission theory was thus his own positive intromission doctrine.

Alhazen on the Anatomy and Physiology of the Eye

One of the principal merits of Alhazen's theory of vision, as I have suggested above, is that it successfully integrated the anatomical, physical, and mathematical approaches to sight. If then we are to appreciate Alhazen's achievement, it is essential to make a brief survey of his views on ocular anatomy and physiology.

Alhazen admits that his discussion of ocular anatomy is drawn from the works of others,[56] and it is apparent from a consideration of content that he was familiar with the anatomical tradition issuing from Galen. According to Alhazen, the optic nerves, each consisting of two tunics, issue from the anterior part of the brain.[57] They meet to form the optic chiasma or common nerve, a uniting of the optic nerves into a single hollow nerve, before again dividing and extending to the orbits of the eyes. On the extremity of each nerve is constructed an eye, consisting of four tunics and three humors. The first and largest tunic is the *consolidativa* or white fat. Within it is the uvea or grapelike tunic, entirely surrounded by the *consolidativa* except in front. The uvea is perforated by an aperture (the pupil) in front, directly opposite the extremity of the optic nerve.[58] The eye is covered in front by the cornea, a strong, uncolored, transparent tunic. The fourth and final tunic is the *aranea* or arachnoid membrane, resembling a cobweb, which surrounds the combined anterior and posterior glacial humors.

Within the uvea is the glacial humor, consisting of two parts. The anterior part, the crystalline humor, is not excessively transparent but somewhat dense, having a transparency resembling that of ice. Its anterior surface is flattened, its surface as a whole is lenticular, and it is situated directly opposite the extremity of the optic nerve. The interior or posterior part of the glacial humor has a transparency resembling that of glass, and it is therefore called the vitreous humor.[59] The vitreous humor and the anterior glacial humor together constitute a sphere, and they are surrrounded by the arachnoid membrane.[60] The third humor of the eye is the albugineous (our aqueous), resembling the white of an egg, which fills the hollow of the uvea in front of the glacial humor and the space between the uvea and the cornea.

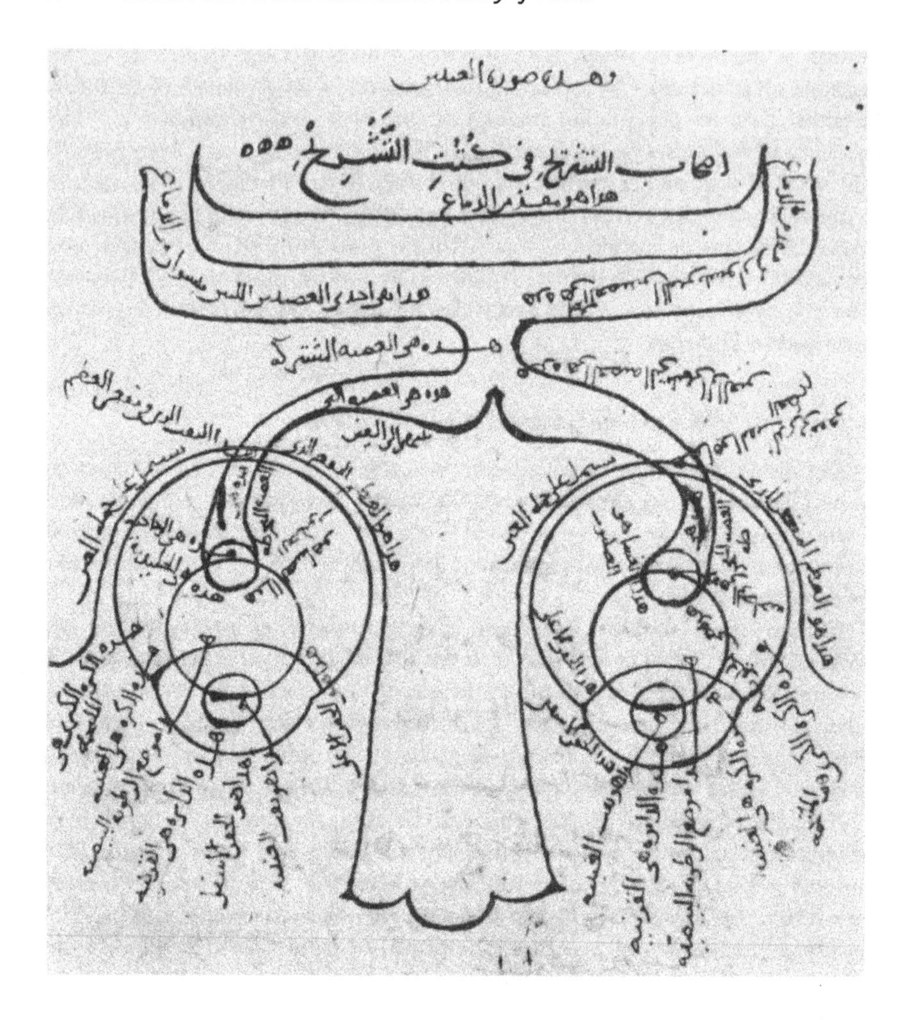

Fig. 7 The visual system according to Alhazen. From a copy of the *Kitab al-Manazir* (*De aspectibus*) made in 1083 A.D. MS Fatih 3212, vol 1, fol. 81b, of the Süleymaniye Library, Istanbul. For identification of the various parts represented, see Stephen L. Polyak, *The Retina* (Chicago, 1941), figure 8 and accompanying text.

Alhazen's account of ocular anatomy is thus far quite traditional. If he defines the tunics and humors somewhat differently from Galen, Hunain, and 'Ali ibn 'Isa, he nevertheless remains well within the mainstream of medieval Galenism.[61] However, Alhazen introduces an element of some novelty when he attempts to impose a geometrical model on ocular anatomy.[62] He argues that all parts of the eye are of spherical surface and that all ocular surfaces have their centers on the single line that passes through the center of the pupil and terminates at the center of the extremity of the optic nerve. Morever, all centers except those of the uvea and posterior surface of the (anterior) glacial humor are at a single point, with the result that all surfaces through which light must pass before emerging from the posterior surface of the glacial humor are concentric.[63] Finally, since the common center of these surfaces is also the center of the eye, its position is not affected by rotation of the eye. This is a highly idealized anatomical scheme, and it is clear that it was dictated not by the results of dissection or any other experimental technique, but by the theoretical necessities of Alhazen's intromission doctrine.[64]

When we turn to Alhazen's physiology of sight, we again discover traditional conclusions. The sensitive ocular organ is the glacial humor, as Alhazen claims to demonstrate by the following argument:

> And we would say in the first place that sight occurs only by means of the glacial humor [i.e., anterior glacial humor or crystalline lens], whether sight takes place through forms coming from the visible object to the eye or in some other way. Sight does not occur through one of the other tunics in front of it, since those tunics are merely instruments of the glacial humor; for if injury should befall the glacial humor, the other tunics remaining sound, sight is destroyed; if the other tunics should be corrupted, their transparency and the health of the glacial humor being retained, sight is not destroyed.[65]

Moreover, sight is prevented by the insertion of an opaque body in the pupil or in the albugineous humor on a direct line between the pupil and glacial humor, and when the opaque body is removed sight returns. This is an extension of an argument presented previously by Galen and Hunain, whose extramission theories Alhazen undoubtedly had in mind when he wrote that the glacial humor must be regarded as the principal organ of sight "whether sight takes place through forms coming from the visible object to the eye *or in some other way.*" Hunain had written: "Further proof that [the power of] vision is in this [crystalline] humor, and not in any other part of the eye, lies in the circumstance that . . . vision ceases when cataract intervenes between it [the lens] and the perceptible object, and that . . . vision returns when the cataract is removed from it by couching [operation]."[66] Now Alhazen improves on this argument by defending the sensitivity of the lens on the grounds that injury to it is the crucial thing in destroying sight. The origin of this sensitivity Alhazen attributes to visual spirit sent through the optic nerves from the brain—the traditional Galenic theory—though he expresses this view with caution:

> And *it is said* that visual spirit is emitted from the anterior part of the brain and fills the channels of the two primary nerves connected to the brain and reaches the common nerve and fills its channel and comes to the two secondary optic nerves and fills them, and it extends to the glacial humor and confers the power of vision on it.[67]

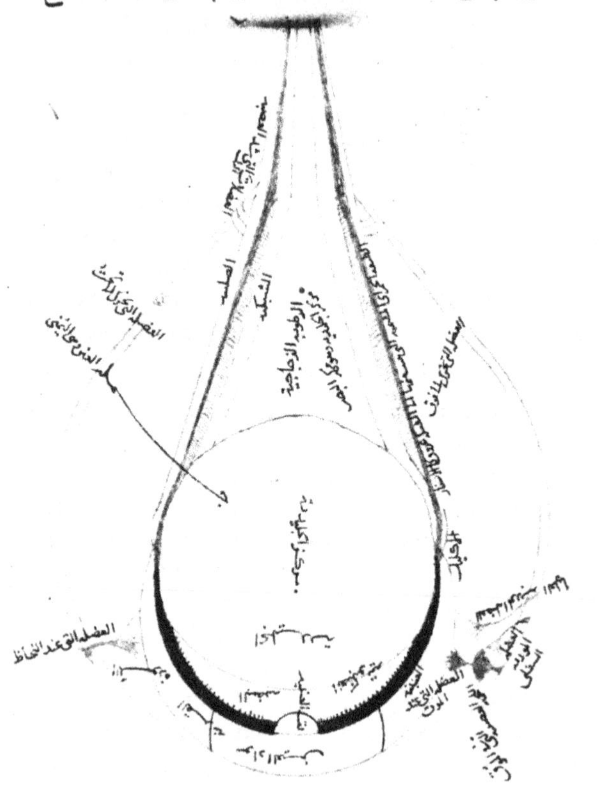

Fig. 8. The eye according to Alhazen and Kamal al-Din al-Farisi. From Kamal al-Din's *Tanqih al-Manazir,* a commentary on Alhazen's *Kitab al-Manazir*. It is usually assumed that the figure was copied from Alhazen. MS III.A.3340, fol. 25b, of the Topkapi Palace Museum, Istanbul (dated 1316 A.D.). For identification of the parts, see Polyak, *The Retina*, figure 9 and accompanying text.

We come then to the question of how the glacial humor perceives the forms of light and color incident upon it. Alhazen presents an extension of the Aristotelian view that the transparent substance of the eye assumes the qualities of the visible object.[68] He argues that because the glacial humor is transparent it receives forms, and yet because it possesses a certain density the forms do not pass freely through it; as a result, the "forms are fixed in its surface and body, although weakly."[69] The result of this "fixation" must clearly be to endow the glacial humor with the qualities of light and color of the visible object. Alhazen states this conclusion quite explicitly later in *De aspectibus*, when he writes: "Essential light (*lux essentialis*) is perceived by the sensitive body from the illumination of the sensitive body, and color is perceived by the sentient body from the alteration of the form of the sentient body and from its coloration."[70] But Alhazen goes beyond this Aristotelian position, arguing that when the form of light

> reaches the surface of the glacial humor, it acts on it, and the glacial humor suffers because of the form, since it is a property of light to act on the eye and a property of the eye to suffer because of the light. And this effect, which light produces in the glacial humor, passes through the glacial body. . . . And the glacial humor perceives on account of this action and suffering, because of the forms of visible things that are on its surface and pass through its whole body; and it perceives through the ordering of the parts of the form on its surface and throughout its body.[71]

Two new elements are here added to the Aristotelian physiology of vision. First, visual sensation is a species of pain.[72] Second, and more important for Alhazen's doctrine of vision, the parts of the form must be properly arranged on the surface of, and within, the glacial humor—arranged, that is, in the same order as the parts of the visible object from which the form originated. The remainder of Alhazen's theory of vision is directed principally toward explaining how this condition is achieved.

Alhazen's Intromission Scheme

When Alhazen rejected the extramission theories of Euclid and Galen, it was not in order to defend one of their ancient rivals—the intromission theory of Aristotle or that of the atomists. Alhazen's intromission theory was a fresh creation, addressed to problems that neither Aristotle nor the atomists had attempted to solve and based on several fundamentally new conceptions of the nature of the visual process.

The foundation of Alhazen's new intromission theory was the claim, first

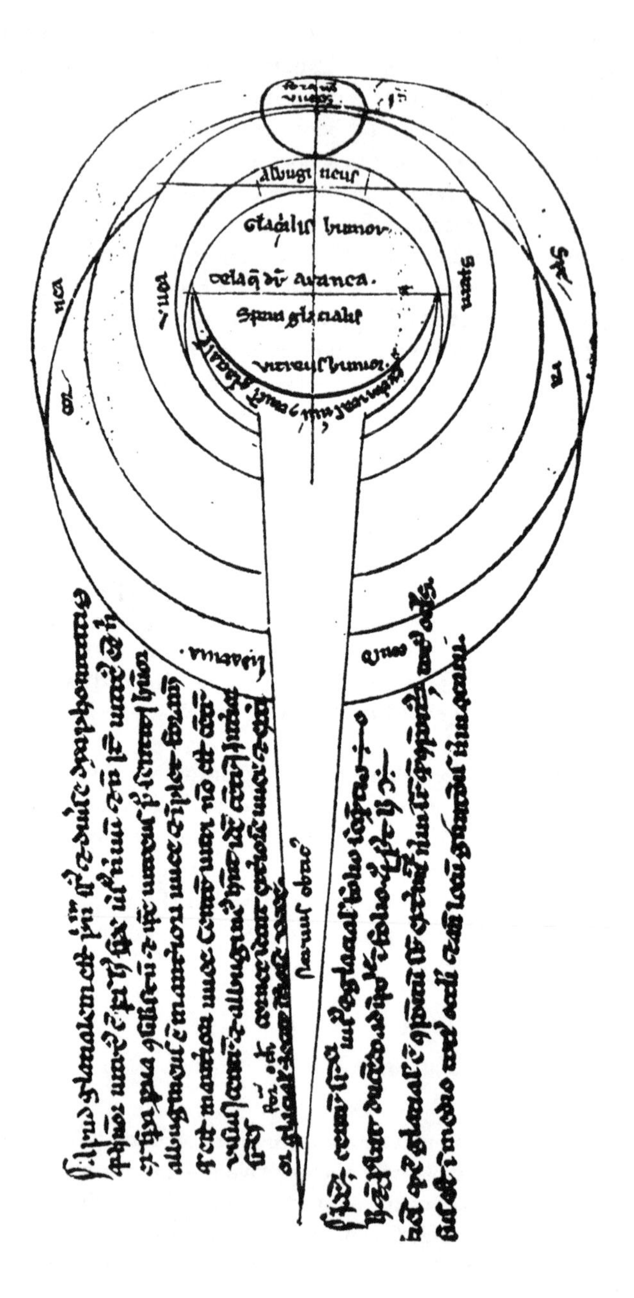

Fig. 9. The eye in an early Latin manuscript of Alhazen's *De aspectibus*. MS 9-11-3 (20), fol. 4v, of the Crawford Library, Royal Observatory, Edinburgh (dated 1269 A.D.).

clearly articulated by al-Kindi, that each point or small region on the surface of a body radiates in all directions. The body is thus submitted to punctiform analysis, for it is not the whole body, but each of its points, from which forms issue. Alhazen is explicit on the matter, noting that "from each point of every colored body, illuminated by any light, issue light and color along every straight line that can be drawn from that point."[73] Alhazen thus adopts as his starting point a position that would surely have astonished his intromissionist predecessors, namely, that a coherent visual impression is to be explained by means of independent or incoherent sources of radiation.

The difficulties of this position appear immediately, for it follows from the punctiform analysis of the visible object that each point on the object sends light and color to each point of the eye. Alhazen fully appreciated the problem thus raised:

> In the same object there can be various colors, and from each part of the object issue light and color along all straight lines extending through the continuous air. Therefore when the parts of one visible object are of various colors, the forms of light and color will come to the whole surface of the eye from each of them, and consequently the colors of those parts will be mixed on the surface of the eye. Therefore sight will either perceive them intermixed or not perceive them at all. If it perceives them intermixed, they are not distinguished, and the parts [of the object] or the colors of the parts will not be [properly] arranged. If the forms are not perceived at all, neither are the parts [of the object] perceived; and if the parts are not perceived, neither is the object perceived.[74]

Alhazen thus recognizes that if he wishes to base his intromission theory on a punctiform analysis of the visible object, he must explain why the presence at every point in the eye of radiation from every point in the visual field does not result in total confusion.[75] The problem is a superfluity of rays, for if vision is to be explained, it is required that each point on the surface of the eye (or, more exactly, each point on the surface of the glacial humor, as Alhazen makes clear later) receive a ray from only one point in the visual field. In short, it is necessary to establish a one-to-one correspondence between points in the visual field and points in the eye.

But how can this be achieved? Alhazen finds the key in the phenomenon of refraction. The refraction of light had been analyzed in considerable detail by Ptolemy in the second century, and if a satisfactory mathematical law of refraction had not been formulated, at least there was no doubt about the qualitative features of refraction.[76] Thus it was thoroughly understood that rays passing obliquely between two transparent media of different densities are bent so as to lie closer to the perpendicular[77] in the denser of the media. Above all, it was clear that only light incident perpendicularly on the interface between two media passes through without refraction.[78]

If, now, we take a point on the surface of the eye, we notice that although it receives a form from each point in the visual field, only one of these forms is incident upon it perpendicularly; this form alone passes into the eye without refraction. The same holds, of course, for every point on the surface of the eye:

> Through each point on the surface of the eye pass simultaneously the forms of all points in the visual field, but the form of only one point passes directly [i.e., without refraction] through the transparency of the tunics of the eye, and that is the point located at the extremity of the perpendicular issuing from the point on the surface of the eye. The forms of all the remaining points [in the visual field] are refracted at that point on the surface of the eye and pass through the transparency of the tunics of the eye along oblique lines.[79]

Consequently, each point of the cornea is the recipient of a single perpendicular ray, which is passed through to the glacial humor without refraction. The collection of all such unrefracted rays constitutes a pyramid or cone, with the visual field as base and the center of the eye as apex.[80] Because the rays are rectilinear and converge toward a single apex, they maintain a fixed arrangement and fall on the surface of the glacial humor in precisely the same order as the points in the visual field from which they originated.[81] If it can be successfully maintained that vision is produced by rectilinear rays alone, the required one-to-one correspondence will have been established.

But on what grounds can refracted rays be ignored? Alhazen argues, in the first place, that what is required to explain the facts of vision is a single ray incident on each point of the surface of the glacial humor and that only perpendicular (and hence unrefracted) rays constitute such a unique set, for the number of refracted rays falling on each point is beyond number:

> If one point of the glacial humor perceives all forms incident upon it along all lines, it will perceive at every point forms produced by a mixture of many different forms and of the many colors of the visible objects opposite the eye at that moment. Thus the points on the surface of visible objects will not be distinguished, and the forms of the points coming to that point [of the glacial humor] will not be [properly] arranged. But if one point of the glacial humor perceives only that which is incident upon it along one line, the points on the surfaces of visible objects will be distinguished. And no point from which forms come to the glacial humor along refracted lines is more deserving than another, and no refracted line of incidence is more deserving than another; and the number of refracted forms [received] at one point of the glacial humor at one time is indeterminately large. However, there is but one point whose form is incident with the rectilinearity of the perpendicular. . . . Therefore the form incident along the perpendicular is distinguished from the others.[82]

Therefore perpendicular rays alone must be responsible for sight. The argument is quite ad hoc, but it may well reflect the line of thought that brought Alhazen to his new intromission theory: a unique set of rays is required to provide the necessary one-to-one correspondence between points in the visual field and points in the eye, and perpendicular rays constitute the only unique set.

However, Alhazen also has a second argument to justify restricting attention to perpendicular and unrefracted rays: "The effect of light incident along the perpendicular is *stronger* than the effect of light incident along oblique lines. It is therefore proper that a point of the glacial humor should perceive only the form incident upon it with the rectilinearity of the perpendicular."[83] The claim that light incident along a perpendicular line is stronger may seem gratuitous, and it is true that in book 1 of *De aspectibus* Alhazen merely asserts it without additional comment; however, in book 7, where refraction is discussed in much greater detail, Alhazen attempts to defend the principle by means of two mechanical analogies:

> If one takes a thin board and fastens it over a wide opening, and if he stands opposite the board and throws an iron ball at it forcefully and observes that the ball moves along the perpendicular to the surface of the board, the board will yield to the ball; or if the board is thin and the force moving the ball is powerful, the board will be broken [by the ball]. And if he then stands in a position oblique with respect to the board and at the same distance as before and throws the ball at the same board, the ball will be deflected by the board (unless the latter should be unduly delicate) and will no longer be moved in its original direction, but will deviate toward some other direction.
>
> Similarly, if one takes a sword and places a rod before him and strikes the rod with the sword in such a way that the sword is perpendicular to the surface of the rod, the rod will be cut considerably; and if the sword is oblique and strikes the rod obliquely, the rod will not be cut completely, but perhaps partially, or perhaps the sword will be deflected. And the more oblique the [motion of the] sword, the less forcefully it acts on the rod. And there are many other similar things, from which it is evident that motion along the perpendicular is stronger and easier and that the oblique motion which approaches the perpendicular is [stronger and] easier than that which is more remote from the perpendicular.[84]

If swords and iron balls penetrate more readily when they strike a surface perpendicularly, then light too must be more efficacious when perpendicularly incident on the surface of the glacial humor.[85]

But it is one thing to claim that perpendicular light is more efficacious than oblique light, another to claim that only perpendicular light has any efficacy at all. Why don't oblique rays make a weaker, but still perceptible, impression on the eye? As Alhazen himself points out, the more closely a motion approaches the perpendicular, the stronger it is, and there is no reason to suppose that the efficacy of light in stimulating the glacial humor falls off abruptly as the light diverges from the perpendicular. Moreover, when, at the beginning of *De aspectibus*, Alhazen analyzes the effects of strong and weak lights acting on the eye simultaneously, he maintains only that strong lights conceal weak lights[86]—which is far from the claim that a light is concealed by any light stronger than itself.

Alhazen seems to have realized the precariousness of his position, for in book 7 of *De aspectibus* he returns to the effect of nonperpendicular rays on the eye; that on this occasion he presents a theory wholly at variance with that of book 1 and that he allows the inconsistency to stand suggests that he was not able to resolve the problem to his own satisfaction. He admits in book 7 to having argued earlier that if each point of the glacial humor were to perceive all forms incident upon it, total confusion would prevail, and therefore that only perpendicular rays are able to stimulate the visual power. But this, he now admits, is not exactly the case.[87] Actually, all forms incident on the glacial humor are perceived, but those that are *not* incident along perpendiculars (dropped from their points of origin) are perceived as though they *were* incident along such perpendiculars! How is this possible? Alhazen explains that

> in this treatise [i.e., book 7] we have demonstrated that refracted forms are perceived only along the perpendiculars issuing from the visible objects to the surfaces of the transparent bodies [in which the objects are situated]. Therefore the forms refracted in the tunics of the eye are perceived by the eye only along the perpendiculars issuing from the visible objects to the surfaces of the tunics of the eye, and these perpendicular lines issue from the center of the eye. Therefore all forms refracted in the ocular tunics are perceived by sight [as though] along straight lines issuing from the center of the eye.[88]

This is an astonishing argument, which requires clarification. Earlier in book 7, Alhazen demonstrated that objects observed by refraction appear to be situated along the perpendicular dropped from the object to the refracting interface.[89] That is, if an object at *O* (fig. 10) is observed by eye *E* through refracting interface *SP* (the eye being situated in the denser medium), ray *OS* is refracted toward perpendicular *CS* and falls into the eye. However, the eye interprets ray *SE* to have issued from *I*, where the backward extension of the ray incident on the eye intersects the perpendicular *OC* drawn from the object to the refracting interface. This principle had been firmly established in antiquity,[90] and it makes perfect sense, for it requires simply that the eye be unaware of the break in the ray at *S* and therefore that it project the image backward along the incident ray. However, if we can trust the Latin text, Alhazen employs this principle in a most unreasonable way. In simplest terms his position appears to be that rays refracted upon entering the eye behave in the same way as rays received by the eye through a refracting interface, and therefore that the forms carried by such rays seem to have issued from a point on the perpendicular dropped from the object to the surface of the eye and, indeed, to have reached the eye along that perpendicular; consequently they reinforce the impression made by perpendicular rays.[91] There is a superficial plausibility to this scheme, but only until we consider more closely what it entails. In effect, the analogy that Alhazen seems to urge, between image-formation by a single refracting interface and vision by rays refracted at the surface of the eye, requires us to grant that a ray incident on the surface of the eye *AD* (fig. 11) along line *OA* can stimulate the visual power in the glacial humor in such a way that it seems to have been incident along line *OB*. Clearly this is impossible: the visual power at *E* on the surface of the glacial humor, observing refracted ray *AE*, must (according to Alhazen's own principle of

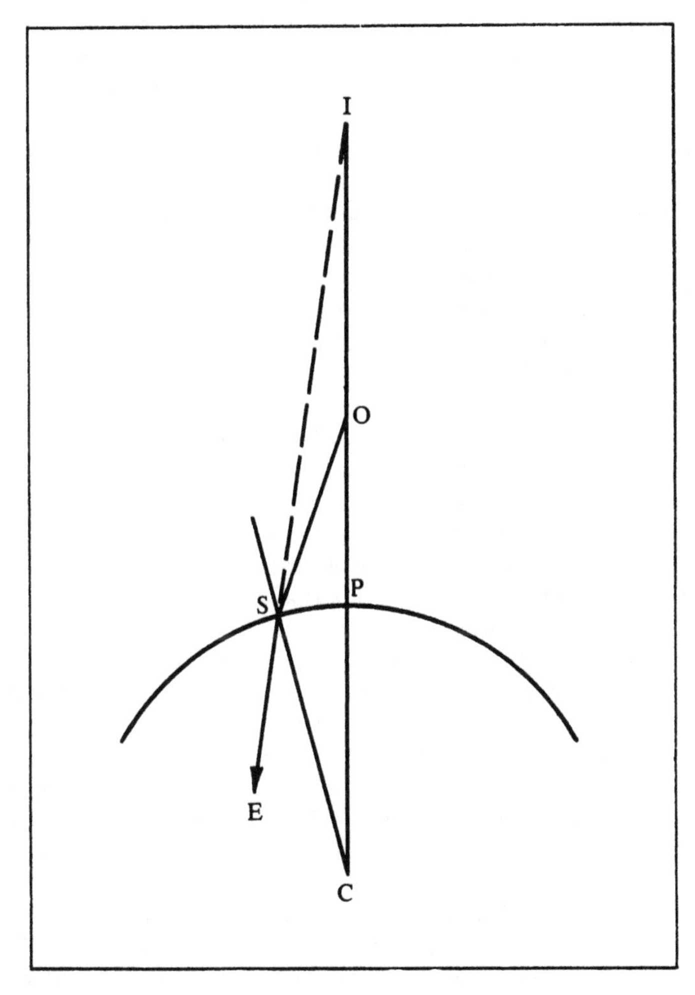

Fig. 10

image-formation) perceive it to have originated from a point along its backward extension *AG*, which may or may not intersect perpendicular

OB;[92] and in any case there is obviously no means by which the ray can seem to reach point *A* or point *E* by means of ray *OB*. Moreover, there is a second refraction at the surface *E* of the glacial humor, which not only raises all the same problems again, but presents the additional difficulty that the visual power extends right up to the refracting surface and hence is the recipient not simply of the refracted ray, but of the incident ray as well; it is not, therefore, analogous to an observer situated at some distance behind a refracting interface. If my interpretation is correct, this is undoubtedly the weakest point in Alhazen's theory of vision, and it was not to be rectified until Kepler devised the theory of the retinal image; indeed, Kepler's principal innovations were a response to precisely this problem of nonperpendicular rays and the necessity of establishing a one-to-one correspondence.[93]

We must not permit the several weaknesses of Alhazen's theory to obscure its importance. Alhazen has taken the intromission theory, which had always been defended on physical and physiological grounds alone,[94] and has revealed its mathematical potentialities. By confining his attention to the perpendicular rays, Alhazen has managed to appropriate the visual pyramid of Euclidean and Ptolemaic optics,[95] thereby transforming the intromission theory into a mathematical theory of vision and negating the principal argument in favor of the extramission theory; the return of nonperpendicular rays in book 7 in no way weakens Alhazen's position if (as he claims) they behave exactly like perpendicular rays. Alhazen's theory of vision is thus able to accomplish mathematically everything that had traditionally been achievable only with the mathematical theories of Euclid, Ptolemy, and al-Kindi.[96] For the first time an intromission theory of vision has become a viable alternative, adequate to compete on geometrical as well as physical and psychological terms with the theory of visual rays.

There remains the physical question of the nature of the radiation responsible for vision. The one certainty is that Alhazen gives no systematic answer to this question. In his treatise *On Light*, he notes that according to philosophers the light of a self-luminous body is a substantial form, whereas the light of an opaque body illuminated by a self-luminous body is an accidental form; according to mathematicians, on the other hand, the light issuing from a self-luminous body (the substantial form of that body) is the heat of fire.[97] What is most noteworthy about this account is that Alhazen makes no attempt to resolve the controversy between the philosophers and mathematicians; in R. Rashed's words, "he renounces from the beginning any wish to define the nature of light or of the luminous ray."[98] This is no doubt true, but we need not therefore give up all hope of understanding Alhazen's implicit position on the matter. In the first place, both the philosophers and the mathematicians, as Alhazen points out, associate the light of a self-luminous body with substantial form, and there are no grounds for supposing that Alhazen broke out of this basically Aristotelian framework.

Second, the terminology of both the *Kitab al-Manazir* and its Latin translation, *De aspectibus*, reflects this same Aristotelian natural philosophy. As Alhazen says repeatedly, it is the *form* of light and color that is transmitted from the visible object to the eye, where it endows the glacial humor with the qualities of the visible object.[99] Thus, luminous radiation is not a substantial entity, a particle or corpuscle, but a quality of bodies that can be propagated in straight lines through a transparent medium to produce visual sensation.[100]

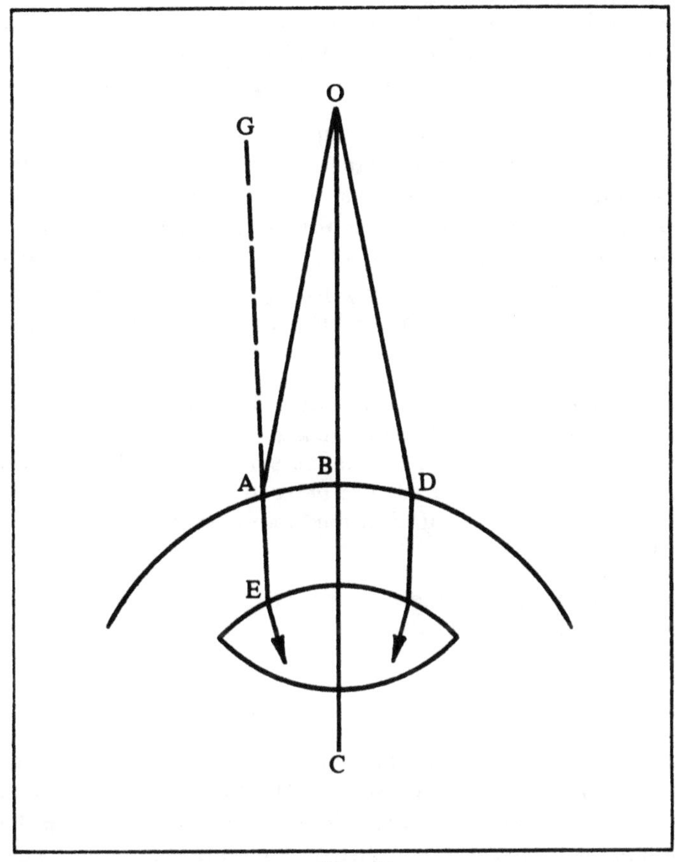

Fig. 11

What, then, are we to make of Alhazen's several comparisons of the propagation of light to the motion of a projectile? Alhazen argues, in the case of reflection, that

> light is reflected along a line having the same slope as the line by which the light approaches the mirror because light is moved very swiftly, and when it falls on a mirror it is not allowed to penetrate but is denied entrance into that

> body. And since the original force and nature of motion still remain in it, the light is reflected in the direction from which it came, along a line having the same slope as the original ray. We can see the same thing in natural and accidental motion, for if we allow a heavy spherical body to descend perpendicularly onto a smooth body from a certain height, we will see it reflected along the same perpendicular by which it descended. In accidental motion, if a mirror is raised to the height of a man and firmly fastened to a wall and a sphere is fixed to the tip of an arrow, and if the arrow is [then] projected by a bow at the mirror in such a way that the arrow and mirror have equal elevations and the arrow is horizontal, it is evident that the arrow approaches the mirror perpendicularly; and it will be seen to recede from the mirror along the same perpendicular.[101]

Alhazen treats refraction similarly, comparing the refraction of light to the deviation of a sphere passing through a piece of thin slate.[102] If we are to understand Alhazen's position, we must recognize that these comparisons are not literal statements of identity, but similes, meant to elucidate the geometry and causes of reflection and refraction but not the nature of the reflected or refracted entity. Indeed, in the very passage on the cause of reflection, Alhazen is careful to distinguish the behavior of light from that of heavy spheres:

> But in the rebound of a heavy body, when the motion of repulsion ceases, the body descends because of its nature and tends toward the center [of the world]. However light, which has the same nature of reflecting [as a heavy body], does not by nature ascend or descend; therefore in reflection it is moved along its initial line until it meets an obstacle which terminates its motion.[103]

Although light and a heavy sphere are similarly reflected, the two entities do not share the same nature.

The Completion and Certification of Visual Perception

Alhazen's doctrine of vision does not end with the reception of forms in the glacial humor. With Galen and Hunain,[104] Alhazen admits that the final perception and interpretation of visual impressions are achieved in the brain, as the result of the propagation of a form or an impression from the eye to the brain. We are thus led to the problems of the connection between the eye and the brain and of the interpretation of visual impressions.

It will be well to begin with a general sketch of Alhazen's view before considering in detail some of its difficulties and ambiguities. The forms responsible for vision are those that fall perpendicularly on the glacial humor and enter it without refraction. Because of the transparency of the glacial humor, these continue their course toward an apex at the center of the eye. However, it is essential that they be prevented from achieving this apex; for if they should reach an apex and be terminated there, they would be reduced to a point, and a point is incapable of communicating the order of the parts of the visible body to the common nerve and brain; if, on the other hand, they were to achieve an apex and continue beyond, they would be inverted and reversed and would consequently communicate an inverted and reversed impression of the visible object to the brain.[105] To avert both of these eventualities, the forms must be refracted before they reach the center of the eye and deflected away from that center; thus diverted from their virtual apex, they can proceed through the vitreous humor and optic nerve in their proper order. All of this can be achieved, Alhazen argues, by placing the interface that separates the glacial and vitreous humors before the center of the eye (see fig. 12) and supposing that the densities of the glacial and vitreous humors are such as to produce refraction in the appropriate direction and of the required degree.[106]

Having been properly refracted, the forms are directed through the vitreous humor to the optic nerve, which is filled with visual spirit (or the "sentient body")[107] and conducts the forms to the optic chiasma or common nerve.[108] There identical forms coming from the two eyes unite. However, the proper uniting of the forms to produce a single image requires that there be a fixed correspondence between points on the glacial humor and points in the common nerve (i.e., that forms arriving at a particular point on the surface of the glacial humor are always transmitted to the same point in the common nerve), that this correspondence be identical for the two eyes, and that the visible point from which the forms originate be similarly situated with respect to both eyes.[109] Upon being united in the common nerve, the forms are perceived by the *ultimum sentiens*, the ultimate sentient power, and the act of sight is thus completed.[110]

Perhaps the most significant feature of this scheme is its curious quasi-optical character. The radiation of forms according to the laws of optics does not cease at the crystalline lens (or anywhere else in the eye), to be followed by a nonoptical transmission of nervous impulses. Rather, forms proceed according to the laws of optics (though with some important modifications—hence the expression "quasi-optical") until they encounter the final sentient power in the optic chiasma. The optical character of the transmission is plainly evident in Alhazen's observation that forms must be refracted at the interface between the glacial and vitreous humors and that the only reason forms are not refracted when passing from the vitreous humor into the optic nerve is that the vitreous humor and the visual spirit contained in the optic nerve are of similar density.[111] Moreover, as Alhazen stresses repeatedly, it is required that the forms maintain their proper spatial arrangement until they reach the *ultimum sentiens*, just as forms must maintain their proper order between the visible object and the glacial humor.[112]

And yet there are many occasions when Alhazen admits that the post-

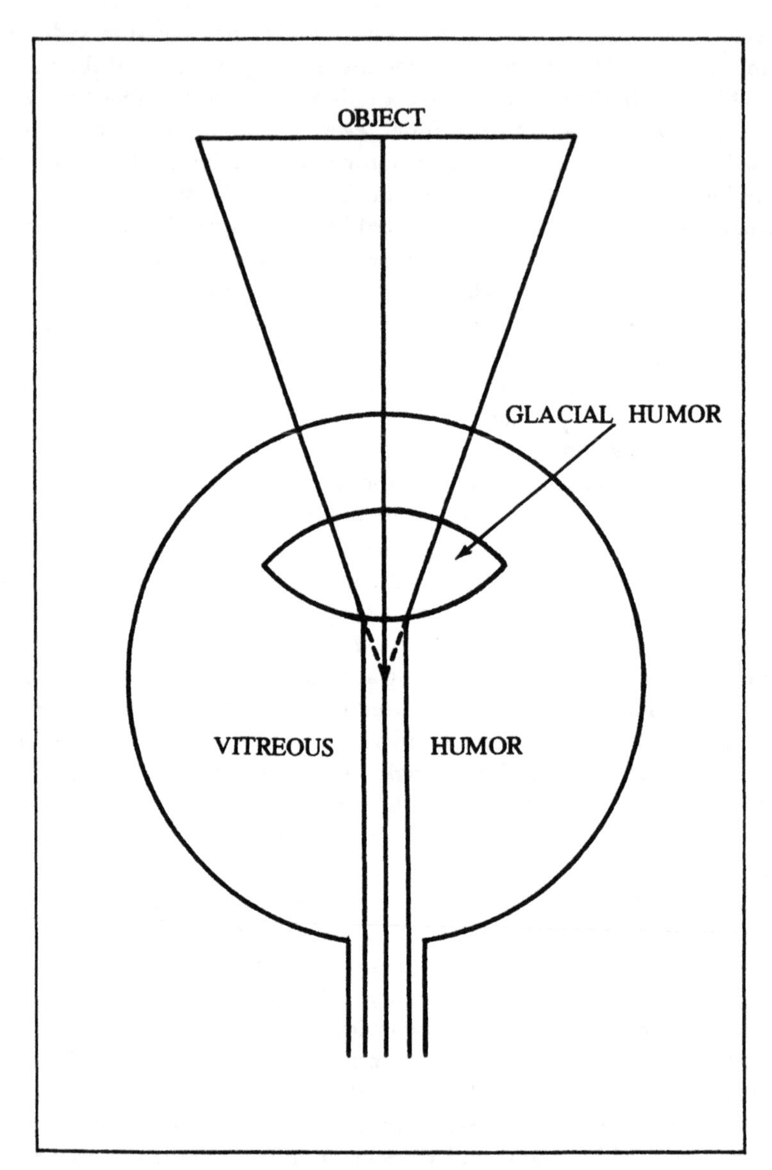

Fig. 12. The geometry of sight according to Alhazen.

crystalline transmission is not strictly optical, or at least does not abide by the ordinary laws of inorganic optics.[113] This is apparent in Alhazen's statement that "the sentient member, namely the glacial humor, does not receive the forms of light and color as air and other nonsentient transparent media receive them, but in a different way."[114] The difference is evident in the fact that the glacial humor does not simply admit the forms, but is affected by them and perceives them. Again, when discussing the refraction of forms as they pass from the glacial to the vitreous humor, Alhazen points out that there are two causes of refraction—the difference in transparency or density between the two bodies (which functions in ordinary inorganic refraction) and the difference in what he calls the "sensitive power" or "sensitivity" or the "receptivity of sense" or simply the "quality of receptivity":

> Therefore forms are refracted at the vitreous humor by two causes, of which one is the difference of transparency between the two bodies and the other the difference between the quality of the receptivity of sense between these two bodies. And if the transparencies of these two bodies were the same [while the receptivities were different], the form would be extended in the vitreous body along the rectitude of radial lines on account of the similarity of transparency; but it would be refracted on account of the difference in the quality of sensitivity. And thus, because of refraction, the form would be monstrous, or it would be two forms.[115]

Fortunately, however, both the transparency and the quality of sensitive receptivity are such as to refract the form identically, and it proceeds through the vitreous humor and optic nerve as a single form.[116] Finally, Alhazen admits the inapplicability of the law of rectilinear propagation to transmission within the vitreous humor or optic nerves. Only as far as the glacial humor are straight lines required, for by means of the visual pyramid they assist in the arrangement of forms on the surface and through the interior of the glacial humor in the same order as the points of the visual field from which the forms originated. Thereafter, all that is required is the preservation of the proper arrangement of forms, and the "arrival of forms at the *ultimum sentiens* does not require extension along straight lines."[117] Indeed, it is clear that forms issuing from a single point in the visual field cannot pass through the two eyes and reunite in the optic chiasma without deviating from rectilinearity.[118]

What one must conclude from the presence of optical and seemingly nonoptical elements in Alhazen's theory—what I have referred to as its "quasi-optical character"—is not that Alhazen was ambivalent on the nature of the post-crystalline transmission. Rather, we must recognize the need to extend our categories to include at least two kinds of optical process—the organic and the physical or inorganic. The propagation of forms to the *ultimum sentiens* is essentially optical because it depends on the transparency of the ocular humors, permits refraction, and requires the maintenance of the proper arrangement of forms; however, there are features of the transmission, such as its ability to depart from rectilinearity, that are unique to organic substances, and we must recognize it as an optical process that is not strictly physical.

Several ambiguities and difficulties still remain in Alhazen's theory. The principal one concerns the nature of the entity propagated from the glacial humor to the common nerve and its relationship to the radiation propagated from the visible object to the glacial humor. Most often Alhazen refers to this entity as the form of light and color, thus indicating that what passes from the glacial humor to the *ultimum sentiens* is the same thing that comes from the object to the glacial humor.[119] But if this is so, we are again faced with the same fundamental ambiguity that we observed in Galen's and Averroes's theory over the function of the glacial humor and the relationship between impressions made there and impressions made in the brain.[120] If the forms of light and color are themselves transmitted to the *ultimum sentiens* for interpretation, what is the function of impressions made in the glacial humor, and in what sense is the latter the primary photoreceptor? Alhazen does not answer these questions or remove the ambiguity; he does, however, equivocate on the identity of the transmitted entity. At one point he seems to argue that what passes to the *ultimum sentiens* is not in fact the form, but its impression or effect: "And this effect, which light produces in the glacial humor, passes through the glacial body only according to the straightness of radial lines."[121] More explicitly, but still with apparent caution, Alhazen writes that "*one may say* that the forms coming to the eye do not extend to the common nerve, but a sensation (*sensus*) is extended from the eye to the common nerve, just as the sensations of pain and touch are extended, and the *ultimum sentiens* then perceives the sensible object."[122] Finally, Alhazen argues in book 2 that the forms of light and color and the effects of those forms (i.e., the sensations that they provoke) proceed together from the eye to the *ultimum sentiens*:

> Therefore the forms [of light and color] arrive at the vitreous humor arranged as on the surface of the visible object, and this body [the vitreous humor] receives them and perceives them. . . . Then this sensation and these forms pass through this body until they reach the *ultimum sentiens;* and the extension of the sensation and of the forms in the vitreous body and in the sentient body occupying the hollow [optic] nerve to the *ultimum sentiens* is like the extension of the sensations of touch and pain to the *ultimum sentiens.*[123]

This seems clear enough if taken by itself, but it remains that except for these few passages Alhazen couches his theory almost entirely in terms of forms propagated from the glacial humor to the *ultimum sentiens*, and he does not satisfactorily explain the relationship between the forms and the sensations or impressions that apparently accompany them. I believe that his lack of clarity on this point reveals his basic indecision over the relative functions of the various sensitive elements in the visual pathway—the crystalline lens, the vitreous humor (the sensitivity of which Alhazen reveals in the paragraph quoted just above), the visual spirits,[124] and the *ultimum sentiens*. This ambiguity would remain until the seventeenth century.

We come at last to the problem of certification.[125] Although Alhazen has argued that clear perception is obtained through all the rays of the visual pyramid—that is, the rays that fall from each point of the visual field toward the center of the eye and hence reach the glacial humor without refraction[126]—he admits in book 2 that in fact there are variations in the clarity of perception. Indeed, there are two categories of visual perception—*aspectus*, a first glance that yields only a superficial perception of the object, and *intuitio*, which provides a "certified" impression.[127] Interestingly, the difference between the two is explained in optical terms. The visual pyramid, the collection of rays by which vision is chiefly brought about, consists only of rays that fall perpendicularly on the eye and proceed without refraction to the glacial humor. However, at the interface between the glacial and vitreous humors, all rays of the visual pyramid are refracted in such a way as to avoid the achievement of an apex—all rays, that is, except the central one, the axis of the visual pyramid. This ray alone falls perpendicularly on the interface between the glacial and vitreous humors and passes into the optic nerve without any refraction whatsoever.[128] Because refraction weakens, this single unrefracted ray is stronger than all others, and points seen through it are perceived with greatest clarity and, indeed, without any possibility of error.[129] The certification of a perception (i.e., the verification of the visible qualities of a body) occurs, then, through a rapid motion of the eye, which carries the axis of the visual pyramid over the visible body so that each of its points is perceived through this central ray.[130]

Conclusion

Occupied as we have been with the details of Alhazen's theory of vision, we must be careful not to lose sight of its significance. Optics before Alhazen was a competition among Aristotelians, Galenists, and Euclideans; every member of the Islamic optical tradition before Alhazen can be assigned to one or another of these ancient schools— al-Kindi to the Euclidean, Hunain and 'Ali ibn 'Isa to the Galenic, and Avicenna to the Aristotelian. Their thought may have been independent, but the framework within which they worked and the battle lines that divided them were Greek. Alhazen's essential achievement, it appears to me, was to obliterate the old battle lines. Alhazen was neither Euclidean nor Galenist nor Aristotelian—or else he was all of them. Employing physical and physiological argument, he convincingly demolished the extramission theory; but the intromission theory that he erected in its place, while satisfying physical and physiological criteria, also incorporated the entire mathematical framework of Euclid, Ptolemy, and al-Kindi.[131] Alhazen thus drew together the mathematical, medical, and physical traditions and created a single comprehensive theory. Although containing ancient materials at every point, the resulting edifice was a fresh Islamic creation.

Alhazen's theory of vision was enormously influential. It was powerfully defended and had the additional advantage of being essentially without competitors, for no other theory could pretend to the same universality. We know little about Islamic optics after Alhazen, although the commentary on Alhazen's *Kitab al-Manazir* (*De aspectibus*) by Kamal al-Din al-Farisi (d. ca. 1320) surely contributed to its dissemination and the development of its theories.[132] It was in the West, however, that Alhazen had his greatest influence. Directly or indirectly, his *De aspectibus* inspired much of the activity in optics that occurred between the thirteenth and seventeenth centuries, and among his followers we must number Roger Bacon, Witelo, John Pecham, Henry of Langenstein, Blasius of Parma, Francesco Maurolico, and Giambattista della Porta.[133]

But Alhazen's theory of vision was more than influential; it was the source of fundamental conceptions on which Kepler based his theory of the retinal image.[134] Lest this seem an exorbitant claim, I must briefly elaborate. I am not arguing that Kepler's theory of vision is to be credited to Alhazen; Kepler's new theory of the retinal image, surprisingly enough, was first formulated by Kepler. But I do claim that Alhazen provided many of the fundamental conceptual materials out of which the new Keplerian theory was built. For example, Alhazen's scheme of incoherent radiation from point sources (an idea appropriated from al-Kindi, but first applied to vision by Alhazen) is one of the most basic principles of the Keplerian theory of vision. A recognition of the need for a one-to-one correspondence between points on the object and points in the eye is another fundamental conception of Keplerian optics that was first clearly articulated by Alhazen; and, indeed, it could be argued that from Alhazen to Kepler one of the principal aims of visual theory (within the mathematical tradition) was to explain how this one-to-one correspondence is brought about. Finally, Alhazen's commitment to a theory of vision that combines the physical, the physiological, and the mathematical has defined the scope and the goals of optical theory from his day to the present. These conceptions and principles may now seem elementary, but we must not permit their present self-evidence to obscure the great intellectual effort that went into their creation or their determining influence on the subsequent course of optical theory.[135] If the modern theory of vision (in broad outline) is to be credited to Kepler, it remains that the questions to which Kepler's theory was addressed and the basic conceptual framework employed in their solution were in large measure supplied by Alhazen.

5 THE ORIGINS OF OPTICS IN THE WEST

The Encyclopedic Tradition of the Early Middle Ages

The origins of optics in Western Christendom must be sought in those chief repositories of Roman theoretical knowledge—the handbooks and encyclopedias designed to communicate the essentials of Greek learning to the Roman upper class.[1] A modicum of optical knowledge passed through Greek handbooks into their Roman counterparts, with varying degrees of accuracy and sophistication, and from there into the handbooks and encyclopedias of the early Middle Ages. The limited optical knowledge available in the early Middle Ages was thus contained in the writings of Seneca, Pliny, Solinus, Chalcidius, and their imitators. Because of the almost negligible input of new knowledge before the twelfth century (Chalcidius's translation of Plato's *Timaeus* excepted), these encyclopedias were not a medium for the communication of an expanding knowledge; rather, they were repositories for a static body of ancient wisdom. Our principal task in this chapter, therefore, is to discover what was preserved and to understand how this preserved knowledge was communicated to subsequent ages.

We get our first taste of optics in the Latin world in the *Quaestiones naturales* of Lucius Annaeus Seneca (4 B.C.?-65 A.D.), tutor and courtier to the Emperor Nero and a leading Stoic philosopher.[2] Relying primarily on Chrysippus and Posidonius,[3] Seneca presents an analysis of meteorological phenomena that is quite the best to be found in the West before the translations of the twelfth century. An account of Aristotle's theory of the rainbow leads to a discussion of mirrors, and here Seneca raises what was to become one of the central questions of the science of mirrors for at least the next fourteen centuries—whether the image observed in a mirror has an objective existence there or is merely the object itself perceived outside its true place.[4] Seneca does not develop a theory of vision in the *Quaestiones naturales* but appears to assume an extramission theory. He quotes Aristotle on the reflection of sight by dense air and notes that a person of weak sight may perceive his own image before him because his sight is unable to penetrate even the closest layers of air.[5]

Seneca's younger contemporary, Pliny the Elder (Gaius Plinius Secundus, 23/24-79 A.D.), was less interested in communicating the wisdom of the Greeks and more interested in entertaining his readers with anecdotal accounts of marvels.[6] On the subject of vision, his *Natural History* recounts such wonders as Tiberius Caesar's ability (unique in all of human history) to see perfectly in darkness and the emission of light from the eyes of cats, wolves, and wild goats.[7] But it would be unfair to create the impression that Pliny had no interest in the ordinary. He supplies a very brief discussion of the anatomy of the eye, noting that it is formed of thin membranes, the outermost of which is perfectly transparent while the inner ones are perforated by an aperture. He points out that the narrowness of this aperture "canalizes" man's gaze—terminology suggestive of an extramission theory of vision, a view supported by other passages as well.[8] The eye is connected to the brain, Pliny reports, by a blood vessel, and it is perhaps similarly connected to the stomach, for "it is unquestionable that a man never has an eye knocked out without vomiting."[9] Pliny also refers briefly to mirrors and their fabrication and explains the reflection of images by "the repercussion of the air which is thrown back into the eyes."[10]

Pliny's *Natural History* became immensely popular in subsequent centuries and determined both the tone and the content of many later encyclopedias. The *Collectanea rerum memorabilium* of Gaius Julius Solinus (fl. third century?) is a good example; this catalog of marvels was derived largely from Pliny, although Solinus succeeded in eliminating many of the duller portions and gathering novelties from a number of other sources as well.[11] In the optical realm, Solinus reports that the most powerful sense of sight (presumably in the history of mankind) was that of Strabo, who could see a distance of 135 miles, and that certain women in Scythia "have two pupils in each eye and kill people by sight if they happen to look at them when angry."[12] Here we have an account of the evil eye—indeed, of the Scythian "double whammy"!

The major event in the history of late Roman and early medieval optics was undoubtedly the translation of the first half of Plato's *Timaeus* by Chalcidius (fl. early fourth century?).[13] The translated portion of the *Timaeus* includes the sections on vision and mirrors, and the Middle Ages thus gained access to a full and reasonably accurate rendering of Plato's theories of light and vision. The wide circulation of Chalcidius's translation is attested by the exceptionally large number of extant manuscripts (including some seventy dating from the twelfth century or earlier),[14] and it is apparent that the primacy and ubiquity of the Platonic theory of vision during the early Middle Ages resulted principally from Chalcidius's efforts. However, almost as important as the translation of the *Timaeus* was Chalcidius's commentary upon it, for here Chalcidius provided not only a further exposition of Plato's theory of vision, but also arguments in favor of it and a discussion of some of the alternatives. Thus most of what early medieval scholars knew about non-Platonic theories of vision also came through Chalcidius.

In his *Commentary* Chalcidius briefly describes the visual theories of the atomists, Heraclitus, the Stoics, the geometers, and the peripatetics.[15] In reporting that the geometers and peripatetics classified vision according to the three modes of radiation—direct, reflected, and refracted—Chalcidius supplies what would become the standard organizing principle of later medieval optics.[16] Finally he turns to the Platonic theory, noting that whereas many ancient philosophers furnished partial explanations of vision (frequently taken from Plato's writings), Plato himself provided a full and complete account. Plato taught that from our eyes flows a light similar to the light of the sun. An exterior light is united with the inner light flowing from the eyes, strengthening it and making it capable of drawing from visible objects their colors:

> Therefore when these three [conditions] concur, sight occurs, and the cause of sight is threefold: the light of the innate heat passing through the eyes, which is the principal cause, the exterior light kindred to our own light, which both acts and assists, and the light that flows from the visible bodies, flame or color . . . ; without these [three causal factors] the proposed effect cannot occur.[17]

To this Platonic teaching Chalcidius adds the anatomical findings of the physicians. Two narrow pathways lead from the brain, in which the principal power of the soul resides, to the orbits of the eyes; the pathways join briefly in the interior regions of the forehead and conduct natural spirit to the eyes. The eyes themselves, Chalcidius reports, consist of four membranes or tunics having different properties and densities.[18]

Chalcidius thus defends the visual theory of Plato, into which he has incorporated elements of the Galenic theory. Largely through his influence, this "Galenized" Platonism would predominate until the thirteenth century.

Before leaving the early medieval period and the encyclopedic tradition in which so much of its intellectual life was embodied, we must take note of the theory of vision expressed by that most influential of all early medieval writers, Saint Augustine of Hippo (354–430). Augustine did not, of course, compose a treatise on vision, and his views on the subject must be discovered in discussions devoted primarily to other matters. Nevertheless, because of his immense authority, Augustine came to be consulted on all sorts of matters to which he had addressed himself only incidentally; on the theory of vision in particular, later medieval writers frequently quoted Augustine when his view paralleled their own.

In book 11 of *De trinitate* Augustine makes one of the fuller presentations of his psychology; there, after distinguishing three factors in sight (the visible object, the act of vision, and the attention of the mind), he comments briefly on the phenomenon of double vision:

> Why, even when the little flame of a lamp is in some way, as it were,

> doubled by the divergent rays of the eyes, a twofold vision comes to pass, although the thing which is seen is one. For the same rays, as they shoot forth each from its own eye, are affected severally, in that they are not allowed to meet evenly and conjointly, in regarding that corporeal thing, so that one combined view might be formed from both.[19]

Here we appear to have a clear reference to the emission of visual rays. There is, however, yet another passage in which Augustine assumes an even more explicit position on the mechanism of vision. In his commentary on Genesis, he remarks that

> surely the emission of rays from our eyes is an emission of a certain light. And it can be gathered that this [light] is emitted, since when we look into the air adjacent to our eyes we observe, along the same line, things situated far away. Nor does this light sensibly fail, since it is judged to discern fully objects that are at a distance, though surely more obscurely than if the power of sight should [itself] be sent to them.[20] Nevertheless, this light that is in vision is shown to be so scanty that unless it is assisted by an exterior light, we cannot see anything.[21]

Although avoiding a detailed analysis, Augustine has clearly opted for an extramission theory of vision.

The Twelfth-Century Revival of Learning

There can be no doubt that the winds of change were blowing over Western Europe during the late eleventh and early twelfth centuries, sweeping in a new naturalism (or "desacralization of nature," to employ Chenu's phrase)[22] which emphasized the capacity of the human mind to discover and comprehend the natural causes of things. Accompanying this renewed naturalism was a stress on quadrivial studies[23] and a zealous elaboration of the Platonic conception of the origin and operation of the material world. But if we expect to find within this new philosophical milieu a new or dramatically more sophisticated theory of vision, we are destined to disappointment. A new conception of nature, or even a new conception of how one must proceed if he wishes to understand nature, does not necessarily transform man's analysis of the particular.[24] Most often (though exceptions can be found) the description and explanation of specific natural phenomena are products not of the fresh application of a natural philosophy or of a scientific methodology to those phenomena, but of earlier theory on the same subject, modified ever so slightly in its rearticulation. In short, the philosophers of the twelfth century did not create a theory of vision on the basis of a new naturalism or a new confidence in human reason; rather, they selected a theory of vision on the basis of the literary sources available to them—which by and large were the same sources available in the West since the time of Chalcidius. To be sure, there were a few new sources, rendered into Latin late in the eleventh century or very early in the twelfth, but the impact of these on visual theory was simply to enhance and emphasize the Galenic elements already present. Theories of vision remained largely Platonic—though centuries of commentary had by this time produced a somewhat garbled version of the Platonic doctrine.

One of the most important scholars of the twelfth century was William of Conches. Born about 1080 in Conches (in Normandy), William studied we know not where (doubt has recently been cast on traditional attempts to link him closely to the school of Chartres) and eventually became a professor at Paris; in 1147 he became tutor to the son of Geoffrey Plantagenet (later to be King Henry II), and he died about 1150 or 1154.[25] In his scholarly career, William was a leader in the revival of Platonism and, above all, in the movement to require that observed phenomena be explained by natural causes. William presented his natural philosophy in a variety of works, including his *Philosophia mundi, Dragmaticon,* and *Glossae super Timaeum Platonis*. William discussed vision in each of these works, but since few significant differences in visual doctrine appear, a single exposition of this theory will be possible.

According to William, three entities are required for the occurrence of sight: an interior ray, exterior light, and an opaque object or obstacle. If any of these is lacking—as the interior ray in the blind, the exterior light on a dark night, or the opaque object in vision through a body of water—sight fails.[26] Having thus begun with an outline of the visual process that is clearly Platonic, William adds a Galenic description of the origin of the interior ray. A digestion of food (the second such) occurs in the liver; a dense vapor is thus produced, which is transformed into "natural virtue" by passage through various members; part of the natural virtue reaches the heart, where it is refined into "spiritual virtue," which passes through arteries to the brain. In the rete the spiritual virtue is further refined until it becomes luminous airy substance, which is sent through nerves to the organs of sense. When the soul wishes to see, therefore, it emits the subtler part of the airy substance through the optic nerve to the eye; from there it issues through the pupil as the "interior ray."[27]

But how does the soul receive impressions from visible objects? When the interior ray emerges from the eye, it mingles with the external light and extends to the opaque object. By its natural mobility it is diffused over the surface of the object and assumes the object's shape (*figura*) and color; thus informed and colored, the ray returns to the soul through the same apertures, carrying the shape and color of the object to the soul.[28] Proof of this theory (against some who argue that the interior ray finds the shape and color of the object already in the eye, and others who maintain that the interior ray encounters the shape and color of the object in the intervening air) is found in the fact that looking at an inflamed eye corrupts the eye of the observer: for this could occur only if the interior ray were to touch the diseased eye and return the blight to the eye of the observer.[29] Another proof is found in the phenomenon of fascination (the evil eye): the ray issuing from a man carries his qualities; when it falls on a man of contrary disposition, it harms him, for contraries are mutually injurious.[30]

In the *Dragmaticon*, one of William's later works, written as a dialogue between the Duke of Normandy (Geoffrey Plantagenet) and his philosopher-teacher, William adds to the above argument a series of objections and replies. In the first place, since we observe the stars the moment we open our eyes, how is the substance that issues from the eyes able to travel so swiftly to the stars and back? The reply is that the substance is so subtle and swift that it is "instantly here and instantly there."[31] Second, if somebody gazes steadily at an object, is the visual substance stable and immobile, or does it run back and forth without intermission? The latter, it is replied. But if it runs continuously back and forth, are there not collisions? The answer is that the substance is so swift that no interval can be noticed between its departure and its return, so subtle that in the meeting of two visual currents there is no interference. Finally, it is inquired how the visual substance is distributed during observation of a star. Is the whole substance in the eye or in the star? If part is in each, is it continuous or interrupted? If the whole is in both, then a single substance is in two different places, which is of course impossible. If part is here and part there, but continuous, how can a substance which is extended continuously from the earth to the heavens be contained within the eye? If, on the other hand, the substance is discontinuous, how can vision occur? The reply to all of these questions is that "no corporeal substance can be so subtle and swift as this. Thus it is instantly here and instantly there, and although some interval elapses between its departure and return, the interval is imperceptible."[32] Thus William dismisses the apparent difficulties simply by pointing out that all of them evaporate if one grasps the extreme subtlety and swiftness of the visual current.

William goes on to give a description of ocular anatomy that represents a slight advance over earlier accounts—undoubtedly because of the availability of anatomical works newly translated from the Arabic.[33] He also discusses the three modes of vision—by direct, reflected, and refracted rays—which he terms *contuitio*, *intuitio*, and *detuitio*.[34] But there is no need to go into these matters; what is important is that throughout his optical discussions, William is a consistent partisan of what he takes to be the Platonic theory—though his own theory of vision, with its stress on the return of the visual ray to the eye, is a bastardized version of the Platonic doctrine.

If William of Conches symbolizes the revival of Platonism in the twelfth century, his contemporary Adelard of Bath (fl. 1116-42) symbolizes another facet of the twelfth-century renaissance—the beginnings of the translation and assimilation of Arabic learning.[35] Adelard was born in Bath toward the end of the eleventh century, studied at Tours, taught at Laon, and traveled throughout Europe and the Middle East. In the course of his travels, he learned Arabic and translated a number of Arabic astronomical and mathematical treatises into Latin.[36] Adelard also composed an encyclopedia of natural philosophy, *Questiones naturales*, in which theories of vision are discussed. Although the *Questiones naturales* claims to present new knowledge acquired by Adelard through the study of Arabic learning, there is little in his theory of vision that was not traditional to the West.

The *Questiones naturales* is in the form of a dialogue between Adelard, who has just returned from extensive travels, and his nephew, who in Adelard's absence has diligently applied himself to the learning indigenous to Europe. After discussing and dismissing three theories of vision, Adelard and his nephew agree on a fourth: in the brain is generated a certain subtle air, having the same nature as fire, which passes through the optic nerve and exits through the pupil; this "visual spirit" or "fiery virtue" then passes with marvelous swiftness to the visible body, where it is impressed with the form of the body. Returning to its place, the visual spirit communicates this form to the observer's soul.[37] As evidence for this fourth theory (aside from the fact that the alternatives have already been proved deficient) Adelard points out that it alone can account for vision of oneself in a mirror: the visual spirit is reflected by the mirror back to the observer's face, receives the form of the latter, and returns it via the mirror to the eye and ultimately to the soul.[38] Precisely how the other three theories of vision would treat the phenomena of reflection and how these treatments are defective, Adelard does not reveal.

Adelard identifies this fourth theory as the "divine opinion" presented by Plato in the *Timaeus* and supports it with several long verbatim quotations from Chalcidius's version of the *Timaeus*.[39] It is by no means clear that Plato would have acknowledged Adelard's version of his theory in all of its details—particularly in its failure to stress the external illumination and its claim that a visual substance proceeds to the object and back again—but by and large Adelard does only minor violence to the Platonic theory. At any rate, this theory, now established to the satisfaction of Adelard and his nephew, gives rise to a series of further questions and objections—some of which are remarkably similar to the objections raised by William of Conches in the *Dragmaticon*.[40] First, inquires the nephew, is the visual spirit substance or accident?[41] Corporeal substance, replies Adelard. In that case, insists the nephew, it is madness to suppose that when we open our eyes, the visual spirit can travel to the fixed stars and back in the instant required for the act of vision to be performed, for even the diameter of the earth is as a point compared to the "infinite" diameter of the stellar sphere.[42] Adelard defends himself by noting, in the first place, that one must not underestimate the "exceedingly swift motion of the visual spirit," which is "more subtly perfected than all things compounded of the elements by the marvelous means of the Creator's power,"[43] and second that it is not certain that the fixed stars are quite so far away as some think. The nephew then raises a second objection, perhaps even more serious than the first, noting the possibility of the eyelid being closed while the visual spirit is out on mission. Adelard's response, in addition to a further reminder that the spirit's swiftness must not be underestimated, is that the visual spirit is the soul's agent, and the soul (which controls the opening and closing of the eye) will surely not close the door in such a way as to injure itself or its agent.[44] Third, Adelard's nephew objects that if we behold an object continuously for some time, since our perception of the object is continuous, the visual spirit must simultaneously go and come and must therefore interfere with itself. Adelard again replies, in a now familiar vein, that the visual spirit moves so swiftly that the interval is insensible, and therefore vision seems continuous even though it is not.[45] Finally, the nephew inquires how one is to understand the impressing of forms on the soul when the visual spirit returns to the observer. Assuming an Augustinian stance, Adelard assures his nephew that the sensible forms cannot act upon the soul, which is incorporeal. Rather, "when the visual spirit is stamped by a hitherto unknown form, the soul, displaying not that form but one similar to it, exhibiting not a corporeal form, but an intellectual one, at once expresses the quality of its own abundance and is then provoked to an act of judgment."[46] The forms are not themselves impressed on the soul, but trigger the creation of similar (but incorporeal) forms out of the soul's own inner resources.

Robert Grosseteste and the Philosophy of Light

Among recent scholars, it has been customary to regard Robert Grosseteste as the herald of a new era in Western optics, the founder of the optical tradition that would finally culminate in Johannes Kepler. There is a sense in which this is true. Surely in Grosseteste's work we can see a greater impact of optical literature newly translated from Greek and Arabic than in the work of William of Conches or Adelard of Bath. Grosseteste was familiar with Euclid's *De speculis*, Avicenna's *Liber canonis*, Aristotle's *Meteorologica* and *De generatione animalium*, and in all probability with Euclid's *De visu* and al-Kindi's *De aspectibus*.[47] And there can be no doubt that this wider knowledge of the Greek and Islamic optical heritage raised the level of Grosseteste's discourse on optical subjects above that of William and Adelard.

However, Grosseteste did not have access to the most important optical works of the past: he does not cite Ptolemy's *Optica* or Alhazen's *De aspectibus*, and there is no evidence that these works exerted any influence upon his optical thought.[48] Grosseteste is thus a transitional figure. He represents the initial stages of the process of assimilating the Greek and

Islamic achievement in optics, before the full impact of Greek and Arabic literature on the subject was felt. His knowledge of this literature may have been noticeably greater than that of his predecessors William of Conches and Adelard of Bath, but it was far less than that of his successors Albert the Great and Roger Bacon. And in the realm of visual theory, Grosseteste must still be classified with the Platonists of the early Middle Ages.

Grosseteste was born about 1168 and probably educated at Oxford and Paris. In 1214 or soon thereafter, he was elected chancellor of Oxford University, and from 1229 to 1235 he lectured in the Franciscan *studium* at Oxford. In 1235 he became bishop of Lincoln, a post he held until his death in 1253.[49] Grosseteste exerted a powerful influence on the intellectual life of the thirteenth century, especially at Oxford and especially among Franciscans.[50] The reasons for his great influence were not only his crucial position near the beginnings of the assimilative process,[51] but also his notable scholarly capacities, his saintly character, and his formative influence on the English Franciscan order[52]—through which he helped to shape the thought of such men as Roger Bacon and John Pecham.

Grosseteste's achievement in optics must be seen against the background of the philosophy (or metaphysics) of light—an ingredient in the complex mix of early medieval optical thought that I have thus far ignored. There has been much discussion of Grosseteste's "metaphysics of light" (for which I prefer to substitute the expression "philosophy of light," since much of it has nothing to do with metaphysics), but this discussion has frequently suffered from a failure to make several indispensable distinctions among differing bodies of ideas.[53] Within Grosseteste's philosophy of light, there are at least four distinct strands, each employing optical analogies and metaphors: (1) the epistemology of light, in which the process of acquiring knowledge of unchanging Platonic forms is considered analogous to corporeal vision through the eye; (2) the metaphysics or cosmogony of light, in which light is regarded as the first corporeal form and the material world as the product of the self-propagation of a primeval point of light; (3) the etiology or physics of light, according to which all causation in the material world operates on the analogy of the radiation of light; and (4) the theology of light, which employs light metaphors to elucidate theological truths. I do not claim that Grosseteste made explicit such a schema, but it is such that we must employ if we are to understand his achievement and its historical ancestry.

Much of the light-imagery of Western thought can be traced ultimately to Plato's *Republic*, where Plato points out that knowledge of the eternal forms is acquired by a process analogous to vision of the imperfect material world. In McKeon's words, "the Form of the Good is to the intelligible world as the sun is to the visible world."[54] The same theme appears repeatedly in the writings of Augustine, here of course given a Christian interpretation. God is the infinite uncreated light, the true light, and the source of all other light.[55]

As John Pecham was later to point out, Augustine held that "God is . . . light not figuratively, but properly";[56] God is the archetypal light, and sensible light is the imitation. Now according to Augustine, God functions in man's acquisition of knowledge exactly as the "form of the good" functioned for Plato, supplying the illumination by which the human soul is enabled to grasp intelligible things:

> But we ought . . . to believe that the intellectual mind is so formed in its nature as to see those things, which by the disposition of the Creator are subjoined to things intelligible in a natural order, by a sort of incorporeal light of an unique kind; as the eye of the flesh sees things adjacent to itself in this bodily light, of which light it is made to be receptive, and adapted to it.[57]

Thus Augustine, like Plato, employs vision as an analogy by which to comprehend the process of cognition.

This epistemological strand of light-philosophy remained an essential part of the Augustinian tradition throughout the Middle Ages, and it was under the aegis of Saint Augustine that it appeared in the writings of Robert Grosseteste.[58] In his *Commentary on the Posterior Analytics*, Grosseteste described divine illumination as a "spiritual light which is shed upon intelligible things and the eye of [the] mind and which has the same relation to the interior eye and to intelligible things as the corporeal sun has to the bodily eye and to visible things."[59] He elaborated on this analogy between the phenomena of physical light and sight and the process of divine illumination in *De veritate*:

> Therefore it is true, as Augustine attests, that no truth is perceived except in the light of the supreme truth. But just as infirm corporeal eyes do not see colored bodies unless they are illuminated by the light of the sun (however, they cannot gaze on the light of the sun itself, but only as radiated onto colored bodies), so the infirm eyes of the mind do not perceive truths themselves except in the light of the supreme truth; however, they cannot gaze on the supreme truth itself, but only in conjunction with and irradiation upon true things.[60]

This epistemological use of light-imagery is but one aspect of Grosseteste's philosophy of light. A second aspect, whose ancestry is considerably more obscure, is the cosmogony or metaphysics of light, according to which light was intimately involved in the very creation of the material world. The origin of Grosseteste's idea is surely to be found in the Neoplatonic doctrine of emanation. According to Plotinus (d. 270), the undefinable transcendent One gives rise to the universe by a process of emanation, just as the sun sends forth its rays. From the One emanates *nous*, from *nous* the world soul, and subsequently individual souls and the material world.[61] Now Grosseteste gives this emanationist metaphysic an important twist; ignoring Plotinus's conception of the emanation of *nous* and the various souls, Grosseteste proceeds directly to the material realm and proposes that God created a primordial point of light, which Grosseteste holds to be the "form of corporeity." Light is by nature self-diffusive, and this point therefore instantly extended itself in all directions, thus giving rise to the material world:

> The first corporeal form which some call corporeity is in my opinion light. For light of its very nature diffuses itself in every direction in such a way that a point of light will produce instantaneously a sphere of light of any size whatsoever, unless some opaque object stands in the way. Now the extension of matter in three dimensions is a necessary concomitant of corporeity, and this despite the fact that both corporeity and matter are in themselves simple substances lacking all dimension. But a form that is in itself simple and without dimension could not introduce dimension in every direction into matter, which is likewise simple and without dimension, except by multiplying itself and diffusing itself instantaneously in every direction and thus extending matter in its own diffusion.[62]

Grosseteste goes on to describe the generation of the firmament, which then diffused its light (*lumen*) back toward the center of the universe to produce the celestial spheres and ultimately the infinite variety of the terrestrial world. What is significant about this scheme is its acknowledgment that the study of physical light permits one to comprehend the origin and structure of the material universe. Optics has thus been given a place at the very heart of natural philosophy.

But the study of light does more than this; it also reveals the very workings of nature, the operation of causes and the production of effects. I come here to the third strand of Grosseteste's philosophy of light, the etiology or physics of light. Once again the idea is Neoplatonic in origin. In the *Enneads* Plotinus wrote:

> All existences, as long as they retain their character, produce—about themselves, from their essence, in virtue of the power which must be in them—some necessary, outward-facing hypostasis continuously attached to them and representing in image the engendering archetypes: thus fire gives out its heat; snow is cold not merely to itself; fragrant substances are a notable instance; for, as long as they last, something is diffused from them and perceived wherever they are present.[63]

Light emanating from a luminous body is, of course, another example of the same effect. In the same vein, Avicebron (d. ca. 1058) argued that powers and rays emanate from all simple substances on the analogy of the emanation of light from the sun: "The essences of simple substances do not issue forth; it is rather their powers and rays that flow forth and spread abroad. . . . Just as light flows from the sun into the air, . . . so every simple substance extends its ray and its light and diffuses them into that which is inferior."[64] Similar ideas had been expressed by al-Kindi in *De radiis stellarum*, and this may well have been Grosseteste's immediate source.[65]

In Grosseteste, these ideas become the doctrine of the multiplication of species.[66] Grosseteste summarizes the doctrine in *De lineis, angulis et figuris*:

> A natural agent multiplies its power from itself to the recipient, whether it acts on sense or on matter. This power is sometimes called species, sometimes a likeness, and it is the same thing whatever it may be called; and the agent sends the same power into sense and into matter, or into its own contrary, as heat sends the same thing into the sense of touch and into a cold body. For it does not act by deliberation and choice, and therefore it acts in a single manner whatever it encounters, whether sense or something insensitive, whether something animate or inanimate. But the effects are diversified by the diversity of the recipient, for when this power is received by the senses, it produces an effect that is somehow spiritual and noble; on the other hand, when it is received by matter, it produces a material effect.[67]

Every natural agent propagates its power from itself to surrounding bodies, and this propagation explains all natural causation. Therefore if one is to engage in natural philosophy, he must study the propagation of light and the lines, angles, and figures thus formed. Grosseteste makes this point explicitly: "Now, all causes of natural effects must be expressed by means of lines, angles, and figures, for otherwise it is impossible to grasp their explanation."[68] And again:

> Hence these rules and principles and fundamentals having been given by the power of geometry, the careful observer of natural things can give the causes of all natural effects by this method. And it will be impossible otherwise, as is already clear in respect of the universal, since every natural action is varied in strength and weakness through variation of lines, angles, and figures.[69]

Thus to understand optics (especially geometrical optics) is to understand nature.

The final aspect of Grosseteste's philosophy of light is his use of light metaphors to express theological and moral truths. The metaphorical use of light in theology can be traced at least as far back as John's gospel, where Christ is referred to as "the Light."[70] The tradition was continued by Augustine and other early church fathers, reaching a zenith in Pseudo-Dionysius's *On the Divine Names*. Here we read, for example:

> And so that Good which is above all light is called a Spiritual Light because It is an Originating Beam and an Overflowing Radiance, illuminating with its fullness every Mind above the world, around it, or within it, and renewing all their spiritual powers, embracing them all by Its transcendent compass and exceeding them all by Its transcendent elevation.[71]

Grosseteste is only moderately representative of this theology of light. He points out that "the divine essence is a light of most perfect lucidity"[72] and attempts to clarify the relationship between members of the Trinity and the operation of grace and free will by the use of light-imagery,[73] but he achieves none of the raptures of Pseudo-Dionysius. Roger Bacon, who came a generation after Grosseteste, is actually a better example of this tradition, though he succeeded in descending from the sublime metaphors of Pseudo-Dionysius to pure banality:

> Since the infusion of grace is supremely illustrated by the multiplication of light, it is useful in every way that by the multiplications of corporeal light should be revealed to us the properties of grace in the good and the rejection of it by the wicked. For the infusion of grace in perfectly good men is analogous to light incident directly and perpendicularly, since they do not reflect grace from themselves nor refract it from the direct path that extends along the way of perfection of life. But the infusion of grace in imperfect, but good, men is analogous to refracted light; for on account of their imperfections, grace does not maintain in them an altogether direct course. However, sinners who are in mortal sin reflect and repel God's grace from themselves, and therefore grace in their case is analogous to repelled or reflected light.[74]

I have made no attempt to write a complete history of the medieval philosophy of light; however, I have said enough to indicate the general character of this movement and of Grosseteste's involvement in it. But what is its significance for the history of optics and visual theory? It has been claimed, and rightly so, that Grosseteste's philosophy of light provided (at least in part) the motivation and justification for his study of optics.[75] Because optics could reveal the essential nature of material reality, of cognition, and indeed of God himself, its pursuit became not only legitimate, but obligatory. In a society as thoroughly Christian as that of medieval Europe, justifying one's priorities in theologically acceptable terms was of the utmost importance, and it was by this demonstration of the connection between optics and superior endeavors that Grosseteste justified his study.

To what extent the same claim can be made for Grosseteste's followers and for medieval optics as a whole is more problematic. To be sure, Roger Bacon followed Grosseteste in justifying optical studies on the grounds of their utility to Christendom. He devoted much of the *Opus maius*, part 4, to defending the usefulness of mathematics, and as many have noted he was more skilled in its praise than in its practice. As for optics, Bacon pointed out that "it is necessary for all things to be known through this science, since all actions of things occur according to the multiplication of species and powers from the agents of this world into material recipients; and the laws of these multiplications are known only through perspective."[76] It is difficult, after Bacon, to discover similar *apologiae* for optical studies. When John Pecham and Witelo wrote their treatises on optics later in the century, they plunged right in, with no more than a perfunctory bow to the beauty and utility of their subject, and proceeded to pursue optics for what would seem to be its own sake. The brief justificatory remarks that do appear strike one as the medieval equivalent of those encomia to the practical utility of knowledge that preface so many works on natural philosophy in the seventeenth century, and I do not think we should take the former any more seriously than we take the latter. Disciplines have a way of acquiring their own intrinsic merit so far as the adept are concerned, and optics was no different; having been once satisfactorily justified by the arguments of Grosseteste and Bacon, it became self-sustaining, as successive generations of practitioners found themselves so fully preoccupied with internal questions and problems that they had no time or incentive for wondering whether the entire endeavor was worth the effort. I am suggesting, therefore, that although the philosophy of light may have provided important initial justification and stimulation for optical studies in the West, it did not continue to operate in that capacity. By the end of the thirteenth century and throughout the fourteenth, people devoted themselves to optics simply because it was one of the established disciplines, having a certain intrinsic interest and enough unresolved issues and puzzles to keep one occupied. The justification of the discipline, if it was noticed at all, had ascended to the level of an official theoretical justification, which was available if needed but no longer expressed the personal motivations of the individual practitioner.

We come at last to the content of Grosseteste's optical work. Two works contain the bulk of his specifically optical (as opposed to metaphysical or epistemological) conclusions: *De lineis, angulis et figuris* and *De iride*—with two other works adding material of some significance: *De colore* and *De natura locorum* (the latter of which is treated in many manuscripts as a continuation of *De lineis*).[77] All were written, according to Dales's reconstruction, near the end of Grosseteste's scholarly career, perhaps between 1230 and 1235.[78]

Grosseteste pays scant attention to the theory of vision. His full statement on the subject follows:

> Nor is it to be thought that the emission of visual rays [from the eye] is only imagined and without reality, as those think who consider the part and not the whole. But it should be understood that the visual species [issuing from the eye] is a substance, shining and radiating like the sun, the radiation of which, when coupled with radiation from the exterior shining body, entirely completes vision.
>
> Wherefore natural philosophers, treating that which is natural to vision (and passive), assert that vision is produced by intromission. However, mathematicians and physicists, whose concern is with those things that are

> above nature, treating that which is above the nature of vision (and active), maintain that vision is produced by extramission. Aristotle clearly expresses this part of vision that occurs by extramission in the last book of *De animalibus*, saying: "A deep-set eye sees from a distance; for its motion is neither divided nor destroyed, but a visual power leaves it and goes directly to the objects seen." Again in the same book [Aristotle writes]: "The three senses referred to, namely vision, hearing, and smell, issue from the organs [of perception] as water issues from pipes, and for this reason long noses have a strong power of smell." Therefore, true perspective is concerned with rays emitted [by the eye].[79]

This passage is basically Platonic, though with curious overtones deriving from other traditions. The visual species, a substantial thing "shining and radiating like the sun," is clearly the old visual fire of the Platonic tradition, clothed in the terminology of species. Similarly, the "radiation from the exterior shining body" must be understood, in Platonic terms, as the external light that unites with visual radiation to form a medium capable of returning impressions to the soul; it is by no means an emanation from the observed object. Grosseteste is thus to be seen as reflecting early medieval and twelfth-century Platonism.

However, Grosseteste has also felt the impact of some of the newer traditions. He has read Euclid, in all probability al-Kindi, and a certain amount of Aristotle and Avicenna, and in the passage quoted above we catch a glimpse of the struggle to assimilate these newly available authors into the Platonic framework. We find a fresh awareness of the contest between natural philosophers (the Aristotelians) and mathematicians and physicists (who else but Euclid and al-Kindi?) over the question of intromission versus extramission, and in the present passage we have a defense of the extramission doctrine against those "who consider the part and not the whole." The influence of Euclid is also unmistakable in the concluding claim that the extramission theory is proper to the science of perspective. However, Grosseteste also retains the doctrine of intromission; in the passage above, he implies that the intromission view is incomplete, but not incorrect, a position amplified in his *Commentary on the Posterior Analytics*: "For the visual ray is light passing out from the luminous visual spirit to the obstacle, because vision is not completed solely in the reception of the sensible form without matter, but is completed in the reception just mentioned and in the radiant energy going forth from the eye."[80] Thus Grosseteste appears to have felt that he could reconcile all theories of vision: natural philosophers are correct in maintaining that the eye must receive the forms of visible bodies; mathematicians are correct in speaking of an out-going visual radiation; but Plato has seen that vision is both active and passive and has therefore combined extramission and intromission into a single theory. Grosseteste has in a very primitive way foreshadowed Bacon's later synthesis.

What then is Grosseteste's significance for the history of optics? He was the author of no significant and enduring theoretical "discoveries"; nor did he contribute to any dramatic increase in the sophistication of optical investigation and debate. His famous "quantitative" law of refraction was remarkably primitive by comparison with Greek and Islamic theories of refraction, soon to be reproduced in the West;[81] much has been made of Grosseteste's utilization of refraction in explaining the rainbow, but the luster of that achievement quickly fades when we realize that his theory of the rainbow could not account for even the most basic phenomena and has remained largely unintelligible to the present day.[82] His theory of vision was simply Plato's theory, crudely merged with Aristotelian and Euclidean teachings. Only his doctrine of the multiplication of species had a continuing influence. If we are to appreciate Grosseteste's significance, therefore, we must cease trying to view him as a precursor or the author of striking theoretical novelties; we must look beyond the theoretical content of his work, beyond any possible contributions to "positive knowledge," to his position near the beginning of the Western reception and assimilation of Greek and Arabic optics. Grosseteste is important as a transitional figure, as one of the first to reveal the impact of the new learning on the science of optics. Although Grosseteste's theory of vision remained basically Platonic, we see in his work the beginnings of the attempt to reconcile Platonic doctrine with some of the traditions made newly available through the translating activity of the preceding century. But Grosseteste is not merely a weather vane by which to gauge the appearance of new optical currents; he also provided (again by virtue of his crucial position near the beginning of the assimilative process) a stimulus to optical studies—both by precept, in his defense of the importance of optics, and by example.

The Translations

Before concluding this chapter, I must add a word about the translations. As many scholars have pointed out, the character of the twelfth-century revival of learning was dramatically transformed by a flood of translations from both Greek and Arabic; what was at first chiefly an intensification of interest in ancient Latin sources became a quest for new knowledge, previously unavailable in the West. No adequate history of this translating activity has yet been written, and I do not pretend to write one here.[83] We as yet know very little about the Western perception of Greek and Arabic learning that attracted the translators to their task, the motivations of the translators, the means by which works to be translated were acquired or selected, or even the mechanics of translation.[84] What we do know is that the initiative was almost entirely on the part of the West and that the effect was to revolutionize Western intellectual life. We know, moreover, that the trickle of translations in the eleventh century became a flood in the twelfth and was reduced again to a small but steady flow by the end of the thirteenth.[85] In optics, William of Conches and Adelard of Bath, writing in the first half of the twelfth century, had access to one or two significant works; a century later, Robert Grosseteste had much more available to him; but it was not until the middle of the thirteenth century that the full corpus of Greek and Arabic works on the subject was at hand in the major European centers of learning, able to shape (and indeed revolutionize) the thought of Western scholars.

Although we lack a history of translating activity in general, we are in a much better position concerning the particulars. While it is by no means possible to identify the translator of every optical treatise, or to specify the place and date of its translation, such information can be supplied in a surprisingly large number of cases. We know, for example, the identity of the men who translated Ptolemy's *Optica*, al-Kindi's *De aspectibus*, Hunain's *Ten Treatises*, and Avicenna's *Shifa*. But rather than interrupt the narrative by inserting such data at this point, I relegate it to the Appendix, where the reader will find discussed the translation of the principal optical works.

6 THE OPTICAL SYNTHESIS OF THE THIRTEENTH CENTURY

Introduction

In the thirteenth century the influx of Greek and Arabic literature in translation and the development of the European scholarly tradition produced a great flowering of optical studies. The essential characteristic of this movement was its progressive rediscovery and mastery of Greek and Islamic optical traditions, with concomitant shifts in the loyalties of successive generations of practitioners. We have already seen the beginnings of the movement in Robert Grosseteste, who was influenced by a few of the new works but still managed to fit them into the Platonic framework. Albert the Great, writing a decade or two later, represents a far more advanced stage in the assimilative process: he was familiar with a much wider range of sources, including the full corpus of Aristotelian works (and their commentaries), and from these he managed to construct an intelligent and skillful defense of the Aristotelian theory of vision; he also wrote some ten pages on the subject of optics to every one page supplied by Grosseteste, and with conspicuously greater understanding and sophistication.

In the second half of the century, Alhazen's *De aspectibus* at last made its full influence felt, and optical studies of the mathematical variety gained the ascendancy. The key figure in the reception of Alhazen's theories was Roger Bacon, who attempted to reconcile Alhazen's doctrines with those of Aristotle, Augustine, Grosseteste, and other authorities and thus to establish the mathematical tradition (or *perspectiva,* as it was called) in the West. Bacon was followed by Witelo, John Pecham, and others, who spread the gospel of *perspectiva* to a wide audience. The historically most significant threads of visual theory run through the works of the "perspectivists"[1] to Johannes Kepler in the seventeenth century, and it is to the synthesis developed by these writers (especially Bacon) that I principally devote this chapter. However, I must begin with a brief analysis of Albert the Great and his defense of the Aristotelian theory of vision.

Albert the Great and the Establishment of the Aristotelian Tradition

Albert the Great (Albertus Magnus or Albert of Bollstädt) was born into the Bavarian nobility between about 1193 and 1207. He studied the arts at the University of Padua, where he entered the Dominican order; later he studied and taught in the faculty of theology at Paris; in 1248 he went to Cologne, where Thomas Aquinas became one of his students. The remainder of his life was divided between teaching and administrative duties; he died in Cologne in 1280.[2] Albert's intellectual scope was enormous, as he demonstrated a mastery of virtually the entire learning of his day; he wrote what have been described as paraphrases or commentaries, but which were actually his own fresh compositions, on all known Aristotelian works for the purpose of explaining, correcting, and defending the newly received Aristotelian system. His interest in optics spanned the decades of the 1240s and 1250s, beginning with *De homine* (part 2 of the *Summa de creaturis*), written in the early 1240s, and ending with the two treatises on animals (*De animalibus* and *Questiones de animalibus*), written in the late 1250s or early 1260s.[3] In between, he included long discussions of optical matters in the *Meteora, De anima,* and *De sensu et sensato.*[4]

The problem that dominates the optical parts of Albert's *opera* is vision. This occupies some sixty pages of *De homine,* nineteen of *De anima,* thirty-two of *De sensu,* and nineteen of *De animalibus* (all in Borgnet's edition). One observes a certain degree of development in Albert's thought on the subject as new sources came to his attention, and it would be tempting to undertake a detailed analysis of his theory of vision. Such a study would throw light on Albert's position in the reception of the new learning as well as on his own scholarly methods, and the effort (I believe) would be amply rewarded. However, it is precluded here by limitations of space; moreover, since Albert drew so largely from the very texts upon which I have based the preceding chapters, such an analysis would be repetitious. For the present it must suffice to offer a very brief and general description of his theory of vision.

Albert was thoroughly acquainted with the principal theories of vision formulated by Greek and Arabic authors (except Alhazen, of whose teaching he seems to have had only limited knowledge), and in all of his works he expends much effort in describing each of them and refuting all but Aristotle's. In *De homine,* he discusses and attacks the theories of Plato, Empedocles, Euclid, and al-Kindi.[5] In *De anima* and *De sensu* he adds Democritus to his list of victims.[6] In place of these discredited theories, he attempts to establish the Aristotelian doctrine that vision is caused by an alteration (*immutatio*) of the transparent medium by the visible object and the propagation of this alteration to the watery substance of the eye.[7] However, Albert adds several non-Aristotelian embellishments: following Avicenna, Averroes, and (in his later works) Alhazen, he argues that the visual power resides in the crystalline humor, to which it is conducted by the spirit that fills the optic nerves, and that vision occurs through a pyramidal figure with base on the visible object, apex in the eye, and an axis running through the center.[8] He also provides a full description of ocular anatomy, concluding that there are seven tunics and three humors.[9] Albert argues most forcefully and effectively when opposing the extramission theory of vision, as in the long appendix to question 22 of *De homine.*[10] Here he assaults the doctrines of Euclid, al-Kindi, and Plato with a variety of objections drawn principally from Avicenna and Averroes. Albert admits that all animals (especially the lion, wolf, cat, and snake) have a certain light in their eyes, but he insists repeatedly that this is simply the result of fiery particles in the outer surface of the eye and is not a light that issues forth from the interior to produce vision.[11] On several occasions Albert refers to "certain moderns" who maintain that sight is produced by a combination of extramission and intromission (probably a reference to Grosseteste or Bartholomeus Anglicus), but he scornfully dismisses their teaching.[12]

Albert is thus a loyal defender of Aristotle on the subject of vision, although he adds to Aristotle's teachings a number of doctrines (primarily Galenic) gathered from the Islamic optical tradition. His teachers in this, besides Aristotle himself, were chiefly Avicenna and Averroes. Albert is thus the first in the West to defend Aristotle's theory of vision, and in so doing he raises Western optical theory to roughly the same level it achieved in Islam in the works of Avicenna and Averroes. But Albert's significance is not restricted to his part in establishing the Aristotelian theory of vision in the West. We must also note his position in the midst of the process of assimilating the entire corpus of Greek and Arabic writings on optics. Albert has read more of the newly translated works on optics than any of his predecessors or contemporaries, and he thus mirrors better than anybody else the sources circulating about mid-century.

What sources did Albert employ? He knew the sources traditional to the West: Seneca's *Quaestiones naturales*, Plato's *Timaeus*, Chalcidius's commentary, and the writings of Augustine.[13] He also utilized the full Aristotelian corpus, citing twelve different Aristotelian treatises in the works on vision alone. He was deeply indebted to the Aristotelian commentators, Avicenna and Averroes, whom he knew through Avicenna's *De anima* and Averroes's *De anima* and *De sensu;*[14] indeed, a significant portion of Albert's discussion is little more than a commentary on Averroes's *De sensu.* Of works in the mathematical tradition, Albert cited al-Kindi's *De aspectibus,* Euclid's *De visu, Liber de fallacia visus* (*De radiis visualibus*), and *De speculis,* and the Pseudo-Euclidean *De speculis* (*De speculo et visu*).[15] Albert made no reference to Alhazen in *De homine,* written in the early 1240s, but cited him in *De sensu* and the *Questiones de animalibus,* written in the late 1250s;[16] it is clear, in any case, that Albert made only limited use of Alhazen's *De aspectibus.* Finally, Albert's knowledge of medical literature seems to have been limited: he cited Costa ben Luca,[17] Palemon (presumably Polemon of Laodicea, a Greek medical writer of the second century), Loxus (another Greek medical author),[18] and Isaac Israeli.[19]

What is important about this list is not only its apparent Aristotelian bias and the information it supplies about the availability of particular treatises, but also its magnitude. Grosseteste, writing in the 1230s, could make use of only a handful of newly translated sources; Albert, writing in the 1240s and 1250s, had full command of a vastly larger body of literature. This growth in the number of sources led, of course, to a growth in the sophistication of analysis; whereas Grosseteste presented a brief and primitive discussion of the Platonic theory of vision, superficially reconciled with Euclidean and Aristotelian teachings, Albert offered a long and relatively abstruse defense of the Aristotelian position, enriched with a variety of Islamic accretions. Grosseteste could not yet see beyond the Platonic theory; Albert, on the other hand, recognized other schools of thought and the need to arbitrate among them.

Roger Bacon and the Thirteenth-Century Synthesis

The enthusiastic patronage of learning by the mendicant orders during the thirteenth century is well established. Moreover, a near monopoly on optical studies may be suggested by the fact that three of the most important optical writers of the century (Bartholomeus Anglicus, Roger Bacon, and John Pecham) were Franciscan friars, three (Albert the Great, Vincent of Beauvais, and Theodoric of Freiberg) were Dominican friars, another (Robert Grosseteste) was closely associated with the origins of the English Franciscan order, and two more (Witelo and Henry Bate of Malines) were encouraged in their scientific work by their friend the Dominican philosopher and translator William of Moerbeke. However, one must beware of superficial correlations and of gratuitous causal inferences drawn from them. If this list of optical writers suggests certain traditions of Franciscan and Dominican science (or, better yet, of English Franciscan science and continental Dominican science), a quick reexamination of the list reveals another correlation having equally inviting possibilities: all of the writers studied or taught at the University of Paris. Should we then leap to the conclusion that the crucial thing was a Parisian optical tradition?[20] It is not my purpose, however, to defend the idea of a Parisian optics (of which I am even more skeptical than I am of an English Franciscan optics), but to urge caution. We must avoid simplistic explanation of scientific achievement in terms of birthplace or membership in a religious order or education at a particular university. There were many ways of becoming interested or educated in optics, and we need specific evidence before we can determine which of them was operative in a particular instance. Within the Franciscan order, Bacon has been viewed as Grosseteste's immediate and most distinguished successor, and with some justification. Surely Bacon was thoroughly familiar with Grosseteste's optical works and drew from them certain basic doctrines. But there were many other influences at work as well, and Grosseteste does not appear to have been the major one.

Bacon was born in England early in the thirteenth century, but there is not a shred of evidence regarding the specific place or date. The date traditionally assigned, 1214, is as good a guess as any.[21] We know scarcely more about his education. He studied the arts at Oxford and took the M.A. degree either there or at Paris. We know nothing more of his life until he appears as a member of the arts faculty at Paris sometime before 1245. After lecturing for several years on a variety of Aristotelian and Pseudo-Aristotelian works, Bacon returned to Oxford (between 1247 and 1250) all afire with his new vision of a universal science. He joined the Franciscan order in Oxford at an unknown date, undoubtedly supposing that his studies would thus be promoted. However, he soon found himself in difficulty with his superiors and chafing against numerous restrictions. The eventual outcome was that Bacon was sent to Paris for closer watching (perhaps about 1257), and it was probably there in the early 1260s that he composed his optical works. He died about 1292, after a long and stormy career. Three treatises contain the bulk of Bacon's optics: the *Perspectiva,* which appeared as part 5 of the *Opus maius* and also circulated independently; *De multiplicatione specierum*; and *De speculis comburentibus.*[22] Optical matters are also treated in parts 4 and 6 of the *Opus maius,* the *Opus tertium,* and part 1 of the *Communia naturalium.*

Even the briefest perusal of Bacon's works cannot fail to reveal that optics was one of the central elements in his planned synthesis of all human knowledge, but the sources of his inspiration are less obvious. Bacon's belief that optics is central to natural philosophy probably derived from Grosseteste, who had displayed this view more prominently than anyone else among Bacon's predecessors.[23] There is no evidence that Bacon studied under (or even knew) Grosseteste, though it is conceivable that he and the bishop of Lincoln became acquainted during one of Bacon's sojourns in England. A more likely channel of influence would have been the presence of Grosseteste's books and manuscripts in the Franciscan friary at Oxford.[24] In any case, it is clear that Bacon was familiar with Grosseteste's ideas, including the attempt to construe optics as the most fundamental of all the natural sciences. However, the same conception could have been found in other works, such as al-Kindi's *De radiis stellarum,* and it is very likely that Bacon was acquainted with a variety of treatises that reinforced Grosseteste's point of view.[25]

As for the content of Bacon's optics, Grosseteste again undoubtedly provided important elements, but here he cannot have been the major influence. Another author who may have been influential is the Franciscan Bartholomeus Anglicus; Bartholomeus had lectured in the Franciscan convent in Paris until about 1231 (when he left for Germany), and it is possible that his *De proprietatibus rerum,* with its fairly substantial sections on light and vision, came to Bacon's attention or that his influence lingered in the Parisian friary in some other form.[26] In addition, Bacon could hardly have avoided knowing Albert the Great during his Parisian teaching career (though it is impossible to say how well), and he was certainly acquainted with Albert's reputation and works; and although Bacon expressed annoyance over Albert's popularity and success, he may well have been influenced by Albert's teachings.[27] But above all, one must reckon with the fact that Bacon had at his disposal virtually everything of importance from the Greek and Islamic optical tradition. He knew Aristotle's works thoroughly; he cited Avicenna and Averroes, Euclid, al-Kindi, Tideus, and Constantinus Africanus (to whom Hunain's *Ten Treatises* was attributed); and most important of all, he made full use of Ptolemy's *Optica* and Alhazen's *De aspectibus*, works that had theretofore been little used or totally unknown.[28] Indeed, without disregarding important borrowings from other sources, we may firmly assert that Bacon owed his greatest debt to Alhazen.

The essentials of Bacon's theory of vision are all drawn from Alhazen. Rays or species issue in all directions from every point on the visible object and reach all points on the surface of the eye—or, shifting our perspective and viewing this radiation from the observer's end, each point on the surface of the eye is the apex of a pyramid of radiation emanating from the entire visible object. But if confusion is to be avoided, a one-to-one correspondence must be established between points in the visual field and points in the eye—and that requires some mechanism to eliminate all but a single ray from each visible point. This is a problem to which Alhazen first called attention, and it is of course Alhazen who furnished the solution. Bacon explains the matter as follows:

> Although to every point of the eye and cornea comes the apex of one pyramid [originating] from the whole object, and the species of all parts [of the object] are there mixed, nevertheless to one point of the eye or cornea and the aperture of the uvea [i.e., the pupil] comes a species perpendicularly from only one point of the visible object, although to the same point come an infinity of species inclined at unequal angles [i.e., obliquely]. Therefore since the body of the eye is denser than air, it is necessary, according to the laws of refraction determined above, for all oblique lines to be refracted at the surface of the cornea. And since oblique incidence weakens the species, and similarly refraction, and perpendicular incidence is strong, therefore the perpendicular species conceals the oblique ones, as a bright and strong light conceals many weak lights.[29]

A visual pyramid is thus formed, having its base on the visible object and its apex at the center of curvature of the cornea. Vision occurs as this pyramid of radiation enters the observer's eye and its rays are arranged on the surface of the anterior glacial humor:[30]

> The principal requirement is simply that sight should perceive the object distinctly and sufficiently and with certitude, and this can occur through a pyramid in which there are as many lines as there are parts or points in the

> visible body, along which individual species come from the individual parts [of the object] until they reach the anterior glacial humor in which the visual power resides. And those lines are terminated on individual parts of the glacial humor, so that the species of the parts of the object are arranged on the surface of the sentient organ exactly as are the parts of the visible object [from which they originated].[31]

Bacon continues to unfold Alhazen's scheme, arguing that the species are refracted at the rear surface of the anterior glacial humor (the crystalline lens) in such a way as to avoid intersection; they are then propagated through the vitreous humor and optic nerve to the common nerve, where species from the two eyes are joined and a final judgment is made.[32]

Bacon illustrates his conception of the visual process with several drawings, two of which are reproduced here as figures 13 and 14.[33] In the former, rays from visible object *MP* fall perpendicularly on the anterior surface of the gla-

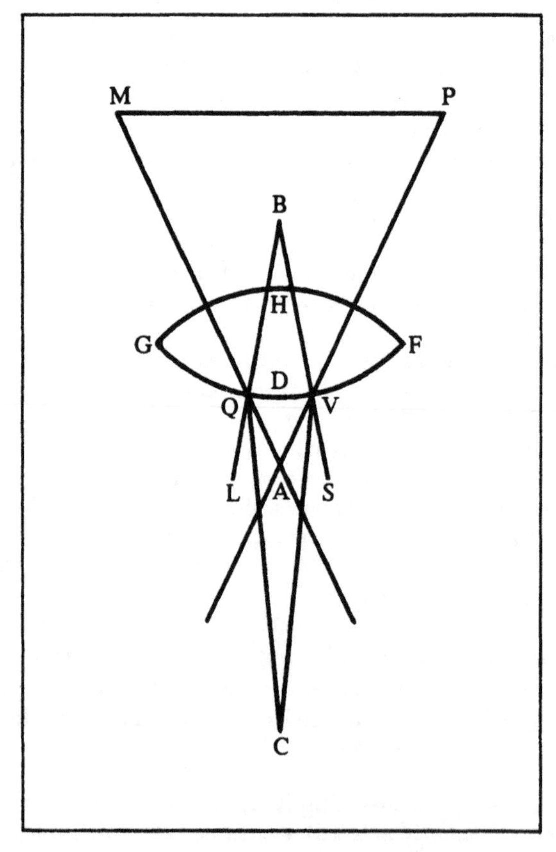

Fig. 13. Radiation through the glacial (or crystalline) humor according to Roger Bacon.

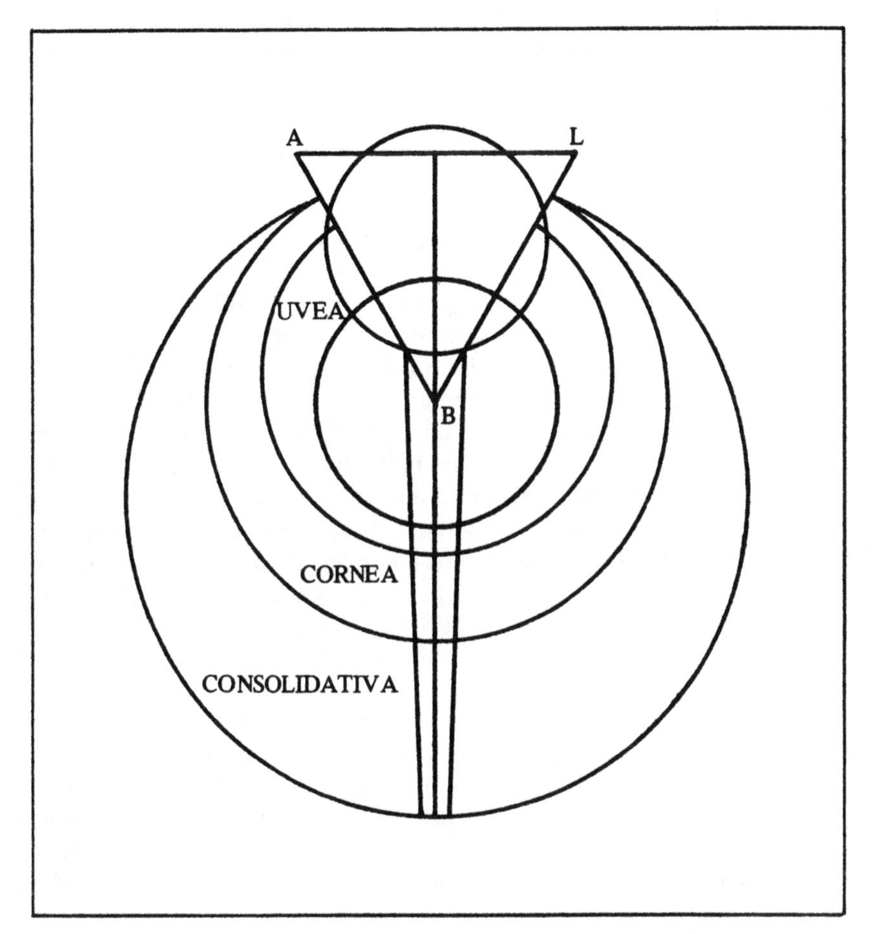

Fig. 14. The path of radiation through the eye according to Bacon.

cial humor *GF* and pass through without refraction, directed toward an apex at the center of the eye, *A*. However, before reaching this apex, the rays are refracted at *Q* and *V* on the posterior surface of the glacial humor and descend through the vitreous humor and optic nerve to an apex at *C* in the common nerve. The same scheme is illustrated in figure 14, but with all of the humors and tunics of the eye included. Rays from visible object *AL* pass into the glacial humor and are diverted by refraction at its posterior surface to an apex behind the eye, in the common nerve.[34]

Bacon follows Alhazen in many other details. Having disposed of all nonperpendicular species early in his account of vision, he later reintroduces them, arguing that the refracted species actually reinforce the unrefracted ones to improve vision.[35] He lists Alhazen's six requirements for vision (separation between the object and eye, light, magnitude of the visible object, transparency of the medium, and density of the visible object), adding two additional requirements (time and health of the eye) which Alhazen had made explicit elsewhere.[36] He repeats Alhazen's list of twenty-two visible intentions: light, color, shape, position, and so forth.[37] Regarding ocular anatomy, Bacon admits to following three authorities (Alhazen, Avicenna, and Constantinus Africanus), and his description of the tunics of the eye is actually closer to Constantine's than to Alhazen's. What is of crucial importance, however, is that he follows Alhazen in imposing a geometrical model on ocular anatomy, arguing that all parts are spherical and all surfaces anterior to the glacial humor are concentric.[38] Bacon's argument for the sensitivity of the glacial humor is identical to Alhazen's: injury to it destroys sight, whereas injury to any other tunic or humor (as long as its transparency is not affected) does not destroy sight.[39] It would be possible to go on almost without end, pointing to additional parallels between the visual theory of Bacon and that of Alhazen. But what we would prove is a point that must already be evident—that Bacon has adopted all the essential details of Alhazen's theory of vision.

But Bacon was more than a follower of Alhazen; he was also, as he saw it, a follower of almost everybody else.[40] Bacon was deeply convinced of the unity of all human knowledge (on account of its divine source) and saw his particular contribution as one of synthesis.[41] Fisher and Unguru achieve precisely the right emphasis when they point out that all of Bacon's works "are directed toward the aim of incorporating all human knowledge into one teachable system."[42] Bacon was quite capable of identifying errors in the inherited philosophical systems, especially in matters of detail, but his usual proclivities were syncretistic. Thus if apparent discrepancies were discovered in the teachings of various authorities, Bacon's first impulse was to attempt a reconciliation. This tendency was not Bacon's alone, of course, but characterized much of the Middle Ages; however, Bacon made it the basis of his life's work, and in his theory of vision it is vividly apparent.

On most points there was little or no direct conflict among the various traditions and therefore no serious obstacle to the achievement of Bacon's program of synthesis. Insofar as the mathematicians had concentrated on the mathematical theory of perspective, they had treated a topic ignored by the other major traditions; by contrast, the natural philosophers (principally Aristotelians) had focused on causal and psychological issues, and the physicians on questions of anatomy and physiology. Thus the major schools of thought were largely complementary, and a simple merger would seem possible. However, in reality none of the traditions was as narrowly defined as this simplified scheme suggests, and on certain issues they unavoidably conflicted. Two such issues were so basic to visual theory that they demanded attention, and to these Bacon applied his conciliatory skills.

The first was the physical nature of the radiation responsible for sight. Here Bacon was faced with Aristotle's theory of the qualitative transformation of the medium, Alhazen's discussion of the forms of light and color, and Grosseteste's doctrine (Neoplatonic in origin) of the multiplication of species; other alternatives, such as the Platonic doctrine of a visual fire and the atomistic theory of *eidola* or *simulacra*, which imply a material emanation, were apparently never taken seriously. Bacon's procedure, then, was simply to gloss over the differences—to take Grosseteste's visual species, endow them with all of the properties of Alhazen's forms, and claim that this was what Aristotle had meant all along. Actually, this was not so crude as it might seem, since Alhazen had restricted himself largely to the geometrical properties of forms, and Aristotle had been vague enough to permit a variety of interpretations.

The resulting doctrine is Neoplatonic in its metaphysical basis. Sounding very much like Grosseteste, Bacon writes:

> Every efficient cause acts through its own power, which it exercises on the adjacent matter, as the light (*lux*) of the sun exercises its power on the air (which power is light [*lumen*] diffused through the whole world from the solar light [*lux*]). And this power is called "likeness," "image," and "species" and is designated by many other names, and it is produced both by substance and by accident, spiritual and corporeal. . . . This species produces every action in the world, for it acts on sense, on the intellect, and on all matter of the world for the generation of things.[43]

There is a general multiplication of power or species throughout the universe, of which visual species is a single instance. Species, Bacon points out, is the "first natural effect of any agent."[44] But he argues emphatically, against Plato and the atomists, that this species is not a material emanation or effluence; rather, an object produces its likeness or species in the adjacent transparent medium, which in turn produces a further likeness in the next part of the medium, and so forth:

> But a species is not body, nor is it moved as a whole from one place to another; but that which is produced [by an object] in the first part of the air is not separated from that part, since form cannot be separated from the matter in which it is unless it should be mind; rather, it produces a likeness to itself in the second part of the air, and so on. Therefore there is no change of place, but a generation multiplied through the different parts of the medium; nor is it body which is generated there, but a corporeal form that does not have dimensions of itself but is produced according to the dimensions of the air; and it is not produced by a flow from the luminous body but by a drawing forth out of the potentiality of the matter of the air.[45]

Now that this is exactly what Aristotle and Alhazen had in mind, Bacon feels, is quite certain. Aristotle had stressed the transformation of the medium from potential to actual transparency by the presence of a luminous body,[46] and

Bacon sees this reproduced in his own stress on species as "a drawing forth out of the potentiality of the matter of the air"; moreover, Bacon explicitly employs the Aristotelian form-matter dichotomy. As for Alhazen, Bacon attributes to visual species all the properties with which Alhazen had endowed his "forms of light and color," including the tendency to radiate in all directions from every point of a visible object,[47] obedience to the established laws of propagation, and ability to stimulate the glacial humor of the eye. Bacon can thus go so far as to affirm that species "is called 'form' by Alhazen, author of the well-known *Perspectiva*."[48] Indeed, Bacon makes the more general claim that all the various expressions used to denote the effect of an agent—*lumen, idolum, phantasma, simulacrum, forma, intentio, similitudo, umbra, virtus, impressio,* and *passio*—are merely synonyms of the word "species" employed in particular contexts.[49] Consequently, almost all who have addressed themselves to the nature of light or radiation are in agreement. The unity of knowledge triumphs!

If reconciliation of opposing views on the nature of the radiation proved relatively easy, the second crucial area of conflict—the direction of radiation—posed far more serious difficulties. Bacon was in possession of virtually all the works discussed so far in this book, and the variety of opinions thus presented to him was enormous, ranging from the visual ray theories of Euclid and al-Kindi to the intromission theories of Avicenna and Alhazen, with various intermediate options. Everybody from the ancient Greeks to Grosseteste and Albert the Great had spoken on the subject, and it would require some very clever argumentation to demonstrate that all were in basic agreement.

Bacon approaches the problem by affirming, with Aristotle and Alhazen, that sight is basically a matter of intromission. He quotes Aristotle to the effect that the act of sensation requires the reception by sense of the species of the sensible thing—that sense participates passively in the visual process, while the species is the agent.[50] He also recounts several of Alhazen's observations on the appearance of an afterimage when the observer shifts his gaze from a bright spot to a dark region or from one color to another and the injurious character of bright light, from which he concludes that the species of light and color make an impression on the eye and that without such an impression there can be no vision.[51] Bacon then proceeds to develop his theory of vision at great length in terms of intromitted rays.

Thus far Bacon has remained faithful to Aristotle, Alhazen, and the rest of the intromissionist school. But although he has demonstrated that species from the visible object are necessary for vision, he has refrained from suggesting that they are sufficient. Moreover, the doctrine of multiplication of species clearly teaches that species issue from everything, substance as well as accident and spiritual things as well as corporeal things; therefore the power of sight must also be a source of species. As proof of this position, Bacon points out that the eye can see itself in a mirror, which would not be so unless a species issued from the eye and were returned to the eye.[52] The confusion here is horrendous—for Bacon has equated the species of the eye (by which the eye is itself observed) with the species of vision or the visual power (by which the eye sees)[53]—but if we disallow Bacon's proof and avoid his confusion, it remains that his doctrine of the multiplication of species requires the visual power itself to serve as a source of species. Moreover, the existence of visual radiation can also be established on the basis of authority. Ptolemy, Tideus, al-Kindi, Euclid, and Augustine all affirm its existence; even Aristotle says in *De animalibus* (according to Bacon's paraphrase) "that seeing is nothing other than the visual power coming to the thing seen."[54]

But what function could visual radiation perform, since vision, by Bacon's own admission, is fundamentally a matter of reception? Bacon explains:

> Every natural thing *completes* its action by its own power and species alone, as the sun and other celestial bodies, through their powers sent into the things of the world, cause the generation and corruption of things. Similarly lower [i.e., terrestrial] things, such as fire, dry and consume and do many things by their own power. And therefore it is necessary that vision effect the process of seeing by its own power. But the process of seeing concerns the cognition of a distant visible object, and therefore sight perceives the visible object through the multiplication of its own power to the object.[55]

The visual power is not only a recipient, but also an agent. Bacon can even argue that this is Aristotle's own position, since in *De anima* Aristotle stresses the passive aspect of vision and in *De generatione animalium* (by virtue of Michael Scot's mistranslation) the active aspect. But this does not yet identify a specific function for the active part of vision; if the species of the visible object produces a qualitative change in the transparent medium and thus makes an impression on the eye, what could the visual radiation do? Bacon replies that the species coming from mundane objects are not suited, owing to their lack of "nobility," to act on the eye and the visual power, and therefore visual species from the eye are required to excite or ennoble them (and also the medium through which they pass), thereby rendering them capable of stimulating sight:

> The species of the things of the world are not suited to act immediately and fully on sight because of the nobility of the latter. Therefore these species must be aided and excited by the species of the eye, which proceeds through the locale of the visual pyramid, altering and ennobling the medium and rendering it commensurate with sight; and thus it prepares for the approach of the species of the visible object [itself], so that it is altogether conformable and commensurate with the nobility of the animated body, i.e., the eye.[56]

Bacon has nicely succeeded in agreeing with everybody. He agrees with Alhazen and the Aristotelians that vision occurs through an impression made on the eye, a position that he does not compromise in the least. But this does not forbid the existence of visual rays that perform some other function, such as ennobling the medium and the species of the visible object. Bacon was acute enough to notice that Alhazen, Avicenna, and Averroes had never disproved the existence of visual radiation; they had, according to him, merely demonstrated the absurdity of maintaining that something material passes from the eye to the visible object, seizes the species of the object, and returns it to the eye.[57] Thus a beautiful compromise is possible. Visual rays exist, as Euclid, Ptolemy, al-Kindi, Augustine, and the Stoics insist, but they have none of the characteristics and perform none of the functions denied them by Aristotle, Avicenna, Alhazen, and Averroes.

The Baconian optical synthesis is thus complete. Roger Bacon has brought all of the optical traditions of the past—Greek, Islamic, and Christian—into a rough unity under the leadership of Alhazen. This unity may now appear superficial and flawed, but it proved sufficient to bring Alhazen's theory of vision, with its Aristotelian and Neoplatonic modifications, to a position of prominence. If the Baconian synthesis lacked depth and reflected an uncritical belief in the unity of knowledge, that by no means interfered with its wide acceptance in the thirteenth century and its continuing influence during the later Middle Ages.

However, I must add two important qualifications. First, I do not wish to suggest that the Baconian theory was always properly distinguished from that of Alhazen. It proved relatively easy, during the later Middle Ages, to overlook the differences and view the two theories as one—for they might indeed seem nearly identical when viewed beside the gulf that separated both of them from the alternatives. I am thus arguing that although Bacon's amalgamation of Alhazen's visual theory with the various other traditions contributed to its favorable reception, this Baconian amalgam was easily (and often) confused with its dominant ingredient. Second, my use of the term "synthesis" to describe Bacon's achievement is not meant to imply that his theory of vision was triumphant—that it was without serious rivals—during the later Middle Ages. On the contrary, the Aristotelian theory of vision rose to a position of clear supremacy in the fourteenth and fifteenth centuries; moreover, physicians clung to their Galen, and defenders of the Platonic and Euclidean theories, although exceedingly scarce, were not altogether lacking. The Baconian theory clearly reached its zenith in the thirteenth century, though it continued thereafter to have its practitioners and to exercise an important influence. But these are matters to be dealt with more fully in the following chapter.

The Dissemination of the Baconian Synthesis: John Pecham and Witelo

The work of unification and synthesis that Roger Bacon began was carried on by two younger contemporaries, John Pecham and Witelo. And yet we must be careful how we describe the relationship between Bacon and the two younger men, for it would be easy to misrepresent the facts. Although both Pecham and Witelo derived important features of their thought from Bacon, neither can be strictly described as his disciple. The problem, of course, is that Pecham and Witelo were exposed to a multiplicity of influences, of which Bacon was only one; and although they followed the broad outlines of Bacon's synthesis, they were not above deserting him for other authorities on occasion or having thoughts of their own. Indeed, it is safe to say that they regarded themselves principally as followers of Alhazen—although it is equally clear that they frequently viewed Alhazen through Baconian glasses.

Because Bacon's works were written in partial secrecy and dispatched directly to the papal curia (and therefore cannot be assumed to have had any immediate local circulation), one might legitimately question whether they influenced Pecham and Witelo at all. But Pecham's and Witelo's familiarity with Bacon's works is easily established on internal grounds. For example, Pecham takes Bacon to task for the animistic terms in which he had couched his theory of refraction, and Witelo mentions Bacon's value of 42° for the sum of the altitudes of the sun and rainbow (a value expressed by no other predecessor).[58] Other examples are also available, but perhaps the most important evidence (despite its diffuse and somewhat ambiguous nature) is simply the broad similarity between their respective theories of vision, to which I have alluded above and to which I shall return. As for possible channels of communication, Pecham and Bacon were fellow inhabitants of the Franciscan convent in Paris during the early 1260s, the period during which Bacon was studying optics most intensely, and there would thus have been ample opportunity for them to discuss optical matters or for Pecham to read Bacon's optical manuscripts.[59] Witelo, on the other hand, appears to have had no opportunity for significant personal contact with Bacon.[60] However, Witelo appeared at the papal curia in Viterbo at the beginning of 1269; since Bacon's works had been dispatched to the curia about a year earlier, there is no difficulty in supposing that they were still available and were placed at Witelo's disposal.[61]

John Pecham[62] relied on a wide variety of sources for his ideas on optics, including Euclid, Pseudo-Euclid, Aristotle, Augustine, al-Kindi, Avicenna, Alhazen, Grosseteste, and Bacon;[63] of these, Alhazen is by far the most significant, and Pecham could speak of his intention to "follow in the footsteps" of the *auctor*.[64] Indeed, as I have argued elsewhere, it is appropriate to conceive of Pecham's *Perspectiva communis* as a compendium of Alhazen's optics, designed to communicate Alhazen's essential achievement to an audience that required an introductory account.[65] Yet when it came to interpreting Alhazen or finding a means of reconciling his views with opposing traditions, Pecham was happy to follow Bacon's lead. Thus the Baconian synthesis was reproduced in Pecham down to all except the finest details. Following Bacon, Pecham provides an effective summary of Alhazen's theory of vision, which he modifies by admitting that visual rays issuing from the eye also play an essential part in the visual process. Pecham makes minimal use of Bacon's doctrine of the multiplication of species, but it is apparent in his work nonetheless.[66]

When Pecham diverges from Bacon, it is usually in order to rectify or improve an argument. For example, Pecham attacks Bacon's attribution of the power of choice to species as they assume a particular direction after refraction.[67] He argues that species from different points in the visual field intersect neither in the eye nor in the common nerve.[68] He avoids Bacon's confusion of the species of the eye (by which the eye is seen) with the visual radiation and also assigns to visual radiation a slightly different function—moderating, rather than ennobling, external light.[69] Indeed, Pecham's entire argument on the existence of visual radiation is a model of clarity and precision, as he maintains that what Alhazen had demonstrated was merely the insufficiency, rather than the nonexistence, of visual rays.[70]

Witelo was dubbed "Alhazen's Ape" by Giambattista Della Porta in the sixteenth century, and the epithet has been repeated by several historians.[71] It is highly misleading, however. Witelo attempted to collect in one massive volume (entitled *Perspectiva*) the teachings of the entire mathematical tradition in optics, including Euclid, Pseudo-Euclid (author of *De speculis*), Hero of Alexandria, Ptolemy, al-Kindi, Abhomadi Malfegeyr, Alhazen, and Bacon. Witelo was sometimes guilty of utilizing or juxtaposing his sources with an appalling lack of comprehension, but he can hardly be charged with slavish adherence to any one of them. Nevertheless, Witelo's finished system is dominated by the teachings of Alhazen (just as Bacon's had been); on the theory of vision Witelo is remarkably faithful to Alhazen, even avoiding any reference to visual radiation (except in refutation thereof) and Baconian species.[72]

If Pecham's debt to Bacon is determined with relative ease, the extent of Witelo's reliance on Bacon is much more problematic. That Witelo knew Bacon's optical works appears certain, but it remains to determine how much Witelo owed to the "Baconian synthesis." Aleksander Birkenmajer has argued, on the basis of the prefatory letter attached to the beginning of Witelo's *Perspectiva* and occasional remarks scattered throughout the work, that he was a close follower of Bacon. In the prefatory letter, addressed to William of Moerbeke, Witelo presents a discussion of light in Neoplatonic terms; a representative passage follows:

> Light is the diffusion of the supreme corporeal forms, applying itself through the nature of corporeal form to the matter [literally: matters] of inferior bodies and impressing the descended forms of the divine and indivisible artificers along with itself on perishable bodies in a divisible manner, and ever producing by its incorporation in them new specific or

individual forms, in which there results through the actuality of light the divine formation of both the moved orbs and the moving powers. Therefore because light has the actuality of corporeal form it makes itself equal to the corporeal dimensions of the bodies into which it flows and extends itself to the limits of capacious bodies, and nonetheless since it always contemplates the source from which it flows according to the origin of its power, it assumes *per accidens* the dimension of distance, which is a straight line, and thus it acquires the name "ray."[73]

In other passages, Witelo refers to light as "the first of all sensible forms" and as illustrative of "the efficient causes of all sensible things."[74] Witelo's remarks on such themes, although exceedingly brief, have persuaded Birkenmajer that Witelo was "profoundly influenced" by Bacon's *De multiplicatione specierum* and that "the frequently quoted introduction to his [Witelo's] *Perspectiva* can hardly be understood at all unless this influence is taken into account."[75] The main thrust of Birkenmajer's argument is that whereas the great host of Neoplatonic authors regarded light as the mediator of all natural influences in the world, Witelo, following Bacon and Grosseteste, regarded light merely as one special case of natural action that reveals (in its laws of propagation) the mode of all other natural actions; light is a particular instance which, by analogy, enables us to comprehend the general law. The main object of *perspectiva*, therefore, "is not to expound the science of optics, but rather the 'mode of action of natural forms.'"[76] Birkenmajer summarizes his view of the relationship between Witelo and Bacon by asserting that "the 'form' of Witelo is identical to the 'species' of Bacon in all the vast breadth of meaning given this term by Bacon."[77] If Birkenmajer is correct, Witelo's decision to employ Alhazen's term "form" to denote the radiant force of the visible object, rather than Bacon's term "species," by no means reflects an escape from Baconian influence; on the contrary, it becomes clear that while accepting Alhazen's teachings on the mathematical and physiological aspects of vision, Witelo supplied them with metaphysical underpinnings derived from Bacon. Although Birkenmajer's argument is persuasively presented and his conclusion likely, the tenuity of the case must also be acknowledged; not only is it based on tiny fragments from Witelo's *Perspectiva*, but there are sources besides the works of Bacon from which Witelo's ideas on the radiation of force might have come—including Grosseteste's writings and, if I understand its contents, al-Kindi's *De radiis*.[78]

I do not, in any case, think that Bacon's influence was limited to Witelo's prefatory letter and the scattered remarks on the radiation of force scattered throughout the remainder of the *Perspectiva*. I think it likely (though unprovable in any strict sense) that Bacon influenced Witelo's theory of vision as well. What makes it likely is simply that we know Witelo had access to Bacon's optical works (because of his mention of Bacon's value of 42° for the sum of the altitudes of the sun and rainbow), and we strongly suspect that

Witelo's prefatory letter reveals Baconian influence; therefore when we observe that Witelo's theory of vision echoes Bacon's attempt to defend Alhazen's optical theories in a form compatible with other traditions in Greek, Islamic, and Christian thought, there is no reason to doubt Bacon's influence here as well. And whereas there are several possible sources for Witelo's views on the radiation of influence, the Baconian synthesis of Alhazen's theory of vision with other optical currents available in the West was the only prior attempt of this kind to which Witelo could have been exposed.[79]

But it is not, after all, the derivation of Witelo's position that really matters. Nor is Witelo's significance (or, for that matter, Pecham's) a function of his faithfulness to (or even improvement upon) the details of the Baconian synthesis, but rather of his enlistment in the larger campaign on behalf of Alhazen's theory of vision, so interpreted as to be at least outwardly compatible with the teachings of Aristotle, Avicenna, Grosseteste, and whatever other authorities could not be ignored or dismissed. The importance of Witelo and Pecham then is primarily related not to their defense of Bacon, but to their defense of Alhazen—though, as I have suggested, Bacon appears to have been among the most important of those who mobilized them for the battle and shaped their strategy.

A rough measure of Pecham's and Witelo's influence on the spread of Alhazen's teachings can be obtained by counting the extant manuscripts and printed editions of their works.[80] Nineteen manuscript copies of Witelo's *Perspectiva* are now known (excluding tiny fragments), eight of Pecham's *Tractatus de perspectiva*, and sixty-two of Pecham's *Perspectiva communis*. By comparison, Alhazen's *De aspectibus* is extant in sixteen known manuscripts, Bacon's *De multiplicatione specierum* (which does not deal with visual theory directly) in twenty-five, and Bacon's *Perspectiva* in thirty-six. I am not arguing, of course, that Alhazen and Bacon ceased to have an influence of their own, but rather that their influence was augmented by the efforts of Witelo and Pecham. The contribution of Witelo and Pecham grew in importance after the invention of printing. Witelo's *Perspectiva* was printed in 1535, 1551, and 1572, and Pecham's *Perspectiva communis* went through ten editions between 1482 and 1593 (including a translation into Italian) and two more editions in the seventeenth century. By comparison, Alhazen's *De aspectibus* appeared in a single printed edition, that of 1572, and none of Bacon's works was published until 1614.

But manuscripts and editions tell only part of the story. One of the more significant events in the history of late medieval optics was the incorporation of optical studies in the university curriculum, frequently under the auspices of Pecham or Witelo (though not uncommonly of Alhazen). I have not made an exhaustive survey of university statutes of the later Middle Ages, but the following data are perhaps representative and illustrative. Pecham's *Perspec-*

tiva communis was intended as an elementary textbook and probably served that purpose from the very beginning; the earliest record of its presence in a university library indicates that it (along with Alhazen's *De aspectibus*, Bacon's *Perspectiva*, and several other optical works) was in the library of the Sorbonne before 1338.[81] By 1390 and continuing for at least the next sixty years, the *Perspectiva communis* was the basis of regular lectures at the University of Vienna, and in 1390 it was required for the M.A. degree at Prague.[82] At Leipzig in the fifteenth century there were regular lectures on "*perspectiva communis*," undoubtedly based on Pecham's work, and Andreas Alexander, who prepared the 1504 edition of the *Perspectiva communis* (published in Leipzig), justified his edition on the grounds that he had been appointed by the faculty of arts at Leipzig to lecture on perspective and needed the book for students.[83] One of the most vigorous optical traditions was found at the University of Cracow, where the *Perspectiva communis* was lectured on by a series of masters, including Sędziwój von Czechel about 1430 and the astronomer Albert Brudzewski in 1489.[84] Many copies of the *Perspectiva communis*, as well as commentaries on it, are found in the Jagellonian Library in Cracow; and in one of these commentaries is a note listing Pecham, Witelo, Bacon, and Alhazen (along with a certain Radanus) as the well-known "perspectivists" and implying that all (but especially Pecham) were in use in the schools of the land.[85] By 1431 the works of Witelo and Alhazen had become part of the curriculum at Oxford, for statutes from that year require Euclid's *Elements* or Alhazen or Witelo for the B.A. degree.[86] In 1472 Witelo's *Perspectiva* was studied at Cambridge, and in the late sixteenth century it could still be substituted for Euclid at Oxford.[87] In Spain, Pedro Ciruelo paraphrased the entire *Perspectiva communis* in his *Cursus quatuor mathematicarum artium liberalium*, published four times in Alcalá (1516, 1523, 1526, and 1528) and used at the universities of Alcalá, Salamanca, and Paris.[88] Finally, the *Perspectiva communis* was lectured upon at the University of Würzburg as late as 1594–95.[89]

Other data could doubtless be uncovered; for example, an examination of manuscripts and early printed copies of the works of Alhazen, Witelo, and Pecham would reveal the names of previous owners and would undoubtedly shed light on the uses to which the works were put. Influence can also be measured through citations: if we examine optical and philosophical works from the fourteenth until well into the seventeenth century, we find an enormous collection of citations, and it becomes apparent that most of those who touched upon optical matters were familiar with the works of Alhazen, Witelo, Pecham, and Bacon.[90] Through the influence of these works, Alhazen's theory of vision became firmly established as an integral part of the Western intellectual tradition, and so it would remain until displaced in the seventeenth century by Kepler's theory of the retinal image.

7 VISUAL THEORY IN THE LATER MIDDLE AGES

INTRODUCTION

A glance at the history of visual theory during the fourteenth and fifteenth centuries might seem to bear out old stereotypes about the later Middle Ages as a barren period in the history of thought. By the close of the thirteenth century the intromission theory was firmly established, and in the Baconian synthesis we find many of the principles on which Kepler would later build his theory of the retinal image—the punctiform analysis of the visible object, the requirement of a one-to-one correspondence between points in the visual field and points in the eye, a stress on mathematical analysis, and a relatively advanced understanding of the propagation of light and its refraction in transparent substances. But in the fourteenth and fifteenth centuries we find little or no additional progress. Why is this so? Why wasn't the Baconian theory quickly corrected and elaborated into the theory of the retinal image? Or why, at least, was there not steady (if slow) progress in that direction? Indeed, as we survey the literature emanating from the fourteenth and fifteenth centuries, why do we discover so few treatises of the perspectivist variety? Did the development of visual theory really come to a halt in the fourteenth and fifteenth centuries? Did late-medieval Scholasticism have a suffocating influence on intellectual life, as the stereotype suggests?

These are important questions, not easily answered. We are vastly ignorant of the late-medieval literature in which optical material might appear, and only the most preliminary and tentative answers will be possible. But we must make the attempt, beginning with an examination of the perspectivist tradition and then shifting our attention to other classes of optical literature.

VISUAL THEORY AMONG THE PERSPECTIVISTS

A considerable portion of the total energy devoted to the science of *perspectiva* during the later Middle Ages was tied up in transmitting the heritage of the past—copying, studying, and lecturing on treatises written in the thirteenth century or before. Among the main objects of attention were the works of Alhazen, Bacon, Witelo, and Pecham, whose influence and circulation were discussed in the preceding chapter, but the entire mathematical tradition was known and studied. To give but a few additional examples, many manuscript copies of Euclid's *De visu* and *De speculis* and

Grosseteste's optical *opuscula* emanate from the fourteenth and fifteenth centuries.[1]

However, the perspectivist tradition also had practitioners capable of producing original treatises. In the fourteenth century there appeared three treatises entitled *Questiones super perspectivam* or *Questiones perspective*: the earliest was written by Dominicus de Clavasio (d. ca. 1357-62), a set of six questions based on Alhazen's *De aspectibus* and other standard works of the perspectivist tradition; the second was by Henry of Langenstein, probably written between 1363 and 1373, consisting of fifteen questions on Pecham's *Perspectiva communis;* finally, in 1390 Blasius of Parma wrote twenty-four questions on the *Perspectiva communis.*[2] But one must not be misled by the titles of these works to suppose that they cover *perspectiva* or optics in comprehensive fashion; the questions, though drawn from many areas of the science of *perspectiva,* are individually quite narrow. Moreover, rather than being confined strictly to optical matters, some of them are primarily an occasion for applying the theory of infinitesimals or some other mathematical or logical development of the fourteenth century.[3] All are in the Scholastic tradition: arguments against the thesis to be defended are ordinarily presented first; these are followed by a lengthy argument (sometimes drawing on previous authorities) in which the author presents his own viewpoint; finally, the author responds to the contrary arguments with which the question began.[4]

A few other treatises were also produced by the perspectivist tradition. Early in the fourteenth century Theodoric of Freiberg (d. ca. 1311) wrote a treatise on the rainbow, *De iride et radialibus impressionibus,* in which the mathematical analysis of the perspectivists was heavily employed, and near the end of the century Wigandus Durnheimer (fl. 1390) wrote a lengthy *Perspectiva* in which he commented upon Pecham's *Perspectiva communis.* Finally, there appeared a number of brief tracts or individual *questiones* bearing such titles as *Questio de reflexione, De perspectiva, De speculo,* and *De speculo comburenti.*[5] Most of these treatises consist of no more than a few leaves, and none except Theodoric's is extant in more than three manuscripts; many appear to be the product of debate and disputation in the medieval university, and although they constitute an important index of what was known and discussed, none had a significant influence.

If we wish to investigate developments in visual theory among the perspectivists, we must examine the *Questiones super perspectivam* of Henry of Langenstein and the work of the same title by Blasius of Parma—the only treatises of the late-medieval perspectivist tradition that had a circulation of any significance, and among the few that deal at any length with visual theory.[6]

Henry of Langenstein (also known as Henry of Hesse) was born about 1325 in the village of Langenstein in Hesse.[7] In 1363 he earned the M.A. degree at the University of Paris and joined the faculty of arts; before the end of the year he was elected Proctor of the English Nation.[8] He appears to have been actively involved in the arts faculty until the early 1370s, when he matriculated in the faculty of theology. He earned the theological doctorate in 1376 and became a member of the College of the Sorbonne; later he became vice-chancellor of the university. In 1383 he was summoned by Albert III (Duke of Austria) to the University of Vienna, where he was asked to reorganize the university and initiate instruction in theology. He became dean of the theological faculty in 1389 and rector of the university in 1393. He died in 1397. Henry wrote a variety of treatises on natural philosophy, including one on Ptolemaic astronomy, another on the comet of 1369, an antiastrological work, two treatises on the nature of cause, and the *Questiones super perspectivam.*[9]

No complete and systematic analysis of vision is to be found anywhere in Henry's *Questiones super perspectivam,* though questions 1, 4, and 9-11 deal with important aspects of visual theory, and scattered remarks found throughout the treatise reveal Henry's intromissionist sentiments.[10] Henry's credentials as a perspectivist are established (and the foundations of his theory of vision laid) in the first two questions, where he argues that light is multiplied by means of rays, which are understood not as mathematical lines but as pencils or pyramids of radiation, and that a luminous body terminates the pyramid of its light at every point of an adjacent transparent medium. The philosophical basis for these claims is found in the doctrine that natural agents act through "instrumental" or "intentional" qualities sent from agent to recipient, which are impermanent and of different kind from the "acting quality" of the natural agent. These are multiplied radially in straight lines from all sensible qualities. In the case of light, the natural agent is *lux* (the luminous quality of the bright body) and the instrumental quality or species issuing from it (through which *lux* is able to act on vision) is *lumen.*[11]

The geometry of vision is first broached in the fourth question, where Henry inquires into the paradoxical fact that a luminous body, such as the flame of a candle, may appear larger from afar than from nearby. One explanation of this phenomenon is refraction in the observer's eye.[12] In figure 15, *DSRC* represents the eye, *DC* its aperture, and *KE* the visible object; triangle *KOE* represents the visual pyramid composed exclusively of perpendicular rays.[13] However, not all radiation remains within this "perpendicular pyramid"; rays *KS* and *ER* fall outside the pyramid and are refracted toward the center of the eye.[14] Points *K* and *E* are therefore perceived as though situated at *K'* and *E'*, and the object appears larger than it really is, because angle *SOR* is larger than angle *KOE.* What is in fact revealed by this relatively primitive demonstration is not that the apparent size of a body increases with remoteness from the eye, but merely that under circumstances which Henry does not define a body may appear larger than it would if viewed by unrefracted rays. Henry does not explain the effect on angle *SOR* of increasing the distance between the eye and the object, except to note that not everything is observed through a pyramid whose base is the object, for if this were so the farther the object the smaller it would have to appear.

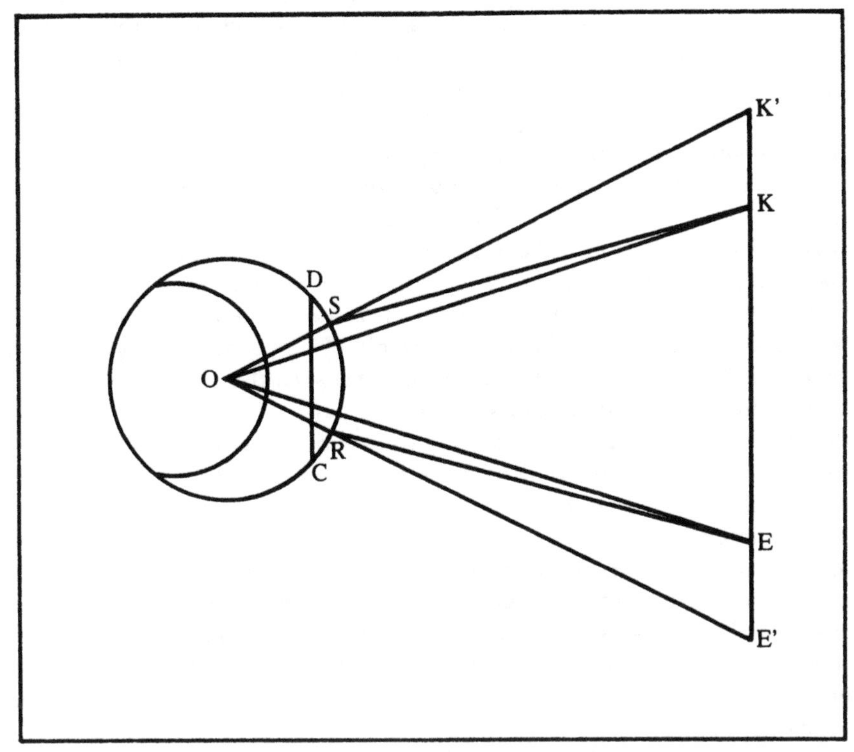

Fig. 15

In question 9 Henry returns to the problem of rays refracted in the eye—specifically, to the question "whether in a uniform medium visible things are seen without refraction."[15] The answer is no, and in elaborating on this theme (through a series of six conclusions) Henry makes a serious attempt to say something about the path of radiation within the eye. In the first conclusion he maintains that no ray is, strictly speaking, perpendicular to the eye because all rays have width, so that if the center of the ray is perpendicular to the curved surface of the eye, the lateral edges (parallel to the central line) must fall on the eye obliquely. Second, all rays having their central lines perpendicular to the cornea are directed toward an apex at the center of the eye. However, different tunics and humors have different centers, and therefore these "perpendicular" rays can be directed toward only one of them—the center of curvature of the cornea or outer surface. The third conclusion returns to the subject of the first, noting that because rays are physical and have width, no ray at all (not even the central axis of the visual pyramid) can pass without refraction through the centers of curvature of all the eye's tunics and humors. This may seem like quibbling, but it illustrates the late-medieval inclination to explore and test the meaning and implications of philosophical claims—in this case, the claim of the perspectivists that vision through the central ray is able to certify visual perception because the central ray passes through the eye without refraction.[16]

In the fourth conclusion of question 9, Henry provides a nice demonstration (marred only by the fact that one of its premises is false) that refraction of rays as they pass from the crystalline humor to the vitreous humor cannot prevent intersection of the rays—thus denying what had been fundamental to the perspectivist tradition since Alhazen; for at all costs intersection and inversion were to be prevented. In figure 16, *O* is the center of curvature of the anterior surface of the crystalline humor, and *K* is the center of the vitreous humor; *ED* is the visible object, and *EOD* is the pyramid of perpendicular rays.[17] When rays from *E* and *D* encounter the interface between the crystalline lens and vitreous humor, they are of course refracted; but no matter what the relative densities of the two media, these rays cannot be refracted beyond perpendiculars *FK* and *CK*. If the vitreous humor is less dense than the crystalline humor, the rays will be refracted away from their respective perpendiculars and will intersect above *O*; if the vitreous humor is denser than the crystalline lens, the rays will be refracted toward their respective perpendiculars and will intersect between *O* and *K;* therefore, according to all possibilities, the rays intersect. The only flaw in the demonstration is that Henry has assumed that surface *FC* is convex (as seen by the incoming ray), whereas in the past it had generally been regarded as concave;[18] and in the latter case, Henry's conclusion does not follow.

The fifth conclusion follows from the fourth. If rays were to advance in the eye according to the ordinary physical laws of propagation, they would necessarily intersect, and the observer would have a reversed and inverted impression of the visual field—unless, of course, cognition were to occur before the point of intersection, "immediately beneath the convex surface of the interior glacial humor."[19] However, having offered this as a possible solution, Henry expresses his preference for another: once within the vitreous humor rays are no longer bound by the law of rectilinear propagation but are multiplied according to the "path of visual spirits, with which that humor happens to be filled."[20] Hence rays do not intersect because within the vitreous humor they are not propagated in straight lines.

Finally, in the sixth conclusion of question 9, Henry deals with rays outside the primary visual pyramid. Rays obliquely incident on the eye reach the visual

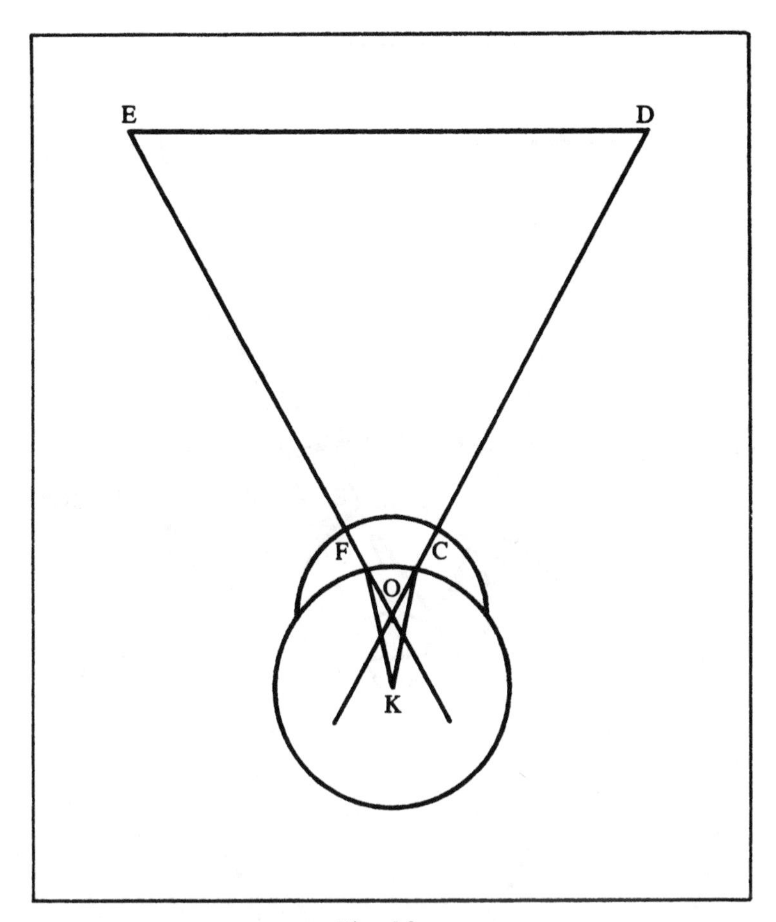

Fig. 16

power by means of four refractions—toward the perpendicular at the surface of the cornea, away from the perpendicular at the surface of the albugineous humor, away from the perpendicular at the surface of the crystalline humor, and away from the perpendicular at the surface of the vitreous humor (see fig. 17). Because three out of four times they are refracted away from their perpendiculars, few of these rays are able to reach the visual power (which Henry apparently locates in or behind the vitreous humor); and if they reach it, they are generally "on the opposite side of the eye" and apparently have little effect because of their small number.[21]

In question 10, which inquires "whether every visual perception takes place under an angle," Henry considers the size of ocular apertures and the geometry of radiation through them.[22] He claims that the eye has two apertures, the

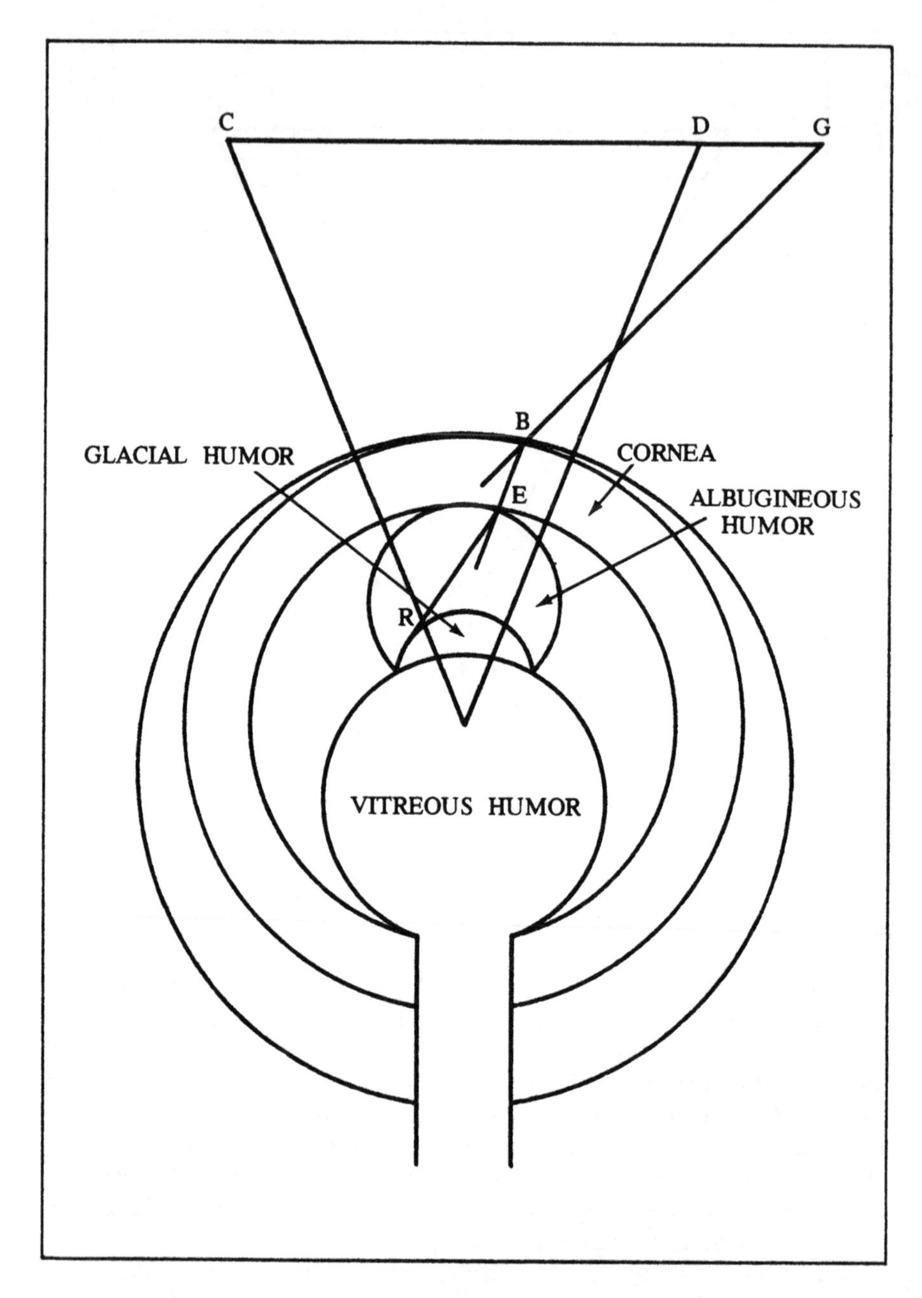

Fig. 17. The refraction of radiation within the eye according to Henry of Langenstein.

innermost being the pupil and the outermost corresponding to the junction between the opaque sclera and the transparent cornea. Both apertures are equal (in diameter) to the side of the square inscribable within their respective tunics; this means that both intercept a right angle at the center of the eye, and the sides of both will be grazed by the outer edges of the maximum visual pyramid.[23]

Henry is aware of the implication of this ocular geometry. The least angle under which the object is just too large to be seen in its entirety is the angle having its sides in contact with the aperture (apparently the outer one) and its apex at the convex surface of the vitreous humor. (Fig. 18 illustrates what he has in mind: *AB* is the aperture, and *C* is a point on the surface of the vitreous humor.)[24] Under this angle, Henry concludes, "vision cannot occur since no part of it enters the interior glacial [i.e., vitreous] humor, in which resides the power of apprehending."[25] What is most remarkable about this argument is Henry's insistence that the visual power is situated in the vitreous humor. I do not know whether this represents a conscious attempt to improve the theory, or a misunderstanding of the past optical tradition, or an awareness of the passage in which Alhazen affirmed the sensitivity of the vitreous humor— though I am disposed to rule out the second alternative on the grounds that Henry could not have misunderstood the unambiguous affirmations by Pecham and the other perspectivists (not to speak of medical writers) of the sensitivity of the crystalline humor.[26] In any case, it is noteworthy indeed to find a medieval author suggesting alternative locations of the visual power within the eye.

Finally, in question 11 Henry delves into the psychology of perception. He argues that the species of color and light are the intermediaries by which all visible intentions (such as shape, size, number, motion, and so forth) are

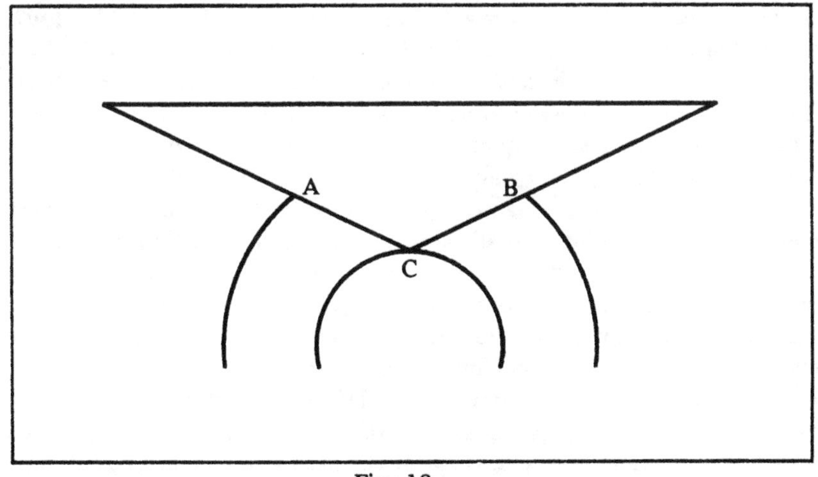

Fig. 18

perceived.[27] The species or similitudes of sensible things produced in the exterior sense organ are of the same kind as those produced outside the organ in the air or transparent medium, and the similitudes received by the internal senses are of the same kind as those received in the external sense organs.[28] This question also makes an additional point of interest for visual theory. Henry argues that species pass from the external to the internal senses either by multiplication or by motion through hollow nerves containing sensitive spirits. He elaborates on these two alternatives in the following passage:

> In the hollow nerves descending from the brain and carrying the sensitive spirits there must be a transparent body suited, when illuminated, to the multiplication of species, which body is terminated at the exterior organ of the senses; or else it is necessary that in those hollows [of the nerves] there be most subtle and clear bodies, namely spirits, flowing continually from the brain to or toward the outside and afterwards flowing back with a certain motion, in which reflowing spirits the received *simulacra* of sensible things are carried to the common sense or the imagination.[29]

The second of the alternatives, according to which the spirits themselves return to the brain bearing the simulacra with them, can perhaps be interpreted as an overly hydrodynamic version of the Galenic theory; its presence here, in any case, is a curious modification of the perspectivist theory of vision.

Blasius of Parma (or Biagio Pelacani da Parma) was somewhat younger than Henry of Langenstein, having received the doctorate in arts from the University of Pavia about 1374.[30] He remained on the faculty of arts at Pavia until at least 1378, then taught at the universities of Bologna, Padua, and Florence. There is evidence to suggest that he also studied at the University of Paris, perhaps receiving a doctorate in theology.[31] In 1389 he returned to Pavia, and in 1407 to Padua. The University of Padua dismissed him in 1411 on the grounds that he was no longer fit to teach, and in 1416 he died in his native Parma. Blasius wrote treatises on a variety of topics, including Aristotelian commentaries, a treatise on weights, a set of questions on the latitude of forms, two or three astonomical works, and the *Questiones super perspectivam.* The latter is dated in the explicit of one of the manuscripts to the year 1390.[32] Since it presents no coherent theory of vision, I shall limit myself to a few brief conclusions.[33]

In the first question, Blasius inquires whether it is necessary to posit the diffusion of species from the object to the eye in order to explain vision. He presents the case for both sides, the negative clearly revealing Ockhamist influence, and appears to conclude (against Ockham) that one must indeed posit species.[34] Blasius argues that one can defend the multiplication of species from the eye to the object as well as from the object to the eye, but he regards the latter as the more probable.[35] As for the nature of species, Blasius argues that there is no contradiction in maintaining either that species are true substances (so that "when I see . . . an ass, that ass multiplies asses from itself through the medium to the eye") or that species are qualities rather than substances;[36] however, he prefers the latter alternative.

Blasius also touches upon the location of the visual power in this first question. He argues that it is more probable that vision occurs in the glacial or crystalline humor than at the junction of the optic nerves: in the first place, the visual power should have its seat in some organ, but in the optic nerves there is nothing but visual spirit flowing from the brain to the eyes; second, the Master of Perspective (John Pecham) and the rest of the perspectivists argue that vision must occur in the glacial humor because injury to it impairs vision.[37] This conclusion is reinforced in the second question, where Blasius appears to assert without qualification that the visual power is situated in the glacial humor; he also explains how vision can continue even after the removal or destruction of the perceived object, arguing that the glacial humor, because of its substance and density, can conserve the species diffused into it from the object for a certain period of time.[38]

The location of the visual power is investigated once again in question 12 of the first book, where Blasius inquires "whether it can be evidently concluded that vision occurs at the junction of the optic nerves."[39] The argument is long and convoluted, and a complete analysis of its contents could be a chapter in itself. However, to summarize very briefly, Blasius does not (as far as I can judge) come down clearly on either side of the question, but rather maintains that either view can be persuasively argued. In support of the view that vision occurs in the glacial humor rather than the optic nerve, Blasius argues that the intromission theory of the perspectivists is built upon the idea of pyramids of radiation consisting of rays perpendicular to the eye and that this scheme assumes that vision occurs in the glacial humor. He also recounts the argument that injury to the glacial humor impairs sight, whereas a healthy glacial humor permits sight. On the other side of the question, Blasius attempts in the main to reveal weaknesses in the opposing arguments. He points out, for example, that the argument about an injured glacial humor impairing sight is not demonstrative, since the glacial humor can be regarded merely as a pathway, whose corruption prevents passage of the species. He illustrates this with an analogy: from the fact that "removal of the teeth impairs digestion, it does not follow that digestion occurs in the mouth where the teeth are located."[40] Blasius also argues that there is no good way of explaining the appearance of a double image when one eye is dislodged from its normal position except by positing that the visual power resides at the junction of the optic nerves and that through elevation or depression of the eye "the united nerves are dislocated or separated from each other and therefore remain distinct in place and subject."[41]

It is clear from this analysis that Henry and Blasius accept at least the broad outlines of the perspectivist theory of vision. They do not doubt (seriously, at any rate) that vision is produced by intromitted rays, and both are prepared to submit this radiation to geometrical analysis—and, indeed, to conceive the vision-producing rays as forming a visual pyramid with base on the visual object and apex at the center of the eye. A fuller analysis of their treatises would generate a much longer list of perspectivist commitments. But despite their perspectivist outlook, neither Henry nor Blasius attempts a comprehensive and coherent statement of the perspectivist theory of vision; rather, they make unsystematic forays into perspectivist visual theory in order to question it on a few very restricted issues. For example, Henry inquires into the relationship of *lux* to *lumen* and explores the implications of the fact that rays are three-dimensional pencils rather than geometrical lines. Blasius inquires whether species are required for causing vision and debates with himself whether the seat of the visual power is in the crystalline humor or the junction of the optic nerves (for in the past, perspectivists had defended both locations). In questions that I have not discussed in this chapter, both Henry and Blasius inquire whether a radiant pyramid from a luminous body is terminated in an apex at every point of the adjacent transparent medium, and why radiation passing through a small triangular aperture gives rise to a circular image; Henry inquires whether perpendicular rays are indeed stronger than oblique rays, and Blasius explores the possibility that bodies no longer in existence are nevertheless perceived.[42] Geometrical analysis is undertaken on occasion, more often by Henry than by Blasius, but with no conspicuous skill. Questions are sometimes left unresolved, more often by Blasius than by Henry, and one has the distinct impression that visual theory was not significantly advanced for the attention they gave it.[43]

What then can we conclude concerning *perspectiva* in the later Middle Ages? The one thing that is perfectly clear is that Henry and Blasius were engaged not in explaining or defending the perspectivist theory of vision, nor in extending and improving it, but in submitting it to criticism a piece at a time. They were not partisans of the theory, but investigators of it. Does this represent a decline of the perspectivist tradition in the fourteenth century? Probably so. But before we explore the matter further, we must consider other bodies of late-medieval literature in which visual theory comes under discussion. Two present themselves immediately: commentaries on Aristotelian works, and commentaries on the *Sentences* of Peter Lombard. Let us consider each of them in turn.

Visual Theory among Aristotelian Commentators

Before the end of the thirteenth century Aristotelian natural philosophy became the heart of the arts curriculum in the universities, and the literary outcome in the fourteenth and fifteenth centuries was an enormous number of commentaries (often in the form of questions) on the works of Aristotle. Three Aristotelian works in particular—the *Meteorologica, De anima,* and *De sensu*—were apt to raise questions about vision, and it is in their commentaries that much of the visual theory of the later Middle Ages is to be found.[44] The emphasis in these commentaries is, of course, somewhat different from that of the perspectivist tradition, for Aristotelian commentators were naturally more interested in the physical, ontological, and psychological concerns of the Aristotelian tradition than in the geometrical issues that preoccupied the perspectivists. Yet we will see that certain Aristotelian commentators incorporated elements borrowed from the perspectivists into their visual theories.

Visual theory within the late-medieval Aristotelian tradition has been so little studied that there is no hope of providing a broad and synthetic summary. I cannot even promise to illustrate the tradition with men who properly represent the full spectrum of opinion. What I will do is consider several Aristotelian commentators whose theories of vision are at least interesting, under the assumption that they will reveal in a general way the kind of question to which the Aristotelian tradition addressed itself.

One such Aristotelian commentator of the first half of the fourteenth century is the Parisian scholar John Buridan. Buridan was born about 1295 in northern France, received the M.A. degree at Paris soon after 1320, and spent most of his remaining years in Paris as regent master in the arts faculty; he died about 1358.[45] Buridan was clearly one of the most distinguished natural philosophers of the fourteenth century; known as a nominalist, he made significant contributions to logic, scientific method, dynamics, and cosmology. But he dealt with many other topics as well, and in his *Questiones* on Aristotle's *De anima* and *De sensu* he touched upon vision.

Buridan's discussion of light and vision reveals how peripheral were the typical interests of fourteenth-century Aristotelian commentators to the issues that have thus far occupied us in this book. Buridan does not undertake a full discussion of vision and provides only a very brief and very general defense of the intromission theory. His purpose is merely to comment upon very restricted problems raised by Aristotle's psychological tracts, and if Aristotle's theory of vision is assumed as a basis for discussion, there is no way of knowing how Buridan saw its relationship to other theories within the intromission tradition.

Most of Buridan's effort is applied to understanding the nature and interrelations of light (both *lux* and *lumen*) and color.[46] In book 2, question 17 of his *Questiones breves super librum de anima,* he inquires whether external illumination (*lumen*) is required for seeing colors on account of the colors or on account of the medium. After attacking a wide variety of alternatives, Buridan concludes that *lumen* is required for seeing colors because colors alone are frequently too weak to affect the medium and sight; however, working together, "*lumen* and color are sufficient for affecting vision and impressing in it species sufficient for vision," just as two candles may make vision possible when one will not.[47] Having thus entangled his readers in the theory of light and color, Buridan proceeds to unravel the relationships among *lux, lumen,* and color. *Lux* and color have fixed existence in the lucid and colored body, respectively, and are moved with it. Indeed, as Buridan points out in question 18, they terminate or bound their subject, making it impossible for the interior to be seen unless (as in the case of wine and colored glass) the subject "is not perfectly and intensely lucid or colored."[48] *Lumen,* by contrast, does not terminate or bound its subject, the transparent, and does not have fixed existence in it. Or expressed in other terms, *lumen* is the species of *lux,* its image or representation in the medium; and *lux* is seen through *lumen,* its species, just as color is seen through the species of color.[49] This analysis is employed in question 18 to determine that in one manner of speaking *lux* is the first object of vision, in another manner of speaking *lux* and color together are the first objects of vision, and in yet a third manner of speaking there is no first object of vision.[50] What is perfectly clear from this discussion is that *lumen* and the species of color are not themselves perceived, but are merely the agents by which *lux* and color are perceived.

Buridan presents a very sketchy account of the visual process in book 2, question 24, where he inquires whether a sensible object placed against or within the eye (or any of the external senses) can be perceived. In the course of proving that sensibles against or within the eye are unobservable, Buridan sets down his "first conclusion":

> A sensible object is never perceived unless it or its species is placed within the organ of sense; for a sensible object is not perceived through something issuing from the sense, since it would be absurd when we see the stars [to suppose] that something issues from vision [and extends] to the stars. Therefore it must be supposed that sensation occurs when sense receives something from the sensible object, and this is either the sensible object itself or a species representing it. Therefore we do not perceive except by reception of the sensible object or its species in sense or the organ of sense.[51]

Subsequently Buridan establishes, in part by what he calls "experimental induction," that sensible objects placed against the eye are unobservable and therefore that only through its species can an object be observed.[52]

One of the stumbling blocks for the intromission theory throughout the ages has been the observation that the eyes of cats and certain other animals shine in darkness, for this suggests that the eye has its own intrinsic light, which might reasonably be supposed to issue forth and cause vision in the absence of external illumination. Buridan turns to this problem in question 4 of his *Questiones super librum de sensu et sensato,* where he inquires "whether the brightness with which the eye appears to shine or scintillate is within the eye."[53] He argues that it is not, on the grounds that we sometimes see this brightness in our own eye, as when the eye is struck; but since it has been proved that a sensible object placed against or within the eye cannot be observed, the brightness must be external. Moreover, if this brightness were within the observer's eye, he would always see it or else never see it, contrary to experience, since we sometimes see it; consequently it must be the result of external *lumen*. But why, then, is it seen only when the eye is struck, and how can this occur in a dark room or when the eyelids are closed? Buridan argues that external *lumen,* too weak to stimulate sight, is able to penetrate closed eyelids[54] and is present in very small quantities even in a dark room. This external *lumen* is reflected[55] in the "blackness" (pupil) of the eye because of its smoothness and is thus united and strengthened so as to affect vision, just as a burning mirror takes light that is not heat-producing and by uniting it through reflection kindles fire. As for the necessity of a blow,

> when the eye is round it reflects the entire [*lumen*] toward the outside and none into itself; therefore it is unable to see [any brightness] on account of this. But when the eye is struck it is bent inward, and thus one part reflects *lumen* into another part and vice versa; therefore one part perceives *lux* through *lumen* reflected by the other part.[56]

Finally, what about brightness observed in eyes other than one's own, as in the eyes of a cat? Buridan replies that this is merely external light reflected from the eye of the cat in a unified way so as to become perceptible. Once again the intromission theory is preserved.

Nicole Oresme, one of the most original and creative natural philosophers of the fourteenth century, is usually assumed to have been a disciple of Buridan at the University of Paris.[57] Oresme was born in Normandy, perhaps near Caen, about 1320-25. He was educated at the University of Paris, where in 1348 he held a scholarship in the College of Navarre for the study of theology. He became master of theology about 1356 and Grand Master of the College of Navarre in that same year. During the final two decades of his life, Oresme held a series of ecclesiastical posts (including the bishopric of Lisieux) and served the French crown in a variety of endeavors. He died in 1382. Oresme had a literary output of enormous significance. He wrote a series of *Questiones* on Aristotle's *Physica, De caelo et mundo, De generatione et corruptione, Meteorologica, De anima,* and *Parva naturalia.* He also prepared *Questiones* on Euclid's *Elements* and on the *Sphere* of Sacrobosco. And finally, he wrote a series of independent works on physical and mathematical subjects, antiastrological tracts, theological works, and a treatise on money.

Oresme touches upon optical subjects in many of his works, but I will restrict myself to the *Questiones super quatuor libros meteororum,* with occasional reference to other treatises. What first strikes one is Oresme's much fuller knowledge (or at least use) of the past optical tradition than that of his master, Buridan. He cites Aristotle, Euclid (*De aspectibus* and *De speculis*), al-Kindi (*De radiis stellarum* only), Alhazen, Grosseteste (*De iride* and *De radiis,* the latter presumably including both *De natura locorum* and *De lineis, angulis et figuris*),[58] Bacon, Pecham, Witelo, Albert the Great, and a certain Robert the Englishman (who wrote a commentary on Aristotle's *Meteorologica*). Nor is this a case of teaching an essentially pure version of Aristotelian theory with a thin overlay of citations to other authors; Oresme's use of the various sources is thorough and confident, and his mastery of the optical works of Alhazen and Witelo is apparent. Indeed, at the close of the first chapter ("On the Causes of Marvels Contingent on Sight") of his *Quodlibeta,* Oresme spells out the nature of his attempt to integrate the ideas of Aristotle with those of Alhazen and Witelo:

> [The notions of] what and where the said species are; and what, how, and how many are the interior faculties; and how they are moved with respect to objects; and how they move, and so on, I assume from the book *De anima* and *De memoria* and *De* [*sensu et*] *sensato* and *De somno et vigilia* [of Aristotle]. And [the notions of] how vision occurs, what things are required, what the visibles are, when one color is altered merely in position, and how error arises—these I take from the *Perspectiva* of Alhazen in the first, second, and third books, and also from the *Perspectiva* of Witelo, in the first four books, where they treat these matters exquisitely.[59]

Oresme confronts visual theory most fully in book 3, question 12, of his *Questiones super quatuor libros meteororum.* There, in a discussion of the refraction of "vision," Oresme makes a distinction between two senses of the term "vision"—visual power and visual ray—which leads to a discussion of whether vision occurs by extramission or by intromission. The ancients, Oresme points out, thought that vision occurred by extramission, and this position is still defensible if it is taken to mean not the emergence of a body from the eye, but "the multiplication of a certain power or spiritual influence in the same manner as color multiplies its species in a medium."[60] The extramission theory can also be defended by the authority of Robert Grosseteste and of Aristotle in *De generatione animalium* (which Oresme, like Grosseteste and Bacon, cites in Michael Scot's faulty translation).[61]

But against the extramission theory we have the authority of Aristotle in book 2 of *De anima,* where he proves that sight is a passive power and therefore cannot emit radiation. A second argument is that "the visual power is not limited by a *maximum a quo non* but by a *minimum quod non*" and that this is characteristic of passive rather than active powers.[62] Finally, Oresme inquires whether it is really probable that vision could produce an effect as far as the heavens. But all of these objections can be answered. To the first, one might respond that the object acts on vision through its species, but that it does so only after vision has multiplied its power to the object; vision could then be referred to as a "passive power" simply because its action is more spiritual than that of the object. The second objection is answered in the same way: vision might be both active and passive. As to the third objection, there is no reason why a spiritual power could not be extended to the heavens, for al-Kindi argues in *De radiis* that every object emits radiation throughout the entire world.

Oresme's own view, though not unambiguously spelled out at this point, is made abundantly clear elsewhere—vision occurs through the reception of species or rays from the visible object. Here he merely notes that the proper reply to those who hold up Aristotle's *De generatione animalium* as evidence for the extramission theory is that Aristotle was there speaking "according to the manner of the ancient mathematicians," whereas in *De anima* he spoke "according to the manner of the moderns and according to truth."[63] The more important point, in Oresme's view, is that for present purposes it makes no difference whether vision occurs by extramission or by intromission, since "according to both opinions similar lines and angles and refractions are imagined, just as for astrology it does not matter whether the motion of the heaven is natural or violent, since either way the circles, spheres, and aspects of the planets are similarly conceived."[64] Indeed, in discussing the rainbow Oresme employs the language of visual rays (following Aristotle's example)—but notes that "a visual ray is not defined as a ray emitted from the eye to the visible object, but [as a ray] emitted from the visible object to the eye."[65]

That Oresme holds to the intromission doctrine is easily established from many other passages. It receives no extended defense, but its truth is assumed as a premise or established fact from which implications can be drawn. In book 3, question 13, Oresme argues that the surface of the eye cannot be wholly reflective of the species of color, for if this were so no species would enter the eye, and then vision would not occur.[66] Again, he notes that if the surfaces of all bodies reflect powers incident upon them, then the eye must reflect the species of the visible object "and not permit it to penetrate to the center of the eye, or to the inner eye where vision occurs."[67] Finally, in this same thirteenth question, Oresme adds important detail, noting that vision requires the arrangement of species within the eye exactly as the points from which those species emanated are arranged outside the eye:

> I suppose . . . that rays of light and color represent the visible object to vision. And for such representation it is necessary that pyramids are continually incident upon the eye, since for vision [to occur] the arrangement of species or rays in the eye must be such that just as that which is seen is disposed outside [the eye], in like manner it is represented inside [the eye].[68]

This, of course, is the theory of Alhazen and Bacon and Witelo.

There is an almost inexhaustible supply of late-medieval commentaries and questions on Aristotle's *Meteorologica* and the psychological works, and one could go on indefinitely probing for accounts of visual theory. However, one more instance, this from the fifteenth century, must suffice. Joannes Versoris became master of arts and later master of theology at the University of Paris in the mid-fifteenth century. He was rector of the university in 1458 and was still in Paris as late as 1482; presumably he died soon thereafter.[69] Versoris wrote questions on many of Aristotle's works on natural philosophy, including the *Parva naturalia* where (in the *Questiones super librum de sensu et sensato*) we find two questions devoted to the visual process.

The first of these pertains to the constitution of the eye and is resolved (closely following the text of Aristotle's *De sensu*) in favor of a watery nature.[70] The second of the questions on visual theory is considerably more interesting because it gives Versoris more freedom to develop his own argument.[71] Here the question is whether sight occurs by extramission or intromission, and Versoris begins with four traditional arguments in favor of the extramission hypothesis. First, if a person wishes to see more clearly, he squints his eyes, whereas if the intromission theory were true he would open his eyes wider so as to let in more radiation from the visible object. Second, animals such as cats and wolves see at night, which would be impossible without the emission of an ocular ray. Third, vision is an action or activity, and the eye is an agent; but every agent (by definition) sends forth a power into the recipient. Finally, if a menstruating woman looks in a mirror or if a basilisk (a mythological beast) looks at a man, the mirror and the man are harmed, which could not occur except through the transmission of visual rays from menstruating women and basilisks.[72]

Aside from the fact that Aristotle disproved this theory in *De sensu,* there are two strong arguments against it. First, if the extramission theory were true, something would pass either from the eye to the visible object, as Empedocles supposed, or from the eye to an encounter with external light, as Plato supposed. Empedocles' opinion is untenable because what issues forth would be either corporeal or incorporeal; it could not be incorporeal, since "to issue forth and to be moved locally is [possible] only for bodies."[73] It could not be corporeal for the following four reasons: (1) If that which issued from the eye were corporeal, the same place would be simultaneously occupied by two or more bodies—the medium, the body issuing from the observer's eye, and the corporeal substance issuing from the eyes of other observers in the vicinity. (2) In the emission of corporeal substance, the emanation is of greater diameter near the origin and subsequently becomes more attenuated, as is evident in flame; but in vision, according to the mathematicians, the base of the visual cone is in the thing seen and the apex in the eye. (3) If corporeal substance issued from the eye, it could not reach the stars, for it could by no means be "infinitely" rarefied, since natural things possess a certain limit of rarefaction. (4) Finally, the substance issuing from the eye would necessarily be either air or fire; not air, since air is already "sufficiently abundant outside, and not fire since we would [then] be unable to see in or through water."[74]

Nor can Plato's opinion be true—that a ray issuing from the eye encounters and is conjoined to external light. In the first place, *lumen* cannot be conjoined to (or separated from) *lumen* except through its subject, since *lumen* is not a corporeal substance but a quality. But even if *lumen* were a corporeal substance, internal and external *lumen* could not be conjoined "since not everything is suited to be conjoined to everything else, but only those that are homogeneous and of the same kind (*rationis*), which is not so in the present case."[75] Finally, if *lumen* were corporeal, internal and external *lumen* could not be conjoined because they are separated by the membrane of the eye.

Therefore it is clear that sight occurs by intromission. However, this conclusion does not merely win by default, for there are positive arguments in its favor. Versoris argues that "color actively diffuses its species through an actually transparent medium to the organ of sight, which is actually illuminated within, and this organ is actually illuminated by the species; and since the visual power exists there, vision occurs; therefore vision occurs by intromission."[76] The antecedent is evident, since if light moves the transparent medium, so must color. That the organ of vision is actually transparent is proved by Aristotle. It follows, therefore, that vision occurs through a radiant pyramid with base on the observed object and apex in the eye, and that the object appears large or small according to the size of the angle at the apex. It follows, moreover, that sight does not occur through the intromission of the small parts of bodies, since that would injure the eye and diminish the visible object, nor through a local motion, but "through a motion of intentional alteration produced by visual species."[77] Versoris thus defends the Aristotelian theory of vision, only slightly modified by elements from other traditions.

Visual Theory in the Theological Tradition

If our knowledge of visual theory within the Aristotelian tradition is limited, knowledge of visual theory within the theological tradition is virtually nonexistent. Optical material might appear in several different kinds of theological treatise, including biblical commentaries—especially on the book of Genesis, where the story of creation often provoked a discussion of light; quodlibetal questions, the outgrowths of semiannual public disputations in which members of the audience could raise any question they wished; and commentaries on the *Sentences* (*Sententiae*) of Peter Lombard, the basic textbook (along with the Bible) of theological study in the medieval university. Treatises of the third kind are undoubtedly the most significant from an optical standpoint; moreover, they are exceedingly numerous and, in my opinion, constitute one of the important collections of unexplored sources for the history of medieval optics and visual theory.[78]

I offer several examples to illustrate the optical content of sentence commentaries. Peter Aureoli (d. 1322) inquires, in his discussion of the creation, "whether light (*lux*) was created on the first day" and proceeds to a lengthy analysis of the relationships among *lux, lumen,* and color.[79] Durandus of Saint-Pourçain (d. 1334) inquires "whether angels know a thing through its essence or through species," and turns in the course of the ensuing discussion to the human act of perception, concluding that qualities are not perceived through their species but through extension of the qualities themselves in the medium.[80] Gregory of Rimini (d. 1358) touches on light and vision at several points in his commentary on the *Sentences,* the former in a question on the intension and remission of forms ("whether any corporeal form would be augmented or intensified continuously, or whether any intensive augmentation of a corporeal form would be a continuous motion"), and the latter in a question on cognition ("whether sensible things are understood by us naturally").[81]

This question of cognition was one of the important concerns of fourteenth-century theologians, and since discussions of cognition almost always included vision, it may be well to dwell briefly on the subject as illustration of the theological contribution to visual theory. I will restrict my attention to William of Ockham, who, as one of the more radical figures in the debate, is of special interest. Ockham was born between 1280 and 1290, probably in the village of Ockham in Surrey.[82] After entering the Franciscan order he matriculated at Oxford University, studying and lecturing in the theological faculty from about 1309 to 1319. Although he fulfilled the requirements for the doctorate in theology, he did not become regent master, but spent the years after 1319 studying and writing, while also becoming involved in a series of controversies. In 1324 he was summoned by the pope to Avignon, and in 1326 a set of Ockhamist propositions was censured by a papal commission. In 1327 William became involved in disputes over the question of evangelical poverty, which led ultimately to his flight from Avignon and excommunication in 1328. He spent the next twenty years at the court of the German emperor, Louis of Bavaria, and died in Munich about 1349, probably of the Black Death. Ockham's literary output includes a variety of writings on logic, natural philosophy, theology, and ecclesiastical politics; chief among the theological works is his *Commentary on the Sentences,* and here we find a discussion of cognition.[83]

Ockham's analysis has its roots in Duns Scotus's doctrine of cognition and the search for the sources of certitude regarding contingent propositions.[84] Since in Ockham's view only individuals have real existence, knowledge can only be of individual objects and their relations. However, individual objects and their relations can be known in two ways—by intuitive cognition and by abstractive cognition. Intuitive cognition, Ockham argues, is that form of cognition which enables one to know with certitude not only an object, but also whether or not it exists, whereas abstractive cognition provides no knowledge regarding the existence of things.[85] Moreover, it must be stressed that one can have intuitive and abstractive cognition of either sensible or intelligible things—that is, not only of physical objects, but also of objects of the intellect, such as sorrow and charity. Ockham develops his notions of intuitive and abstractive cognition at length, and his doctrine, to be fully appreciated, must be seen in comparison with the teaching of Duns Scotus and other predecessors;[86] however, to pursue these matters would take us much further into psychology and epistemology than we can afford to go in this book, and so we must turn to Ockham's view of species.

The relevance of Ockham's theory of cognition to sense perception is seen in the question of species. Does intuitive cognition (or, for that matter, abstractive cognition) of an object require species? Is there, in short, a mediating instrument between the object (whether sensible or intelligible) and the knowing mind? Ockham's answer is that there is not. Intuitive cognition requires nothing except the intellect and the thing known, and abstractive cognition requires, in addition to perfect intuitive cognition, only a *habitus*.[87] The point of Ockham's teaching, then, is that both the senses and the intellect have direct, unmediated apprehension of their objects, and this direct apprehension provides the grounds for certitude and a defense against the possibility of skepticism.[88] But how is this direct apprehension—the nonexistence of species—established? First, the principle of economy ("Ockham's razor") eliminates species on grounds of redundancy: "It is useless to achieve by more things what can equally well be achieved by fewer; but intuitive cognition can occur by means of the intellect and the thing seen, without any species."[89] Moreover, there is no experiential evidence for the existence of species, for we are not aware of the species, which are posited solely to explain our awareness of the object. Finally, if intuitive cognition were to require species, then, "since that species could be conserved in the absence of the object, it could cause, by natural means, intuitive cognition in the absence of the object, which is false and contrary to experience."[90]

This general theory, applicable to all of the external senses and to the intellect, can be applied specifically to the sense of sight. The visible object, Ockham argues, has three effects on sight: First, the object impresses in the eye a quality that soothes or irritates (*confortans vel debilitans*) the eye. Second, the object impresses in the eye a quality or likeness of itself, which lingers for a period of time: "Besides this [first] quality is another, which is capable of affecting sense and can be perceived by sense, and it is of the same kind (*ratio*) as the external object."[91] Finally, the object is the cause (or at least part of the cause) of the act of seeing itself: "And besides these [two] qualities there is in sight the act of seeing, which is in the [visual] power (in contradistinction to the organ) as subject."[92] Thus if the object and the eye (with its visual power) are both present and suitably disposed, vision occurs, for it is the nature of the object to act on the visual power and of the visual power to perceive the object; there is no need, and therefore no room, in this scheme for an intermediary (a species) between the object and the observer.

But does not Ockham's theory require action at a distance? And what happens to the medium, which in fact intervenes between the object and the observer? Does it not undergo some sort of transformation? Ockham devotes an entire question to the subject, arguing that there is nothing absurd or impossible about action at a distance, which in fact occurs.[93] A series of examples establishes this: First, the sun illuminates the earth without illuminating that portion of the medium that intervenes between the sun and the moon. Similarly, when a ray passes through a window it produces a very intense light (*lumen*) in that part of the medium near the facing wall, but not in the rest of the medium; "consequently this intense light is not immediately produced by another illuminated [part of the] medium; therefore it is caused immediately by the sun, and consequently the sun can act immediately at a distance."[94] Moreover, two candles placed on one side of a small aperture illuminate the wall on the other side of the aperture in two distinct spots, and it is apparent that the two spots "are not produced by light (*lumen*) in the aperture, since that light would have rectilinear access to all parts of the wall; . . . therefore those distinct illuminated spots are caused immediately by the candles."[95]

However, the fact of action at a distance in no way forbids transformation of the medium by the object; indeed, the medium itself may be the recipient of action at a distance. In the first example above, although the sun does not illuminate that portion of the medium that falls between the sun and the moon, it apparently illuminates the portion between the moon and the earth. Is not this an instance of species in the medium? Again the answer is no. Consider a colored object, which may "color" the medium in its vicinity. This "coloring" of the medium is the result not of a species issuing from the color of the body, but of the color itself; or in general terms, what is propagated through the medium is the very quality of the object, rather than the likeness of the quality. Nor is this merely a semantic quibble, for Ockham maintains explicitly that this propagation of the quality (color in our example) in the medium does not account for visual perception of the object, but only for visual perception of the colored medium—that is, for the fact that we see the medium to be colored.[96]

Ockham has thus established two things: first, there are no species; second, though a visible object propagates its qualities through the medium, these qualities are not responsible for visual perception, which occurs by action at a distance as an instance of intuitive or abstractive cognition.[97]

Conclusion

What can we learn from this brief and inadequate survey of visual theory during the later Middle Ages? First, I have suggested a scheme of classification by which to interpret the optical enterprise in the later Middle Ages and identify the principal schools of thought on the subject of vision. If I am correct, there were chiefly three traditions: (1) the mathematical or perspectivist tradition based upon Alhazen's *De aspectibus* and the Baconian synthesis, (2) the Aristotelian tradition found in Aristotelian commentaries and inspired by the example of Albert the Great, and (3) a theological tradition represented by sentence commentaries, quodlibeta, and biblical commentaries.[98]

I acknowledge that this scheme of classification is still primitive and surely requires elaboration and refinement. Moreover, it is clear that the three traditions I have identified were not totally distinct, but frequently overlapped or even merged. Between the Aristotelian and perspectivist traditions there was considerable confusion of boundary lines from their very beginnings in the West, for Albert borrowed the visual pyramid of the perspectivists, and Bacon attempted to incorporate the substance of the Aristotelian theory into his synthesis. The same confusion and overlap persisted in the fourteenth and fifteenth centuries, as we find Aristotelian commentators employing ray-geometry in their analyses (especially when commenting upon Aristotle's *Meteorologica*) and perspectivists (who, like everybody else in the period, had learned their Aristotle well) leaning in Aristotelian directions.[99] Indeed, my reading of the late-medieval texts suggest that very few fourteenth- or fifteenth-century scholars, in defending either the visual theory of Aristotle or that of the perspectivists, were consciously repudiating the other. As for the theological tradition, where it touched on optics and vision it was indebted to both the Aristotelian and the perspectivist schools—which should come as no surprise, since theological masters had generally received prior training in the faculty of arts.

But despite these reservations and qualifications, I can see no classification scheme better able to organize the literature than the one I have suggested.[100] If the Aristotelian, perspectivist, and theological traditions held a great deal in common, they nevertheless drew their inspirations in large part from different source materials, appealed to different communities, and spoke to different issues. Aristotelian commentators were chiefly occupied with the problems raised by the Aristotelian text, primarily physical, ontological, and psychological in nature despite the occasional intrusion of ray-geometry and other elements of the perspectivist tradition. The perspectivists, though sometimes sounding like Aristotelians, were responding primarily to problems raised within the Baconian tradition and resorted to mathematics often enough to be regarded as advocates of the mathematical approach. Finally, theologians, though dependent on the Aristotelian and perspectivist traditions for optical content, were interested in optics for its theological implications and turned to visual theory chiefly as it pertained to the problem of cognition; moreover, theologians, no less than Aristotelians and perspectivists, were responding to the difficulties and unresolved issues of a particular literary tradition—that of commentaries on Peter Lombard's *Sentences,* biblical commentaries, and other theological tracts.

With this tripartite division of the optical enterprise in mind, we can begin to understand what happened to visual theory during the fourteenth and fifteenth centuries. In the first place, the perspectivist tradition was quite simply overwhelmed by the Aristotelian tradition. The dominant position assumed by Aristotelianism within the educational system and the resulting influence on the literary output of the medieval master transformed many disciplines; as the works of Aristotle came to dominate the curriculum of the medieval university, discussions of natural philosophy coalesced around the problems and issues found in the Aristotelian text. In the realm of visual theory, this meant that although there were more scholars taking a serious interest in vision than during any previous period of comparable duration in the history of mankind, they were applying themselves to problems of an Aristotelian variety. Perspectivist writings did not altogether disappear, but they were vastly outnumbered by commentaries on Aristotle's *De anima, De sensu,* and *Metorologica*—besides which one must not forget that Aristotelianism had insinuated itself into the perspectivist tradition.

The result of this shift toward Aristotelianism was that questions of ray-geometry and the physiology of the visual process were shoved into the background as attention was focused on the ontology of radiation (Do distinct rays really exist? Are light and color of the same species? What is the relationship between *lux* and *lumen*?) and the psychological aspects of visual perception (What is the proper object of sight? Are all things known by means of vision known equally? What is the relationship between sensation and cognition?). When geometrical or physiological issues were raised, the ensuing discussion was usually elementary and unilluminating by thirteenth-century standards. New philosophical and theological currents of the later Middle Ages reinforced the trend toward ontology and psychology, as nominalism gave rise to new theories of cognition and to discussions of the ontological status of species. The reason, therefore, why we find no progress during the fourteenth and fifteenth centuries along those lines that would lead ultimately to Kepler's theory of the retinal image is not that medieval natural philosophy lost its vitality or medieval thinkers their nerve, but that the focus of attention shifted elsewhere.

But it was not only a shift of focus that influenced late-medieval visual theory; there was also a shift in methodology and style. The evolution of pedagogical technique (and corresponding literary forms) from the *lectio* to the *disputatio* and *questio* has been ably described by Grabmann, Glorieux, and Chenu.[101] Beginning in the twelfth century and continuing into the fourteenth, scholarly priorities gradually shifted from primary concern for the exegesis of authoritative texts and the laying of doctrinal foundations toward the resolution of particular (and sometimes minor) difficulties and even the questioning of matters no longer seriously doubted, for the sake of exploring the implications of a doctrine, revealing the limits of necessity and contingency, or demonstrating one's dialectical skills. Chenu writes:

> It became no longer a simple question of submitting to research those problems already under discussion or still open to debate. Even the points accepted by everybody and set forth in the most certain of terms were brought under scrutiny and subjected, by deliberate artifice, to the now usual processes of research. In brief, they were, literally speaking, "called into question," no longer because there was any real doubt about their truth, but because a deeper understanding of them was sought after. Theologians as well as philosophers asked the question: Does God exist? Is the soul spiritual? Should a person honor his parents? etc. Yet, of the question, only the form remained, with the typical word *Utrum* [Whether] everywhere and over and over again employed.[102]

As an expression of this tendency to "call into question" doctrines of all sorts, the *questio* became the most characteristic literary form of the fourteenth and fifteenth centuries (in the philosophical and theological realm).[103] The purpose of such treatises was not to treat in systematic fashion the whole of a discipline, but to select a handful of disputed (or disputable) questions from the standard *auctores*.

These developments in scholarly methodology explain the absence of comprehensive and systematic discussions of the visual process among late-medieval scholars of all persuasions. To the authors of commentaries on Aristotelian works, sentence commentaries, and *Questiones super perspectivam* it must have seemed wholly unnecessary to provide a complete account of visual theory; the basic issues had surely been settled long before, and the appropriate task was therefore to raise questions, restricted in scope, about unresolved and debatable aspects of the theory. For example, the intromission doctrine was not really in doubt, but its implications required further exploration, the arguments in defense of it could always use improvement and clarification, and lingering problems (such as the apparent luminosity of cats' eyes) could benefit from further discussion by anybody with something new to say. Neither was the perspectivists' theory of radiation and the visual pyramid in doubt, for no other scheme had ever been presented to explain the mathematics of visual perception; but it needed to be reconciled with troublesome physical truths (such as the fact that rays cannot be devoid of width if they are to stimulate the visual power or have real physical existence), and there were always problems of ray-tracing within the humors of the eye on which anybody so inclined could exercise his mathematical talents. It was in this spirit that scholars of the fourteenth and fifteenth centuries adopted a piecemeal approach to visual theory and dealt with the limited range of issues which in their view required further discussion. If to us their endeavor seems futile, this is not because we are intellectually superior, but because we are privy to the future course of visual theory.

8 ARTISTS AND ANATOMISTS OF THE RENAISSANCE

Introduction

If the Schoolmen of the Middle Ages were occupied with stale philosophical issues raised in the Aristotelian corpus, the fresh breezes of the Renaissance began to transform the European climate in the fifteenth century—or so the popular stereotype informs us. The practical, urbane, self-confident man of the Renaissance, we are told, turned his back on the book learning of the Middle Ages and transferred his allegience to the teachings of experience. Not only did the artist set forth on an empirical expedition to capture "visual reality," but the same empirical spirit transformed the whole of human activity, including natural science. Artists and architects, anatomists and engineers vowed not to read the ancients; instead they would read nature.

The simplistic character of this general interpretation begins to appear when we consider its applicability to particulars—as, for example, optics and visual theory. We need mention only Leonardo da Vinci, of whom it has been maintained ad nauseum (and by some of his most reputable interpreters) that he broke with medieval visual theories and launched out boldly in new directions. The truth, of course, is much more complex. Artists and anatomists of the Renaissance did indeed undertake to analyze and apply visual theory in original and creative ways—did indeed present more accurate descriptions of ocular anatomy—but in doing so they rarely deviated from the fundamentals of medieval visual theory. Actually, the artists and anatomists were well instructed in medieval book learning on the subject (without which they could have progressed little, if at all), and only on rare occasions were they able to depart from their medieval teachers in more than small details. The traditional framework, though occasionally questioned, remained basically intact until early in the seventeenth century.

Vision and the Theory of Artistic Perspective

About 1303, a little more than a decade after the deaths of Roger Bacon and John Pecham, Giotto di Bondone (ca. 1266–1337) began work on the frescoes of the Arena Chapel in Padua—paintings that later generations would view as the first statement of a new understanding of the relationship between visual space and its representation on a two-dimensional surface.[1] What Giotto did was to eliminate some of the flat, stylized qualities that had characterized medieval painting by endowing his figures with a more human, three-dimensional, lifelike quality; by introducing oblique views and foreshortening into his architectural representations, thereby creating a sense of depth and solidity; and by adjusting the perspective of the frescoes to the viewpoint of an observer standing at the center of the chapel.[2] This was the beginning of a search for "visual truth," an "endeavor to imitate nature," which would culminate a century later in the theory of linear perspective.[3]

Historians of art are unanimous in crediting the invention of linear perspective to the Florentine Filippo Brunelleschi (1377–1446). Although Brunelleschi left no written record of his achievement, his disciple Antonio Manetti gives us an account in his *Vita di Brunelleschi*. According to Manetti, Brunelleschi painted two panels, one showing the Florentine Baptistery in the Piazza del Duomo and the other the Piazza della Signoria. These panels, Manetti insists, were painted exactly according to the principles of mathematical or linear perspective:

> During the same period he propounded and realized what painters today call perspective, since it forms part of that science which, in effect, consists of setting down properly and rationally the reductions and enlargements of near and distant objects as perceived by the eye of man: buildings, plains, mountains, places of every sort and location, with figures and objects in correct proportion to the distance in which they are shown. He originated the rule that is essential to whatever has been accomplished since his time in that area.[4]

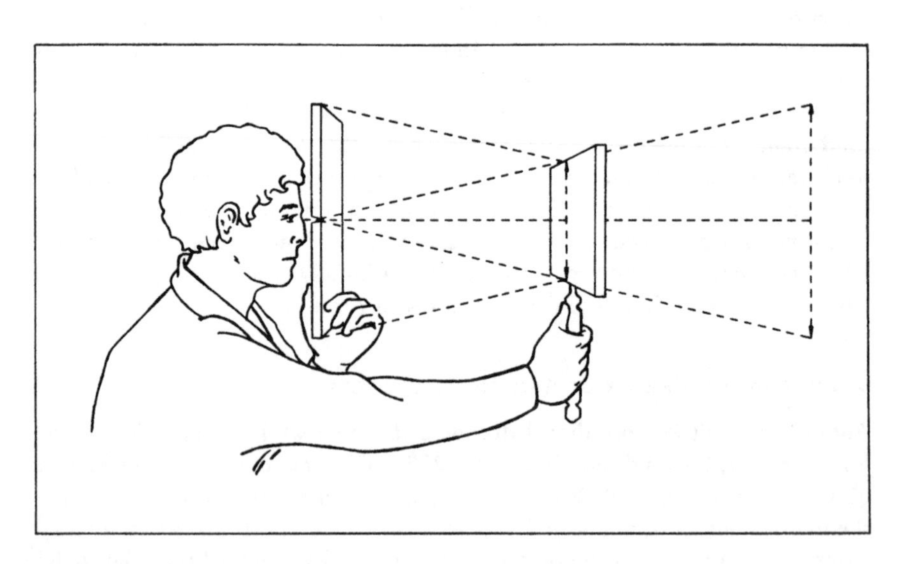

Fig. 19. Brunelleschi's perspective demonstration.

Brunelleschi's painting of the baptistery, according to Manetti, was perforated with a small peephole at its centric point. The viewer was to place his eye behind the peephole and view the painting in a mirror extended at arm's length (illustrated in fig. 19).[5] This may seem like a clumsy procedure, but by providing the peephole and insisting that the picture be viewed through it by reflection, Brunelleschi forced the viewer of his painting to observe it from a point corresponding exactly to the viewpoint from which it had been executed. Manetti reminds us that in a perspective painting the painter must "postulate beforehand a single point from which his painting must be viewed,"[6] and the purpose of Brunelleschi's peephole was to fix that point.

Seen from this fixed viewpoint, Brunelleschi's painting presented a remarkable sight. It appeared so true to life that "the spectator felt he saw the actual scene when he looked at the painting";[7] Brunelleschi had painted the baptistery and its surroundings with such mathematical exactitude that the painting and the painted scene were hardly distinguishable. He had produced a triumph of illusionist painting, of the attempt to imitate nature, which would revolutionize the Renaissance attitude toward the painter's art.

The techniques that Brunelleschi must have used to transfer the dimensions of objects in the visual field to his painted panels were given theoretical expression in a treatise, *Della pittura,* written about 1435 by Leon Battista Alberti (1404-72) and dedicated to Brunelleschi. In this treatise Alberti presents a brief discussion of the visual cone or pyramid,[8] with base on the observed object or scene and apex within the eye. This pyramid includes three classes of rays, differing in strength and function, but all rectilinear: extrinsic or extreme rays, which form the outer surface of the visual pyramid and communicate the size and shape of the object: median rays, which fill the interior of the visual pyramid and, chameleonlike, assume the colors of objects in the visual field and convey them to the eye; and finally, the central and strongest ray of the pyramid, the "prince of rays," through which objects are seen most clearly and which "is the last to abandon the thing seen."[9] Alberti refuses to commit himself to any particular theory of vision, noting that "among the ancients there was no little dispute whether these rays come from the eye or the plane. This dispute is very difficult and is quite useless for us."[10] Several lines later he adds: "Nor is this the place to discuss whether vision ... resides at the juncture of the inner nerve or whether images are formed on the surface of the eye as on a living mirror. The function of the eyes in vision need not be considered in this place."[11] Alberti's point is that the theory of linear perspective, which he is about to develop, requires the visual pyramid, but need not concern itself with the direction of radiation or the functioning of the eye; it requires the mathematics, but not the physics or physiology, of vision.

Having defined the visual pyramid and asserted the rectilinearity of its radiation, Alberti was prepared to explain how the artist imitates nature. The painter's panel is to be thought of as a window through which the painter views the world, and the painting as the intersection of this window with the visual pyramid. When studious painters fill their panels with colors, Alberti writes,

> they should only seek to present the forms of things seen on this plane as if it were of transparent glass. Thus the visual pyramid could pass through it, placed at a definite distance with definite lights and a definite position of centre in space and in a definite place in respect to the observer. . . . He who looks at a picture, done as I have described, will see a certain cross-section of a visual pyramid.[12]

As a further illustration of what he had in mind, and also to provide the painter with practical techniques for executing perspective drawings, Alberti describes a device consisting of a *velo,* or veil:

> Nothing can be found, so I think, which is more useful than that veil which among my friends I call an intersection. It is a thin veil, finely woven, dyed whatever colour pleases you and with larger threads [marking out] as many parallels as you prefer. This veil I place between the eye and the thing seen, so the visual pyramid penetrates through the thinness of the veil.[13]

The veil establishes a set of coordinates for the painter: "Here in this parallel you will see the forehead, in that the nose, in another the cheeks, in this lower one the chin and all outstanding features in their place."[14] When he has located the feature to be painted on the coordinate system provided by the veil, the painter transfers it to his panel, divided into a similar grid.

But this is all in the realm of empirical procedure. What about the geometrical theory that lies behind it? There is no need to delve deeply into the geometrical constructions by which Alberti transfers the dimensions of objects in the visual field, proportionately reduced, to his panel; the details are much debated, and they do not affect the basic scheme.[15] Alberti selects a "centric point" in his picture, where the central ray of the visual pyramid encounters the scene to be depicted. From the centric point he draws straight lines to his base line, which defines the foreground of the picture; these lines represent the orthogonals, horizontal lines perpendicular to the picture plane, perhaps best thought of as parallel lines in a pavement (fig. 20a). However, it is necessary to locate the transverse lines in the pavement, and for this purpose Alberti takes what is in effect a side view of the visual pyramid and its intersection with the picture plane; this is his "distance point operation," illustrated in figure 20b.[16] Where visual rays directed toward the base line intersect the picture plane, horizontal projections are extended to the first drawing, thus marking off the transverse lines of the pavement. Once the pavement has been laid out, other structures and figures may be erected upon it. The most celebrated result of Alberti's construction is that all lines perpendicular to the plane of the picture will recede toward a single vanishing point—Alberti's "centric point."

Alberti was only the first in a long series of artists who attempted to give

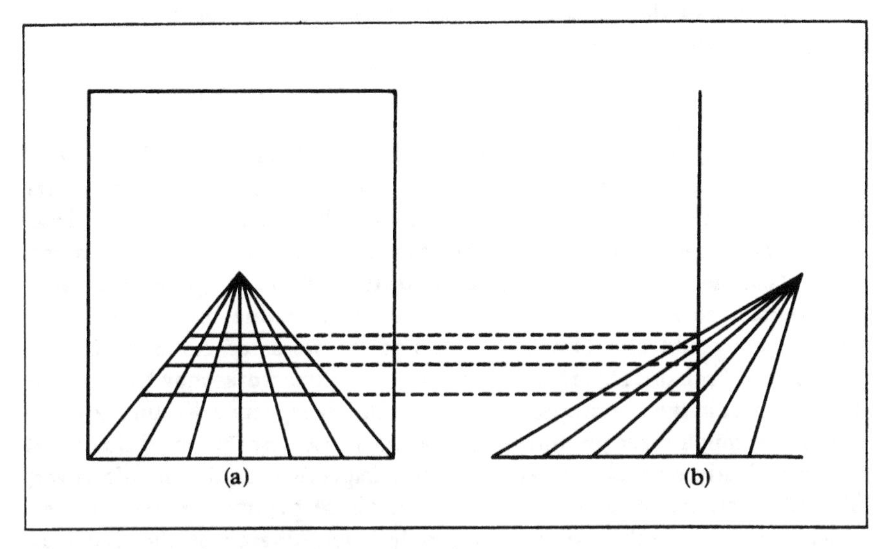

Fig. 20. Alberti's perspective construction.

theoretical expression to the practice of linear perspective. He was followed by Piero della Francesca (ca. 1416–92), Jean Pélerin Viator (ca. 1435–1524), Leonardo da Vinci (1452–1519), Albrecht Dürer (1471–1528), and others—many of whom added important refinements. But most of these were refinements of technique rather than of conception, and they need not concern us. Of far greater importance, within the context of this book, is the question of the relationship between linear perspective and visual theory. Specifically, how much knowledge did quattrocentro artists have of medieval optical theory, and how heavily did they rely upon it in developing the theory and techniques of linear perspective?

We do not know how either Brunelleschi or Alberti came to his theory. Samuel Edgerton has suggested that the newly available flat glass mirror, which began to replace the older flat metal and hemispherical glass mirrors in the fourteenth century, may have had an important influence by allowing people to behold the visual field in a two-dimensional surface. Painters, then, may have been challenged to produce similar results.[17] According to the fourteenth-century chronicler Fillippo Villani, Giotto painted with the aid of a mirror,[18] and we have seen that Brunelleschi used a plane mirror in his perspective demonstration. Alberti too urged the use of mirrors, though for judging rather than composing pictures, when he wrote in *Della pittura:* "A good judge for you to know is the mirror."[19] Finally, Filarete (Antonio Averlino), writing in the middle of the fifteenth century, suggests to his readers:

> If you should desire to portray something in an easier way, take a mirror and hold it in front of the thing you want to do. Look in it and you will see

the outlines of the thing more easily. Whatever is closer or farther will appear foreshortened to you.

And then he adds: "Truly, I think that Pippo di Ser Brunellesco discovered perspective in this way."[20]

Nevertheless, the precise influence of the mirror in the development of linear perspective must remain conjectural. What is beyond conjecture is that the creators of linear perspective knew and utilized ancient and medieval optical theory. Alessandro Parronchi has argued that Brunelleschi's friend Paolo Toscanelli brought a copy of Blasius of Parma's *Questiones super perspectivam* to Florence when he returned from Padua in 1424 and that Brunelleschi could also have had access to the works of Alhazen, Bacon, Witelo, and Pecham. He argues, moreover, that these works may have played a decisive role in the working out of Brunelleschi's perspective demonstration.[21] We are on much surer ground with Alberti, whose description of the visual pyramid clearly reveals knowledge of the perspectivist tradition. Moreover, Alberti's reference to the central ray of the visual pyramid as that through which certainty is achieved can come only from Alhazen or the Baconian tradition.[22] Of even greater significance, if it can be sustained, is Edgerton's suggestion that Alberti's "distance point operation" (the side view of his perspective construction, illustrated in fig. 20b) may have been influenced by proposition 11 of Euclid's *De visu.*[23] Here Euclid argues that "since *GD* [fig. 21] is seen with the rays *AG* and *AD,* and *DE* is seen through the rays *AD* and *AE, GD* appears higher than *DE.* Similarly *DE* will appear higher than *BE,* for objects seen by higher rays appear higher."[24] A comparison of figures 20b and 21 reveals the close analogy between this Euclidean proposition and Alberti's theory: if *EG* is regarded as a section of floor to be portrayed and *IK* as the painter's panel, Euclid's eleventh proposition and Alberti's "distance point operation" are virtually identical.

The most transparent case of the influence of medieval visual theory on a quattrocento artist comes from the work of Alberti's Florentine contemporary, Lorenzo Ghiberti (1381-1455). Toward the end of his life (probably about 1447) Ghiberti began to put together his *Commentarii* from notes gathered during the preceding fifteen or twenty years.[25] In the third and final book of the *Commentarii* (commonly referred to as the *Third Commentary*), he presents a complete survey of the mathematical tradition in optics, consisting mainly of excerpts and paraphrases drawn from the perspectivists: Alhazen, Bacon, Witelo, and Pecham.[26] Ghiberti discusses the various modes of radiation, the nature of species, the anatomy of the eye and visual pathway, vision through the pyramid of radiation, certification, and image-formation by reflection and refraction. But of perhaps greatest significance for the artist is a series of propositions pertaining to the perception of space, which John White gives as follows:

(a) Visible things are not comprehended by means of the visual sense alone. (b) It is only possible to judge the distance of an object by means of an intervening, continuous, series of regular bodies. (c) The visual angle alone is not sufficient for the judgement of size. (d) Knowledge of the size of an object depends upon a comparison of the base of the visual pyramid with the angle at its apex, and with the intervening distance. (e) Distance is most commonly measured by the surface of the ground and the size of the human body.[27]

It is White's contention that these propositions reveal Ghiberti's real intention in composing the *Third Commentary,* namely, to present the optical underpinnings of Alberti's theory of linear perspective:

> The structure of Alberti's artificial perspective reveals at every stage ideas which are the exact reflection of those found in the medieval texts. It seems most likely that just such an application of optical theory to representational problems lies behind the invention of the new system. . . . This close connection between medieval and renaissance ideas reveals Ghiberti's intention in writing the *Third Commentary,* which was to deal with natural perspective or optics. He wished to summarize for his contemporaries the conception of visual reality which was the basis and justification of the new system [of perspective].[28]

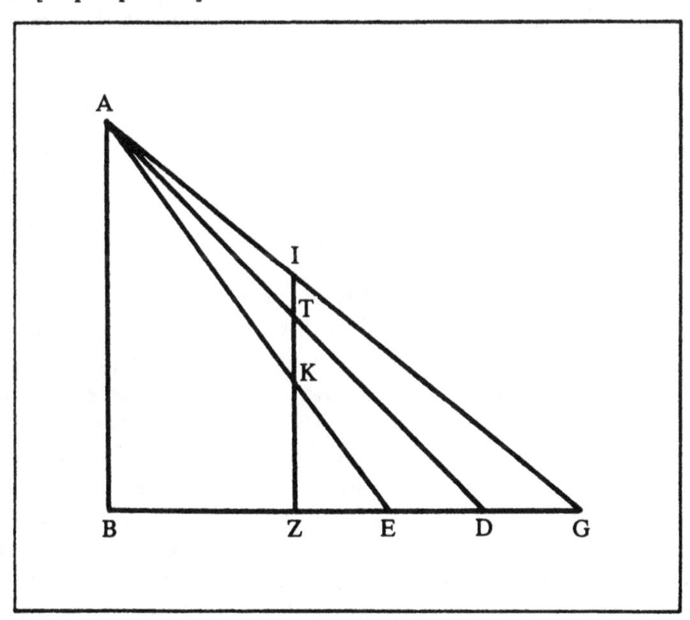

Fig. 21

In view of the fact that the five propositions singled out by White are given no more space or emphasis by Ghiberti than they had received in the medieval works from which he obtained them, we are hardly justified in viewing them as the central features and raison d'être of the *Third Commentary*; moreover, Ghiberti nowhere suggests their relevance to Alberti's theory of perspective.[29] Nevertheless, it is highly probable that Ghiberti's exposition of medieval optical theory as a whole—situated, as it is, in a work devoted to the history and theory of art—was intended to explain and justify, in a very general way, the newly developing representational techniques of Renaissance painting.[30]

What, then, can we conclude about the relationship between theories of vision and the principles of linear perspective? The inventors of linear perspective, we have seen, possessed at least an elementary knowledge of medieval visual theory, and they employed the visual cone or pyramid of the perspectivists to justify the geometrical operations by which the three-dimensional visual field was projected onto the two-dimensional panel. Alberti conceived the panel as a plane intersection of the visual pyramid extended between the fixed eye of the observer and the scene to be portrayed, and his vanishing point was at the intersection of the panel with the central ray of the visual pyramid. Although I have concentrated on Alberti, because he was the first to state the theory of linear perspective, others who followed him were equally conscious of the relationship between their theories of perspective and medieval theories of vision.[31] But it must be emphasized that only the barest framework, the mathematical skeleton, of the perspectivist theory of vision was required. As Alberti pointed out in a passage quoted previously, "Among the ancients there was no little dispute whether these [visual] rays come from the eye or the plane. This dispute . . . is quite useless for us."[32]

Leonardo da Vinci

If it has been established that the artist of the Renaissance required only a modicum of visual theory in order to make his perspective drawings and grasp their geometrical basis, it is nonetheless tempting to inquire how much additional visual theory he may have known; once his appetite was whetted, did he press on to a fuller understanding of the visual process? Ghiberti, of course, copied long excerpts regarding vision from medieval authors, and he undoubtedly grasped at least a good bit of what he copied. Others alluded knowingly to controversies over visual theory, suggesting that they too were well informed on the subject.[33] But no artist of the Renaissance plunged more energetically into investigations of visual theory than Leonardo da Vinci. Of course, Leonardo was more than an artist; though self-taught, he was a man of remarkably wide practical and theoretical interests. Indeed, he was a skillful anatomist, and one might easily suppose that anatomical investigation performed with the practical empiricism of the artist-engineer would lead quickly to a new and more accurate theory of vision.[34] Let us therefore consider Leonardo's work on the subject.

Leonardo was born in Tuscany, near the village of Vinci, in 1452.[35] His father, Piero, moved the family to Florence in the 1460s, and there Leonardo was apprenticed to the artist Andrea del Verrocchio. In 1472 he was admitted to the painters' guild and during the next decade executed a number of commissions. He emigrated to Milan about 1483, and during the next sixteen years undertook an important series of artistic, architectural, and engineering projects for the Sforza court. He returned to Florence for the period 1500-1506, then resided variously in Milan, Rome, and finally Amboise (in France), where he died in 1519 after a stroke.

The image of Leonardo as "a man without letters," who had been instructed by experience rather than books, was encouraged by Leonardo himself when he wrote:

> They [certain arrogant people] will say that because of my lack of book-learning, I cannot properly express what I desire to treat of. Do they not know that my subjects require for their exposition experience rather than the words of others? And since experience has been the mistress of whoever has written well, I take her as my mistress, and to her in all points make my appeal.[36]

And again:

> If indeed I have no power to quote from authors as they have, it is a far bigger and more worthy thing to read by the light of experience, which is the instructress of their masters. They strut about puffed up and pompous, decked out and adorned not with their own labours but by those of others, and they will not even allow me my own. And if they despise me who am an inventor how much more should blame be given to themselves, who are not inventors but trumpeters and reciters of the works of others?[37]

Remarks such as these have led many to suppose that Leonardo (who, of course, had no university education) was virtually isolated from the Latin scholarship of the past—so that George Sarton could write: "Leonardo would have been able to read a book had he wished, but he never wished with sufficient persistence,"[38] and Giorgio de Santillana could conclude:

> To see that for Leonardo Latin was a foreign tongue, it is sufficient to examine his lists of words and his simple grammatical exercises from that time [after he had begun to study Latin]. . . . His Latin remains uncertain and rudimentary; he loses himself in the syntax and confuses the subject and the object in his exercises. This man could never decipher a text without the assistance of a friend.[39]

Doubtless it is true that Leonardo did not devour scholarly treatises written in Latin, but Sarton and de Santillana exaggerate. Marshall Clagett has demonstrated, for example, that in his work on statics Leonardo relied heavily on the *Elementa de ponderibus* of Jordanus de Nemore and the *Liber de ratione ponderis* attributed to Jordanus, and that he also knew Blasius of Parma's *Tractatus de ponderibus* and probably Pseudo-Archimedes' *De canonio* and Thabit ibn Qurra's *Liber karastonis.*[40] In optics, Leonardo quotes the entire proemium of John Pecham's *Perspectiva communis,* which he has accurately rendered into Italian, and he cites Witelo by name on a number of occasions.[41] Yet, as we shall see, Leonardo's knowledge of the past optical tradition was very imperfect.

By whatever means, Leonardo was engaged throughout his life in a process of self-education. His curriculum included studies in mathematics, optics, acoustics, mechanics, natural history, anatomy, physical geography, and painting. Optics was a persistent interest, at least partly because of its relevance to painting, and notes on the subject are sprinkled throughout his manuscripts.[42] Two of the manuscripts can be regarded as preliminary versions of connected treatises on optics: MS C, *On Light and Shade,* and MS D, *On the Eye.*[43] There are many analyses, both scholarly and popular, of Leonardo's work in optics, but most are undependable.[44] I cannot pretend to say the final word on the subject, but I hope at least to bring some order to Leonardo's remarks on vision.

The foundation of Leonardo's theory of vision was his theory of radiation and the radiant pyramid. Objects, according to Leonardo, send their species or likenesses in all directions into a surrounding transparent medium.[45] These species converge along straight lines, thus forming pyramids with base on the object and apex at every point in the medium (see fig. 22):[46]

> Necessity causes that nature ordains or has ordained that in all points of the air all the images of the things opposite to them converge, by the pyramidal concourse of the rays that have emanated from these things; and if it were not so the eye would not discern in every point of the air that is between it and the thing seen, the shape and quality of the thing facing it.[47]

Thus "each body by itself alone fills with its images the air around it, and . . . the same air is able, at the same time, to receive the species of countless other bodies which are in it."[48] Leonardo repeats, until it almost becomes a litany, that the species are "all throughout the whole and all in each smallest part; each in all and all in the part."[49] As proof of the point, he introduces the phenomenon of pinhole images. Three objects *A, C,* and *E* (fig. 23) all produce images through the same pinhole, *N* or *P,* and each object produces its image through both holes *N* and *P.*[50] Clearly the species or images intersect in the apertures without interference or intermingling, and clearly each object sends its species in all directions to converge at each point of the medium.

Never does Leonardo explicitly affirm al-Kindi's principle of punctiform analysis, preferring instead to visualize the image or species as radiating wholistically from the visible object as base to a point in the medium. However, there are scattered hints that Leonardo also entertained (or would have, if pressed) the idea of radiation from point sources. For example, in the

Fig. 22. Pyramids of radiation surrounding a spherical body according to Leonardo da Vinci.

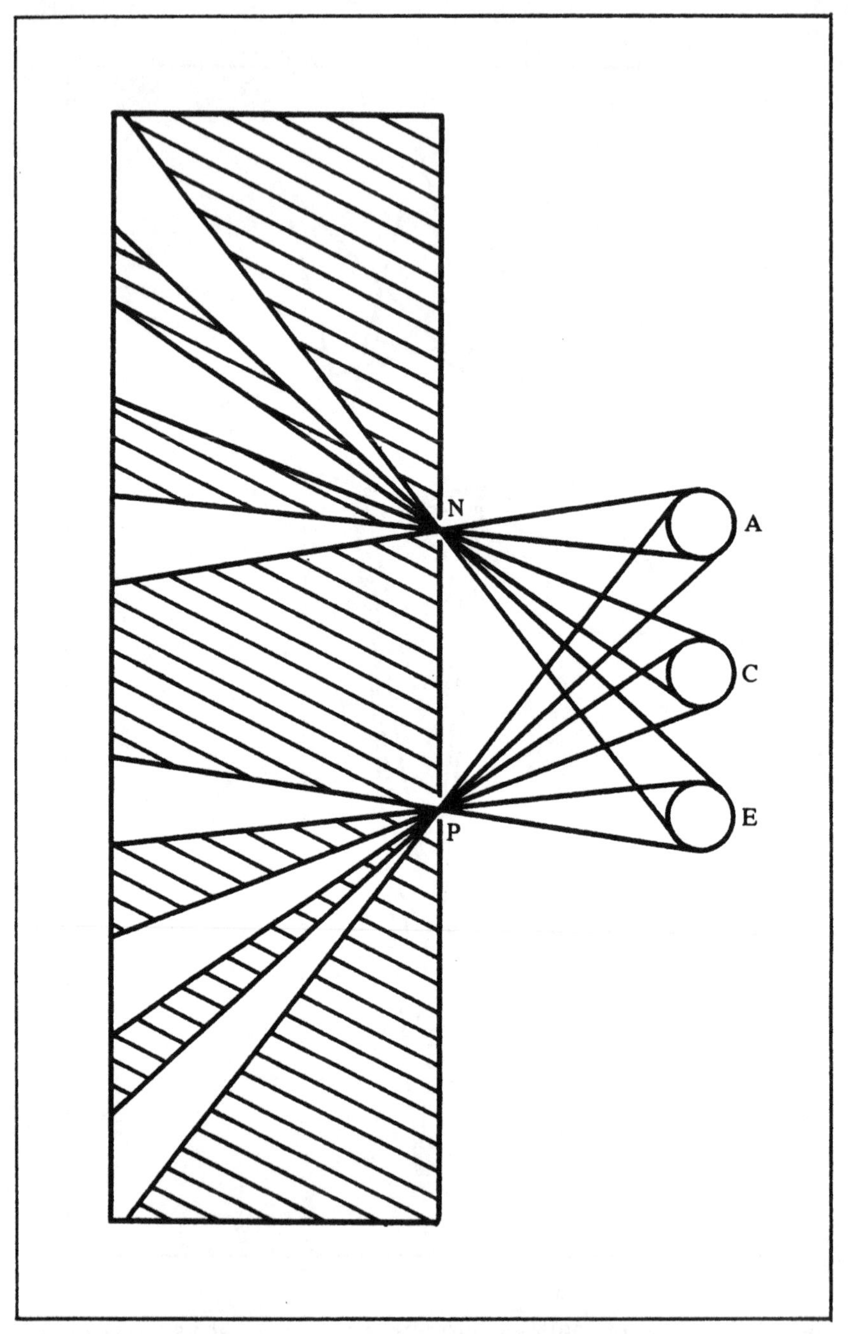

Fig. 23. Radiation through apertures according to Leonardo.

treatise *On the Eye* Leonardo writes that "each point of the pupil sees the whole object, and each point of the object is seen by the whole pupil."[51] Elsewhere he writes that "every point is the termination of an infinite number of lines, which diverge to form a base"[52] and that "where there is a smaller luminous angle there is less light, because the pyramid of this angle has a smaller base, and therefore from this smaller base a lesser number of luminous rays converge at its point."[53]

Having thus defined and established the existence of *radiant* pyramids emanating from all visible objects, Leonardo easily converts them into *visual* pyramids:

> Perspective is a rational demonstration, whereby experience confirms that all objects transmit their similitudes to the eye by a pyramid of lines. By a pyramid of lines I understand those lines which start from the edges of the surface of bodies, and converging from a distance, meet in a single point; and this point, in this case, I will show to be situated in the eye, which is the universal judge of all objects.[54]

Leonardo, demonstrating his usual gift of finding homely illustrations, continues:

> As regards the point in the eye; it is made more intelligible by this: If you look into the eye of another person you will see your own image. Now imagine 2 lines starting from your ears and going to the ears of that image which you see in the other man's eye; you will understand that these lines converge in such a way that they would meet in a point a little way beyond your own image mirrored in the eye.[55]

Vision is strongest, Leonardo points out, along the central line of this pyramid and weakens the farther the object deviates from the center.[56] This can be explained in terms of purpose (nature does not wish to "break the law given to all other powers"),[57] but also through a mechanical analogy:

> Now the objects which are over against the eyes act with the rays of their images after the manner of many archers who wish to shoot through the bore of a carbine, for the one among them who finds himself in a straight line with the direction of the bore of the carbine will be most likely to touch the bottom of this bore with his arrow; so the objects opposite to the eye will be more transferred to the sense when they are more in the line of the transfixing nerve.[58]

Everything said thus far would suggest that Leonardo was a staunch defender of the intromission theory of vision, and it is true that he never denies that species pass from the object to the observer.[59] But in his notebooks we find a variety of opinions on the question whether radiation also passes from the observer to the observed object. In a leaf of the Codex

Atlanticus written no later than 1491, Leonardo argues that visual radiation does indeed exist: "I say that the power of vision extends by means of the visual rays as far as the surface of bodies which are not transparent, and that the power possessed by these bodies extends up to the power of vision."[60] In defense of this claim, Leonardo refers to the ability of certain animals—basilisk, wolf, ostrich, spider, the snake called "lamia," and the fish called "linno"—to kill, produce hoarse voice, hatch eggs, attract nightingales, and illuminate large quantities of water at night by the power of sight alone. In addition, "maidens are said to have power in their eyes to attract to themselves the love of men."[61] Some mathematicians, Leonardo admits, hold that if radiation were to issue from the eye, then "though the eye were as great as the body of the earth it would of necessity be consumed in beholding the stars."[62] But in reply to such people it is noted that the odor of musk, "if it be carried a thousand miles, will permeate a thousand miies with that thickness of atmosphere without any diminution of itself."[63] As for the function of the visual radiation, Leonardo says in one passage that it continues its rectilinear path until intercepted by an opaque object, whereupon it carries the image of the intercepting object back to the eye.[64] In another passage he indicates that "the eye projects an infinite number of lines, and these attach themselves to or mingle with those that come towards it which emanate from the things seen."[65]

In other places, however, Leonardo argues vigorously against any emission of visual radiation. The fullest of these is in Paris, Bibliothèque Nationale, MS Ital. 2038 (written about 1492), where we read:

> It is impossible that the eye should project from itself, by visual rays, the visual power, since, as soon as it opens, that front portion [of the visual power?] which would give rise to this emanation would have to go forth to the object and this it could not do without time. And this being so, it could not travel so high as the sun in a month's time when the eye wanted to see it. And if it could reach the sun it would necessarily follow that it should perpetually remain in a continuous line from the eye to the sun and should always diverge in such a way as to form between the sun and the eye the base and the apex of a pyramid. This being the case, if the eye consisted of a million worlds, it would not prevent its being consumed in the projection of its power; and if this power would have to travel through the air as perfumes do, the winds would bend it and carry it into another place. But we do [in fact] see the mass of the sun with the same rapidity as [an object] at the distance of a braccio, and the power of sight is not disturbed by the blowing of the winds nor by any other accident.[66]

In the past, one of the standard proofs of the extramission theory had been the ability of certain nocturnal animals to see at night. To this argument Leonardo replies that such animals use the small amount of light present, aided by the large diameter of their pupils, and if this fails hunt their prey by sound or odor.[67] The passages in which Leonardo affirms or denies the existence of visual rays can all be dated with considerable confidence, and it appears from these dates that he first adopted and later rejected the visual ray theory: all the affirmations of visual radiation appear in manuscripts dating from the 1480s or very early 1490s, and after about 1492 (if my knowledge of the matter is complete) Leonardo restricted himself entirely to intromitted rays.[68]

Whatever Leonardo's position on the direction of radiation, can we determine his view of the nature of the radiated entity? It has become the standard interpretation among recent commentators that Leonardo took a radically mechanistic approach to radiation, conceiving the propagation of light or images as an instance of wave motion, through which the "power of percussion" is communicated to the sensitive organ.[69] However, in my judgment, this interpretation rests on an unfortunate misunderstanding of a passage in which Leonardo writes:

> Just as a stone flung into the water becomes the centre and cause of many circles, and as sound diffuses itself in circles in the air: so any object placed in the luminous atmosphere diffuses itself in circles, and fills the surrounding air with infinite images of itself.[70]

Now Leonardo asserts here that images (*similitudini*) proceed in circular fashion from an object, just as circles spread out from the point of impact of a stone on a body of water; but neither here nor in any other passage does he ever suggest that images or species or light are *instances* of wave motion. What we have here is merely another of Leonardo's everyday examples to illustrate a theoretical point: the gross geometry of the radiation of images is like that of circular waves in water, but this says nothing whatever about the nature of the radiated entity. This is exactly analogous to Alhazen's use of projectiles to elucidate the geometry of reflection and refraction without intending to say anything about the nature of the reflected or refracted entity.[71] This misunderstanding has led to further misfortunes, as everything Leonardo said about percussion and wave motion in other contexts has been blithely applied to the radiation of light or species.

Leonardo's true view of the nature of the entity responsible for vision, though never spelled out in much detail, harks back to the Aristotelian and perspectivist traditions. The terms most frequently used by Leonardo to denote this entity are *similitudine* and *spetie* (similitudes or species), though occasionally the terms *impressione, forma, eidola,* and *simulacra* are also employed.[72] In one of the few passages that sheds light on the nature of the *similitudine* or *spetie,* Leonardo argues:

> I say that sight is exercised by all animals, by the mediation of light. . . . For it will easily be understood that the senses which receive the images of things do not send forth from themselves any power. On the contrary, the

> air which exists between the object and the sense incorporates in itself the species of things, and by its contact with the sense presents them to it.[73]

This description is strongly reminiscent of Alhazen's and Witelo's forms and of Bacon's and Pecham's species.[74] Perhaps Leonardo is struggling to express the same conception when he writes:

> If the object opposite to the eye were to send its image to the eye, the eye would have to do the same to the object, whence it might seem that these images were an emanation. But, if so, it would be necessary [to admit] that every object became rapidly smaller; because each object appears by its images in the surrounding atmosphere. . . . From this it seems necessary to admit that it is in the nature of the air, which surrounds the objects, to draw to itself like a lodestone the similitudes of the things among which it is placed.[75]

This is a confusing passage, which I do not entirely fathom. It seems clear, however, that the attraction of similitudes by the air, as by a lodestone, is presented as an alternative to the theory of *simulacra,* whereby pieces of the object are thrown off in all directions in such a way that the object must eventually diminish in size. Although I do not understand the precise significance of the analogy of the lodestone, it seems that this must express the replication of similitudes out of the substance and potentialities of the air—in short, something close to the species theory of the medieval perspectivists.

We might expect Leonardo the skilled anatomist to have been at his best regarding the anatomy of the eye; here, if anywhere, observation should be capable of correcting traditional opinion. Indeed, Leonardo himself wrote: "The eye . . . has even down to our own times been defined by countless writers in one way, but I find by experience that it acts in another."[76] Unfortunately, Leonardo's views on ocular anatomy were exceedingly primitive—perhaps more primitive than those of any knowledgeable writer to touch on the matter from Galen to his own day—and it is clear that he would have profited enormously by paying close attention to those "countless writers" of the past. Basically, Leonardo believed that the eye consists of two concentric spheres, the outer of which he referred to as the albugineous sphere and the inner as the vitreous or crystalline sphere. The albugineous sphere is surrounded by the tunic of the uvea, except in front where the pupil opens onto the transparent cornea.[77] At the back of the eye, directly opposite the pupil, is the opening into the optic nerve, though on at least one occasion Leonardo speculated on the possibility that the optic nerve extends through the albugineous sphere and even penetrates a short distance into the crystalline sphere.[78] This scheme differs from the medieval perspectivist (as well as the modern) understanding of ocular anatomy in confusing the vitreous and crystalline humors, maintaining that the albugineous humor surrounds the crystalline humor on all sides, placing the crystalline humor in the center of the eye, and claiming that the crystalline is spherical rather than lenticular.

However, Leonardo made an important contribution to the knowledge of ocular anatomy when he studied the variable diameter of the pupil. This phenomenon had been noted in both Greek antiquity and medieval Islam, but had been largely ignored in medieval Christendom.[79] Leonardo not only gave a fuller description of the circumstances under which the pupil varies its size, but properly understood the cause as the intensity of the incident light:

> I find by experience that the black or almost black fringe of colour which appears round the pupil serves for no other purpose except to increase or diminish the size of this pupil; to increase it when the eye is looking towards a dark place; to diminish it when it is looking at the light or at a luminous thing.[80]

This is fine, but Leonardo had strange ideas about the effect of this variability of the pupil on perception. In notebooks written over a long period of time, he insists that "the larger the pupil the larger will be the appearance of the object it sees."[81] A geometrical explanation of this is given in the treatise *On the Eye,* where Leonardo explains that if the pupil is wide open (circle *MN*, fig. 24) the image of an object placed before the eye will

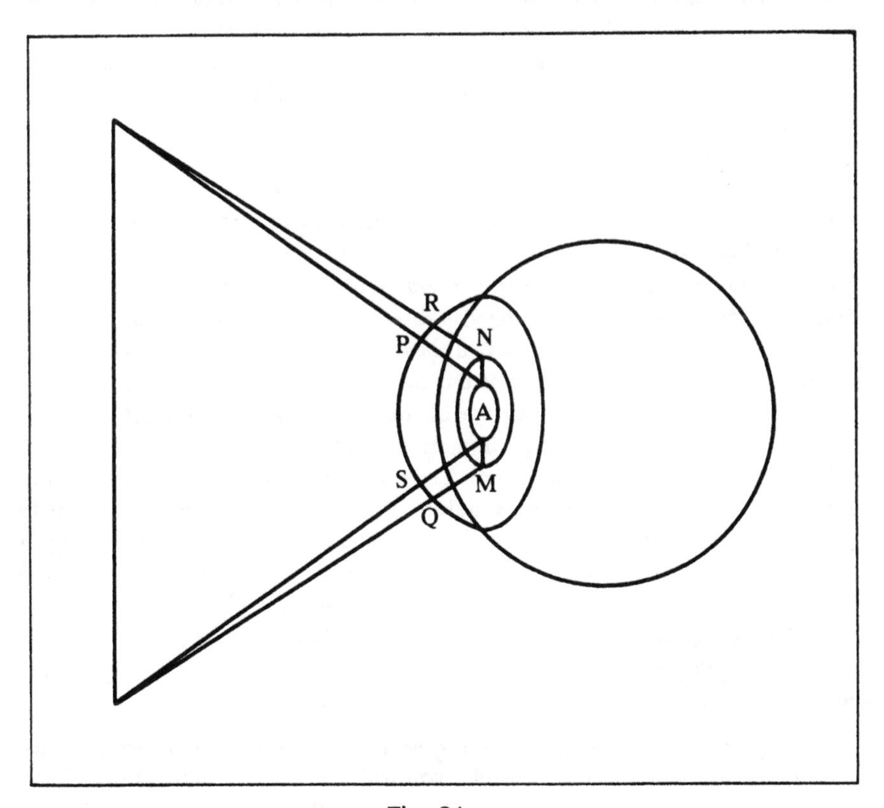

Fig. 24

appear on the cornea between *R* and *Q*, whereas if the pupil is partially closed (circle *A*) the image of the object will appear on the cornea between *P* and *S*.[82] The note breaks off unfinished, but it seems safe to surmise that Leonardo believed that the size of the visible object is judged by the size of the image cast on the cornea. As Leonardo makes clear elsewhere in the same notebook, after the formation of an image on the surface of spectacles or the cornea a "new convergence" is generated, which carries the image into the eye; and from this new convergence the size of the object is judged.[83] This is a remarkably primitive theory: it reveals that Leonardo has not grasped the principles of image-formation in convex surfaces (perfectly understood by both his predecessors and his successors)[84] and that he has failed to perceive the problem of the multiplicity of rays issuing from every point of the visible object which had dominated visual theory since Alhazen and would continue as a problem until solved by Kepler.[85]

Leonardo made a second contribution to the understanding of the eye when he compared it to a *camera obscura*. This comparison, though used sparingly, is explicitly presented. It appears most clearly in the treatise *On the Eye* in a note entitled: "How to make an experiment which demonstrates how the visual virtue employs the instrument of the eye."[86] To perform the experiment, Leonardo tells us, we must construct a pair of concentric spheres filled with water. At the bottom an aperture, simulating the pupil, allows the entrance of light; and at the top the outer sphere is cut away so that the head can be inserted up to the ears (fig. 25 reproduces Leonardo's drawing). "Then," he instructs, "put your face into the water and look into the inner sphere, and you will see how such an instrument dispatches the species of *ST* to the eye just as the eye sends it to the visual virtue."[87] On several other occasions Leonardo refers to the *camera obscura,* but only to demonstrate that rays from various parts of the visual field must intersect within the pupil and thus present an inverted view unless caused to intersect a second time through reflection or refraction.[88] On no occasion does Leonardo assert or imply that the retina (or rear surface of the eye) is a screen, analogous to the back of the *camera,* onto which images are projected.

Where then is the seat of vision? Leonardo argues, first of all, that the visual power is not situated in a point. Thus the perspectivists (Leonardo obviously has in mind Brunelleschi, Alberti, and their followers) are wrong in supposing that the visual power is at the apex of the visual pyramid.[89] Rather, "the pupil of the eye has visual virtue in its entirety and in each of its parts."[90] As experimental proof of this claim, Leonardo points out that a small object placed in front of the eye does not prevent us from seeing objects behind it, as it would if the eye perceived from a point:

> If you place the thickness of a sewing needle of medium size in front of the pupil as near to the eye as possible, you will see that the perception of any object placed behind this needle at however great a distance will not be

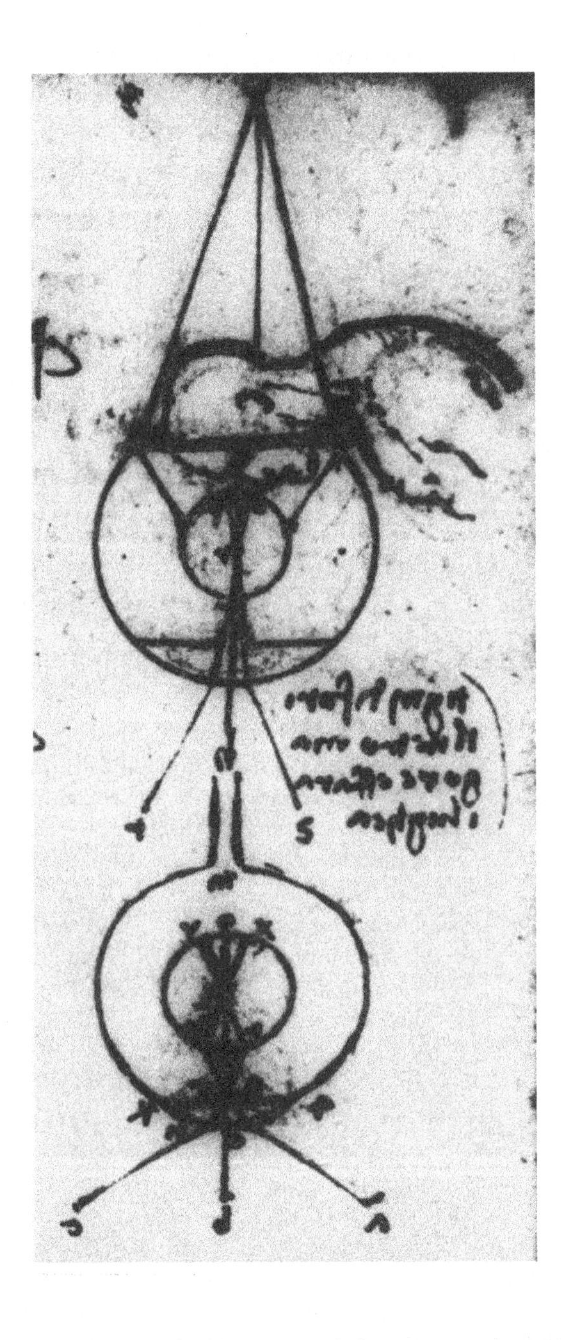

Fig. 25. The eye as a *camera obscura* according to Leonardo da Vinci (*upper drawing*). MS D, fol. 3v, of the Bibliothèque de l'Institut de France, Paris.

> impeded. What I say is entirely borne out by experience and necessity confirms it; for if this visual virtue were reduced to a point every object however small placed in front of such a visual virtue would occupy a great part of the view of the heavens.[91]

Leonardo has thus demolished the theory of the visual cone—strictly defined so as to maintain that vision occurs only through rays emanating from a point. His argument would not, of course, have disturbed the medieval perspectivists, who could have argued that objects behind the small obstacle are visible by nonperpendicular, and hence refracted, rays.

But Leonardo is not to be understood as maintaining that the pupil is the true seat of the visual power—only that radiation falling anywhere within the pupil will be perceived. As for the true location of the visual power, Leonardo sees two alternatives: the crystalline humor and the extremity of the optic nerve. Only on one occasion, so far as I know, does he entertain the possibility of the former, and for the rest he assumes without argument that the visual power is situated at the extremity of the optic nerve.[92] This means, of course, that Leonardo has rejected the standard Galenic and perspectivist opinion on the location of the visual power and that he will consider the crystalline humor strictly as a refracting device.[93]

One of Leonardo's major preoccupations was with the actual path of rays through the eye. His chief concern was to get the rays to the visual power at the end of the optic nerve without inversion, for he must by all means guard against that absurdity. Since the *camera obscura* demonstrates that rays from opposite edges of the visible object intersect in the aperture (and hence in the pupil of the eye), Leonardo must contrive to discover a second intersection within the eye, which will enable him to account for the erect image that must ultimately fall on the end of the optic nerve.[94] He devised a remarkable number of schemes for achieving this goal, employing intersections in front of the eye, within the pupil, at the center of the crystalline humor, and between the crystalline humor and the optic nerve. As mechanisms for producing these intersections, Leonardo called on refraction at the surface of the cornea and the front and rear surfaces of the crystalline humor and reflection from the surface of the crystalline humor and from the uvea, which surrounds the albugineous sphere. I count six such schemes in the treatise *On the Eye,* and I present as illustration the one that Leonardo developed most fully:

> How the species are transmitted to the visual virtue with two intersections by necessity. The object *A* [fig. 26] sends its similitude to the visual virtue through the line *AR* to part *R* of the cornea *EF*. It then enters through the pupil with its intersection at *O,* and passes to the vitreous [i.e., crystalline] sphere at *V* and penetrates this sphere from *V* to *Q*. It then passes through the intersection *N* and terminates in *K* at the head of the optic nerve *KHL* by which it is then referred to the common sense.[95]

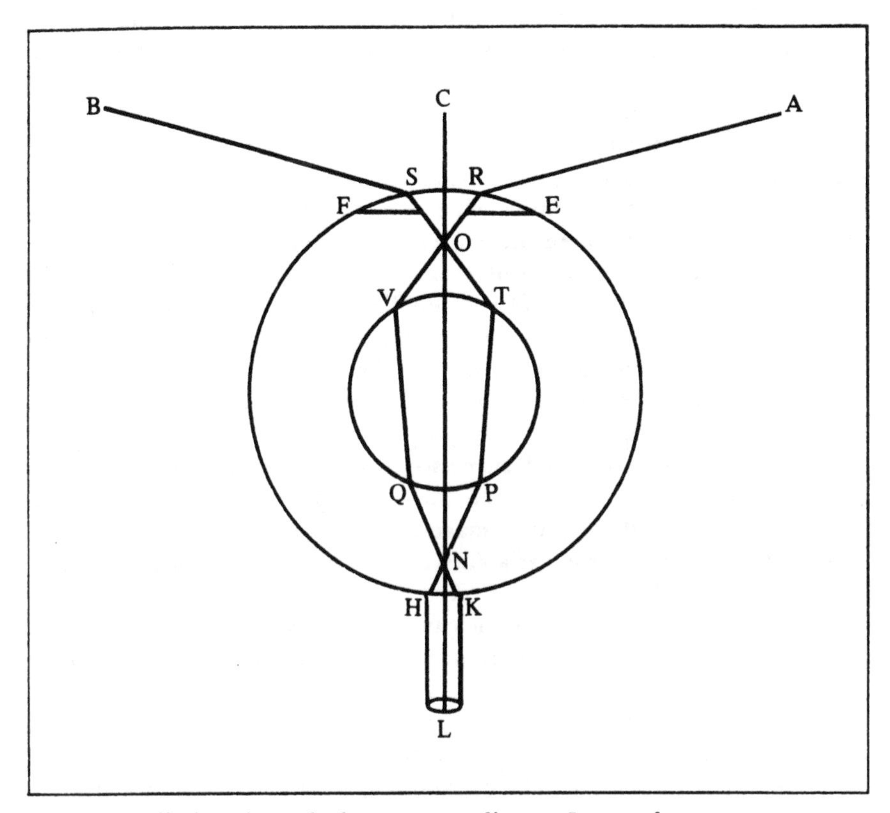

Fig. 26. Radiation through the eye according to Leonardo.

To complete Leonardo's account of visual perception I must mention, finally, his notion of the post-ocular transmission. Leonardo devotes little attention to the problem and merely presents a mélange of medieval ideas. He indicates that the similitudes received in the eye are transmitted through the optic nerve to the *imprensiva,* the faculty of judgment, situated in the anterior or lateral ventricles of the brain. The *imprensiva,* in turn, sends the similitudes to the common sense, which stamps them on the memory.[96] In a slight variation of this, Leonardo argues that "the soul apparently resides in the seat of judgment, and the judgment apparently resides in the place where all the senses meet, which is called the common sense."[97]

How then shall we assess Leonardo's achievement? In the first place, we must recognize his very substantial debt to the past. Much of his visual theory was traditional, ranging from arguments for and against the extramission theory to his view of the radiated entity and the nature of the post-ocular transmission, and no medieval perspectivist could have failed to recognize many familiar elements. However, it was frequently a confused and garbled form of the traditional theory that Leonardo expressed, and our medieval perspectivist would have been horrified by Leonardo's frequent lack of comprehension. Leonardo had a very limited understanding of image-formation by reflection and refraction and no understanding at all of the central issue of traditional optics—the problem of a multiplicity of rays from every point in the visual field influencing all parts of the eye. Instead Leonardo tended to treat radiation in an entirely unsatisfactory wholistic manner. Moreover, his knowledge of ocular anatomy was primitive, and he had strange views regarding the effect of pupillary size on the perceived size of objects in the visual field. These are failures to understand not merely modern optical theory, but also medieval optical theory. We must conclude that Leonardo made use of medieval optical treatises, especially those of the perspectivist tradition, without fully grasping their contents.

On the other hand, few men surpassed Leonardo for native intelligence, ingenuity, independence of mind, and flair for experimental demonstrations. Leonardo employed these gifts to make important original contributions to visual theory. He noted the variability in the size of the pupil and demonstrated that the eye perceives from more than a point. He viewed the eye as analogous, in certain respects, to the *camera obscura,* so that intersection of rays from the visible object must occur within the pupil; and he treated the crystalline humor of the eye strictly as a refracting device. We can only regret that Leonardo's creative talents were not applied to medieval optical theory more fully grasped; for, as Leonardo's various confusions so clearly reveal, the problem of sight was not to be solved through a fresh start by an ingenious empiricist working in an intellectual vacuum, but through careful attention to the past optical tradition—to its doctrines and criteria, its successes and failures. But, alas, it probably would not have mattered anyway, for Leonardo was isolated and without influence. Most of his manuscripts were in private hands until 1636, and they were not seriously studied until the end of the eighteenth century.[98] To examine the ongoing progress of visual theory, we must turn rather to the academic traditions in medicine and mathematics—the first in the section that follows, the second in the final chapter.

Ocular Anatomy in the Renaissance

One of the more important scientific developments of the Renaissance was the advancement of anatomical studies. Many factors contributed to this growth, including improvements in medical education (the increasing frequency of human dissection being the most important improvement),[99] better access to the original texts of Galen, the invention of printing (which made possible the accurate mass production of anatomical drawings), and perhaps the example of anatomical investigations and the skillful representation of human anatomy by such Renaissance artists as Michelangelo, Raphael, and Albrecht Dürer.[100] Let us then consider the contributions to visual theory made by those anatomists who took an interest in the eye.

But first a word about the scope of the inquiry. It would be impossible, in the available space, to attempt a complete history of developments in ocular anatomy during the Renaissance. It would also be useless for our purposes, since much of ocular anatomy in the period before Kepler was only slightly germane to visual theory. For example, knowledge of the muscles of the eye and disputes about the exact structure of the ocular tunics and the precise consistency of the ocular humors contributed nothing to resolving the basic issues of visual theory. We must therefore concentrate our attention on the principal organs of the visual pathway—those that were believed to serve important functions in the reception and transmission of visual impressions or visual spirit. Even here nothing is to be gained from a monotonous recitation of minor variations in description or nomenclature, and I will attempt only to provide a sketch of major developments.

The roots of the anatomical revival can be traced back into the later Middle Ages. It has become customary to grant Mondino dei Luzzi (ca. 1270-1326)[101] much of the credit for planting the seed. His *Anatomia* (written about 1316), though it contained little that was original, became exceedingly popular, and its influence probably justifies Mondino's reputation as "the restorer of anatomy." The *Anatomia* was based on a wide range of ancient and medieval sources—certain of the works of Hippocrates, Galen, and Aristotle, as well as Rhazes's *Liber ad Almansorem*, Haly Abbas's *Liber regalis*, Avicenna's *Liber cannois*, Albucasis's *Chirurgia*, Alcoati's *Congregatio sive liber de oculis*, and Averroes's *Colliget*—and from these same sources came Mondino's views on the anatomy of the eye.[102]

Mondino presented a traditional description of the eye, which he thought comprised seven tunics and three humors.[103] The outermost tunic is the cornea, which is transparent and hard. Next comes the conjunctiva, which covers the forward part of the eye except for the cornea. The posterior extension of the conjunctiva, structurally continuous with it, is the sclera or sclerotic coat. Within the conjunctiva is the uvea or grapelike tunic, with an aperture (the pupil) in front to admit visual species; its posterior continuation is the *secundina*. Finally, the innermost tunic is the *aranea* or arachnoid membrane,[104] with its posterior continuation, the retina. The three humors of the eye are the *albugineus* (so called because it resembles the white of an egg), placed in front of the crystalline humor to moisten it and to separate it from the cornea and the outside air; the crystalline itself, which receives the species of visible things; and the vitreous, which surrounds and nourishes the crystalline humor. Mondino is aware that the crystalline humor is somewhat flattened in front and might seem to assert that it is placed toward the front of the eye, though what he actually states is simply that "this humor is more toward the front than is the vitreous humor in which it is placed."[105] This description differs only in a few details from that presented by Hunain ibn Ishaq four hundred and fifty years earlier.

Mondino touches only incidentally on the visual process, indicating that the "species of the visible object reach the eye in a pyramidal figure, the base of which is the thing seen and . . . the apex of which is in the crystalline humor."[106] The crystalline humor is clearly the seat of the visual power, for Mondino indicates that the purpose of the aperture in the uvea is to permit species to reach the crystalline humor and that the color of an object (a cataract in his example) is able to affect the crystalline humor.[107] Species then pass through the optic nerves to their junction, where species from the two eyes are "returned to oneness."[108] If there is anything novel in Mondino's description, it is his introduction of the visual pyramid of the perspectivist tradition into an anatomical treatise.[109]

The anatomical scheme presented by Mondino was not significantly altered in the next two centuries, at least with regard to the gross features of the eye. Two contemporaries, Lanfranc of Milan (d. 1315) and Henry of Mondeville (1260–1320), wrote surgical treatises containing descriptions of the eye closely resembling Mondino's. Another surgical treatise containing a description of ocular anatomy was that of Guy de Chauliac (1298–1368).[110] By the beginning of the sixteenth century there had as yet been no significant change, as we can see from the work of Gabriele Zerbi (1468–1505), who taught in the medical school at the University of Bologna late in the fifteenth century. Zerbi published a very long and detailed description of the eye in his *Liber anathomie corporis humani et singulorum membrorum illius* (Venice, 1502), where we find some 30,000 words on the subject compared with approximately 1,000 in Mondino's *Anatomia*. Zerbi achieved such length principally because he succeeded in citing all of the ancient and medieval authorities regarding each feature of the eye, and I have not been able to discover any fundamental anatomical innovation. Zerbi still identified the same seven tunics and three humors, though he noted that the tunics can be defined and counted in a variety of ways and that when it comes to the thing itself (as opposed to the name) there are really only four.[111] Nevertheless he described the traditional seven: the conjunctiva, cornea, sclerotic coat, uvea, *secundina,* arachnoid membrane, and retina.[112] The three humors were also described, all in traditional terms. The most important of these, the crystalline humor, was described by Zerbi as being somewhat flattened, indeed lenticular in shape,[113] and situated in the center of the eye; it is the principal organ of vision, and the other parts of the eye are designed to serve it.[114] Finally, it should be noted that Zerbi (following Mondino's lead) introduced the visual pyramid of the perspectivists.[115]

A second illustration of the anatomical scene early in the sixteenth century is Jacopo Berengario da Carpi (ca. 1460–1530?), a student of Zerbi at Bologna and later a teacher there himself.[116] Berengario published a very long and detailed commentary on Mondino: *Commentaria cum amplissimis additionibus super anatomia Mundini cum textu eiusdem* (Bologna, 1521). However, this work was largely without influence, and Berengario's reputation has rested on his much shorter and more readable handbook, *Isagogae breves in anatomiam humani corporis* (Bologna, 1522). The latter work, it has been argued, "inaugurated ... a new era in anatomy and became at once the most authoritative work of its kind until the appearance of Vesalius' *De humani corporis fabrica,*" and "represents the true bridge between the anatomical researches of Mondino and those of Vesalius."[117] Although Berengario claimed to have dissected, by this time, several hundred bodies, and although he displayed a critical attitude toward Mondino (especially in his longer *Commentaria*),[118] the *Isagogae breves* contains essentially a recapitulation of Mondino's scheme of ocular anatomy—the same seven tunics and three humors described in virtually identical terms.[119]

Along with traditional descriptions of ocular anatomy by Zerbi and Berengario went a traditional Galenic physiology of sight. It is true that Zerbi acknowledged the possibility that the *aranea,* rather than the crystalline humor, is the principal locus of the visual power, as Averroes had maintained; but since the *aranea* is the anterior capsule of the lens (in our view) or a tunic attached to the anterior surface of the lens (in the view of Averroes and Zerbi), this is an innovation of small importance.[120] Moreover, Zerbi's own preference was clearly to identify the crystalline humor itself as the seat of the visual power.[121] It is also true that Zerbi repeated the claims of Galen and Averroes to the effect that one of the functions of the retina "is to perceive alterations of the crystalline humor," adding that according to some people "the retina is ... disposed to transmit the forms of visible things from the crystalline humor to the place where the optic nerves intersect, where vision is formed";[122] however, as we have seen, these claims are perfectly compatible with belief in the crystalline humor as the principal organ of vision.[123] Indeed, Zerbi himself explicitly denied that the retina could be the principal organ of sight; after considering Averroes's claim that the *aranea* might be the seat of the visual power, Zerbi continued: "And it cannot be proved in a similar manner that the retina [would] be the principal part of the eye, for although the retina arises from the optic nerves, containing [visual] spirit, nevertheless it does not equal the crystalline humor in lucidity, as does the *aranea.*"[124] In short, the retina is not lucid enough to be the chief sensitive organ of the eye.[125]

Zerbi's only other contribution to the physiology of sight was to add detail to the process of retinal transmission of visual impressions. He argued that impressions do not pass from the crystalline humor to the retina by an optical transmission through the vitreous humor, since the vitreous humor is unsuited for transmission because it is colored, and there is a more suitable medium, the retina, which makes direct contact with the crystalline humor at its periphery or equator.[126] Thus visual impressions pass to the equator of the crystalline humor and from there directly into the retina—though Zerbi did admit that according to some people species pass not through the substance of the retina but through the visual spirit enclosed by the retina.

A far more significant physiological novelty, because it directly challenged the preeminence of the crystalline humor, was expressed by Zerbi's contemporary and colleague Alessandro Achillini (1463-1512) in his posthumously published *De humani corporis anatomia.* In this work Achillini did little more than assemble quotations from ancient and medieval sources and, regarding the location of the visual power, wrote as follows:

> Galen in the book *De iuvamentis anhelitus* has vision occurring not in the crystalline humor of the eye, as it seemed to Aristotle, but in the tunics of the eye. And in the tenth [book] of *De iuvamentis membrorum* [he says that] vision is truly only in these nerves, namely the optic nerves. And in [Book] X, chapter 2, of *De utilitate particularum* [he says that] the first and greatest utility of the retina is to perceive the alterations of the crystalline humor.[127]

Achillini has presented three claims (indeed, they turn out to be verbatim quotations), allegedly from Galen, which challenge the prevailing belief in the crystalline humor as the principal organ of vision. The second and third quotations express ideas that we have seen before, and we need not consider them again;[128] but the first, containing as it does an explicit denial of the primacy of the crystalline humor, must occupy our attention. What is this book, *De iuvamentis anhelitus,* in which Galen allegedly denies that vision occurs in the crystalline humor? It is in fact a Pseudo-Galenic work, which appears under the title *De utilitate respirationis* in a number of early printed editions of Galen's works—included among the *spuria.*[129] I do not know its author, or even the era during which it was written,[130] and therefore I cannot comment on the origin of the ideas expressed in it; what is significant is that in this widely circulated work we have an explicit denial that the crystalline humor is percipient. Since the quotation from Achillini, above, gives only a small portion of the passage on vision in *De iuvamentis anhelitus,* I here provide it in its entirety:

> I say that all the instruments of the senses are contained by tunics; and perception occurs in these tunics, as seeing is in the eye, whose tunics are filled by a watery humor; and these tunics have their origin in the tunics of the brain. And vision occurs not in the crystalline humor, as it seemed to Aristotle, but in the tunics of the eye, two of which proceed from the two tunics of the brain and the third from the optic nerve.[131]

The idea was not further developed, either by the author of *De iuvamentis anhelitus* or by Achillini, but at least the bare claim was now in circulation.

Fuller anatomical investigations of the eye were to be made in the course of the sixteenth century. Andreas Vesalius, in *De humani corporis fabrica* (Basel, 1543), presented a detailed description of the eye which clearly reveals a more intimate firsthand knowledge of ocular anatomy than any of his predecessors could boast; to this description Vesalius added carefully drawn illustrations which in number, accuracy, and agreement with the text surpass any drawings of the eye previously available. *De fabrica* contains significant innovations, such as a fuller and more accurate description of the ciliary body and the assertion that the optic nerves are solid rather than hollow.[132] Nevertheless, Vesalius still placed the crystalline humor (flattened both front and back into lenticular form) in the exact center of the eye and regarded the ciliary body, connected to its equator, as a flat surface bisecting the ocular sphere and separating the aqueous and vitreous humors. Figure 27 reproduces Vesalius's drawing.[133] For the rest, Mondino's anatomical scheme was largely retained.

As for ocular function, Vesalius admitted that he lacked exact knowledge of the process of seeing and therefore would not consider such questions, promising instead to write a special work at some future date on medical and philosophical controversies of this kind.[134] However, he expressed skepticism toward Galen's claim that the crystalline humor is the principal sensitive organ of the eye,[135] and pointed out tantalizingly that "many consider this tunic [the retina] to be the chief organ of sight."[136] It is perhaps noteworthy that Vesalius also discussed the refractive power of the crystalline humor and compared it to a looking glass (*specillum*);[137] although one cannot read into this a denial of the senstivity of the crystalline humor, Vesalius's example may have influenced Felix Platter to view the crystalline humor merely as a looking glass.

Hard on the heels of Vesalius's *De fabrica* appeared other anatomical works containing descriptions of the eye, including some devoted exclusively to the anatomy of the visual system or to vision itself. Among these were Charles Estienne's *De dissectione partium corporis humani libri tres* (Paris, 1545); Realdo Colombo's *De re anatomica libri XV* (Venice, 1559); Constanzo Varolio's *De nervis opticis* (Padua, 1573); George Bartisch's ΟΦΘΑΛΜΟΔΟΥΛΕΙΑ, *Das ist Augendienst* (Dresden, 1583); André Du Laurens's *Discourse de la conservation de la veue* (Paris, 1597); Johannes Jessen's *Anatomia Pragensis* (Wittenberg, 1601); and Hieronymus Fabricius (more accurately, Girolamo Fabrici) of Aquapendente's *Tractatus anatomicus triplex, quorum primus de oculo, visus organo* ... (Venice, 1614). These are a mere fraction of the total output of anatomical and ophthalmological works during the period, but they include the more important sources and can perhaps serve as a representative sample.

Knowledge of the gross anatomy of the eye was little altered by these works. The most important advance was a growing understanding of the location and shape of the crystalline humor. Colombo, who succeeded Vesalius in the

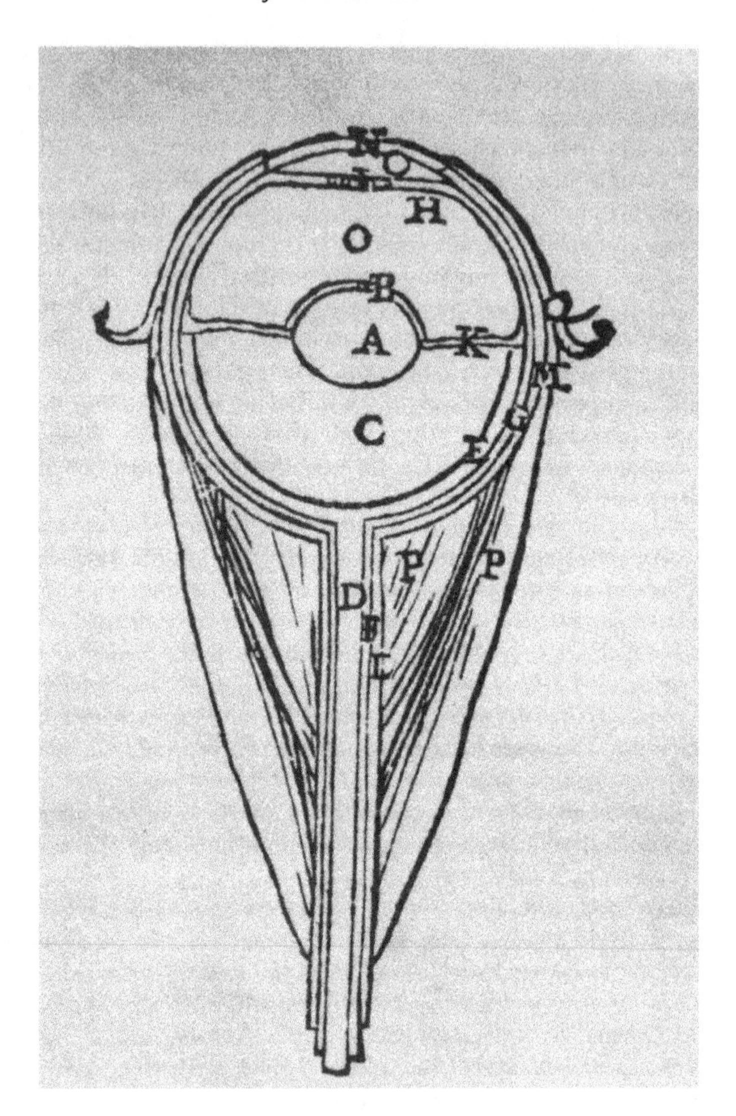

Fig. 27. The eye according to Andreas Vesalius.

chair of anatomy at the University of Padua, attacked his predecessor for "erring totally" in placing the crystalline humor at the center of the eye, for in fact it is "toward the front," though "almost in the center";[138] Colombo also recognized that, contrary to Vesalius's statement the anterior surface of the crystalline humor is more flattened than the posterior surface.[139] Colombo was followed in the former correction by Bartisch and Fabricius and in the latter by Fabricius and Jessen.[140] However, not all innovation was for the better: Charles Estienne admitted that the crystalline humor is flattened in front, but he argued that on the posterior side it comes to a point directly opposite the middle of the optic nerve.[141]

None of the post-Vesalian authors that I have mentioned made significant alterations in visual theory. All maintained that the crystalline humor or the *aranea* (its anterior capsule) is the principal organ of sight, and virtually all admitted that the retina perceives alterations in this principal organ and transmits them to the optic nerve and brain.[142] André Du Laurens (in the English translation of Richard Surphlet, published in 1599) put it charmingly:

> It is now time to make a plaine and open shew of the most precious iewell of the eye, that rich diamond, that beautifull christall, which is of more worth then all the pearles of the East. This is that icelike humour, which is the principall instrument of the sight, the soule of the eye, the inward spectacle: this is that humour which alone is altered by colours, & receiueth whatsoeuer formes of the things that are to be seene. This is that christalline humour, which in more hardie wise then *Hercules,* dares to encounter two at once, namely, the outward and inward light. This is that onely christalline humour, which all the other parts of the eye acknowledge their soueraigne, and themselues the vassals thereof: for the hornie tunicle [cornea] doth the office of a glasse vnto it: the apple [pupil; cf. the German *Augapfel*], the office of a window: the grapelike coate [uvea] is as a fayre flowring garden, to cheare and reioyce the same after wearisome labour: the cobweblike coate [*aranea*] serueth as lead to retaine such formes as are offered: the waterish humour as a warlike foreward, to intercept and breake off the first charge of the objects thereof, assaying all vpon the sudden, and with headlong violence to make breach and entrance: The vitreous humor is his cooke, dressing and setting forth in most fit sort his daily repast: The nerve opticke, one of his ordinary messengers, carrying from the braine thereto, commandement and power to see, and conueying backe againe with all speede whatsoeuer hath been seene.[143]

Regarding the retina, not mentioned in this passage, Laurens had already said: "The vse thereof is to conuey the inward light, which is the animall spirit, vnto the christalline humour, and to carrie backe againe whatsoeuer receiued formes first vnto the nerue optick, and from thence to the braine to iudge thereof."[144]

It is thus apparent that despite an occasional challenge (as by Vesalius and the anonymous author of *De iuvamentis anhelitus*), the essentials of the Galenic theory were widely held among physicians and anatomists of the sixteenth century.[145] To be sure, Galen's idea of a transformation of the medium by visual spirit emerging from the eye had been largely forgotten, but his views on ocular anatomy and physiology (only slightly altered) still prevailed. In this milieu there appeared in 1583 a slender volume by a member of the medical faculty at the University of Basel, Felix Platter (1536-1614), which not only renewed the challenge but also helped to secure the overthrow of the Galenic theory through its influence on Johannes Kepler.[146]

Platter's *De corporis humani structura et usu . . . libri III* was mainly intended to popularize, correct, and supplement the works of Vesalius, Colombo, and Falloppio.[147] It was a short work, consisting of fifty plates and a text, entirely tabular in form (using the method of dichotomy), spread out over approximately two hundred pages. It thus offered no opportunity for extended argumentation or debate, and Platter neither defended nor elaborated upon his scheme of ocular physiology. Of the optic nerve and retina he simply wrote as follows:

> The primary organ of vision, namely the optic nerve, expands when it enters the eye into a hollow retiform hemisphere. It receives and judges the species and colors of external objects, which, along with brightness, fall into the eye through the pupil and are manifest to it through its looking glass, as will be described.[148]

Thus Platter made the optic nerve, together with its expansion in the eye (the retina), the principal organ of vision. As for the crystalline humor, it

> is the looking glass of the optic nerve; and, placed before the nerve and the pupil,[149] it collects the species passing into the eye as rays and, spreading them over the whole of the retiform nerve, presents them enlarged in the manner of an interior looking glass, so that the nerve can more easily perceive them.[150]

The crystalline humor is nothing but an optical lens, through which the retina or retiform nerve views the external world. That the retina is the only seat of the visual power in the eye has by no means been demonstrated, but at least it has been unambiguously claimed.[151]

For this achievement Platter has sometimes been credited with discovering or formulating the theory of the retinal image, but I suggest that such a claim is a considerable overstatement. When considered against the background of sixteenth-century discussions of ocular anatomy, Platter's achievement assumes considerably more modest dimensions. To be sure, his denial of any sensitivity to the crystalline humor was a claim of fundamental importance, which was to become one of the cornerstones of Kepler's visual theory, but it had been anticipated by Leonardo and the author of *De iuvamentis anhelitus* and hinted by Vesalius. Moreover, in endowing the optic nerve and retina with sensitivity, Platter was merely continuing a long tradition. Within the medical tradition it had always been admitted that in some manner the optic nerve and retina are sensitive; it was simply Platter's contribution (and not without precedents) to make them the principal seat of sensitivity.

It must be recognized, moreover, that there were still enormous obstacles to be overcome before we would have anything approaching Kepler's full theory of the retinal image. Platter revealed no understanding of the geometrical

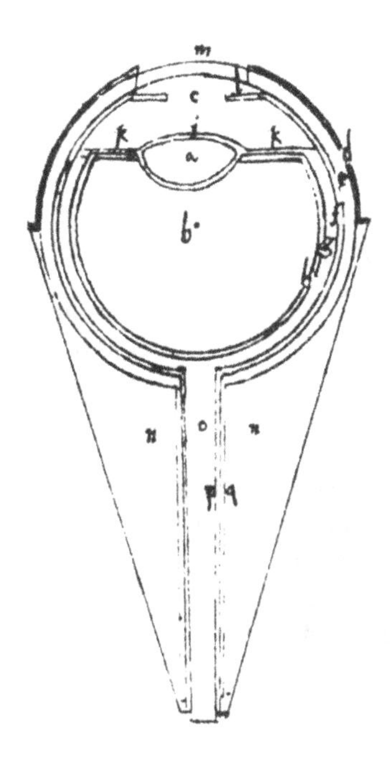

Fig. 28. The eye according to Felix Platter.

issues involved in visual theory, not to speak of a solution to them. There is no reason to assume that he understood the focusing property of lenses or saw how this property could solve the problem associated with the superfluity of rays emanating from each point in the visual field. Nor is there any evidence that Platter was aware of the problem of inversion, which had so vexed Leonardo and which would soon occupy Kepler. Platter's theory simply did not extend to geometrical matters[152]—he claimed merely that the retina views the outside world through its magnifying lens—and we must avoid any suggestion that in Platter's work are to be found solutions to the serious geometrical problems that must be confronted and solved before the theory of retinal sensitivity would be a viable option.

9 JOHANNES KEPLER AND THE THEORY OF THE RETINAL IMAGE

Perspectiva IN THE SIXTEENTH CENTURY

Although the perspectivist tradition was occasionally reduced to a thin stream during the later Middle Ages, its survival was never in doubt: it always had a few practitioners, and it was securely communicated from one generation to the next by the circulation of manuscripts and by a tradition of university teaching.[1] But in the sixteenth century it experienced a revival, manifested by a burst of publication. Printed versions of the works of Euclid, Alhazen, Witelo, and Pecham were issued, numbering some thirty separate editions.[2] Its ablest sixteenth-century practitioners were two Italians, Francesco Maurolico and Giovanni Battista Della Porta, and a German, Friedrich Risner. Although none of the three had a significant influence upon Kepler, a glance at their works will reveal the state of the perspectivist tradition in the sixteenth century, immediately prior to Kepler's own optical investigations.

Maurolico was born in Messina on the island of Sicily in 1494. He became a Benedictine monk and eventually Abbott of Santa Maria del Parto in Messina.[3] He also held civil posts, one as head of the mint in Messina (a post formerly held by his father) and another as director of fortifications for the city under Charles V. In his later years he was charged with writing the history of Sicily (published in 1562) and in 1569 was appointed professor of mathematics at the University of Messina. He died in 1575. Maurolico's most important mathematical writings were published posthumously, including the *Photismi de lumine et umbra* (Naples, 1611), in which he undertook to review and extend the science of *perspectiva*.[4]

In most respects Maurolico's *Photismi* is unremarkable. It begins by setting forth the basic mathematical principles of perspective. Light is held to issue in straight lines from luminous points, and on this foundation Maurolico constructs a series of theorems and corollaries, which describe the illumination of opaque surfaces by luminous objects. Within this framework, however, Maurolico undertakes an analysis of the radiation of light through apertures, and here we find for the first time in the West an adequate geometrical theory of the *camera obscura*.[5] Several theorems contain Maurolico's theory of radiation through apertures, but the nature of his achievement is most clearly revealed in theorem 22, where the following demonstration is presented. Consider light from luminous source *AB* passing through aperture *CD* (fig. 29). *AH*, *AE*, *BF*, and *BK* are rays from the extremes of the luminous object grazing the opposite edges of the aperture, and *FCH* and *KDE* are pyramids of radiation proceeding from vertexes at *C* and *D*, each having a base of the same shape as the luminous source *AB* (but inverted). Now since angles *FCH* and *KDE* are larger than angles *FBK* and *HAE*, the extension of rays *CF, CH, DK*, and *DE* will cause bases *FH* and *KE* to increase more rapidly than bases *FK* and *HE*. "It is therefore possible," Maurolico points out, "for the rays to be extended until spaces *FK* and *HE* become negligible by comparison with *FH* and *KE*."[6] Or to express the same point in terms of visual images, *FH* and *KE* each bear the shape of the luminous source, while *FK* and *HE* represent the separation (owing to the breadth of the aperture) between these two images of the luminous source; therefore, when the ratio of *FH* and *KE* to *FK* and *HE* increases with the extension of the rays, *FH* and *KE* tend to merge, and the image increasingly conforms to the shape of the luminous source. Maurolico thus solved a problem that had perplexed perspectivists in the West for three hundred years.

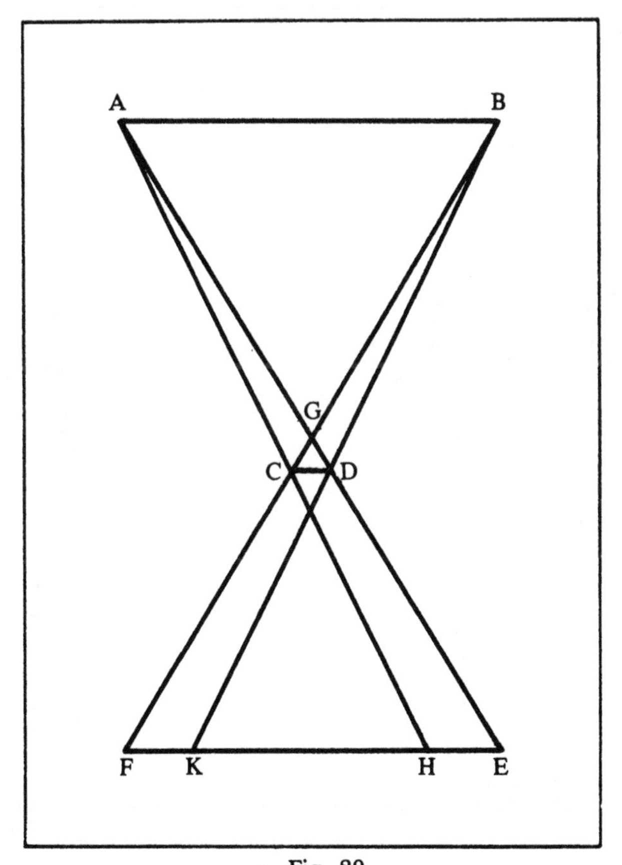

Fig. 29

Maurolico did not, however, apply his conclusions concerning the *camera obscura* to the eye and vision. Rather, he adopted with only small revisions the visual theory of the medieval perspectivists[7]—though the details of ocular anatomy he claimed to have obtained from Vesalius with additions from Bacon and Pecham.[8] He held that the seat of visual power is the crystalline humor, which is nourished by the vitreous humor, which in turn is nourished by the retina.[9] The crystalline humor, he argued, "is enclosed by two convex surfaces, so that by the anterior surface it may receive the species of visible things and by the posterior surface transmit them [in proper order] to the common sense."[10]

However, Maurolico did introduce a significant alteration into the mathematics of the perspectivist theory. An elementary analysis of the geometry of radiation through spectacle lenses (*conspicilia*) led him to the conclusion that double convex lenses produce convergence of the rays, whereas double concave lenses cause the rays to diverge (see fig. 30).[11] Now the crystalline humor of the eye, Maurolico realized, is a double convex lens, which must

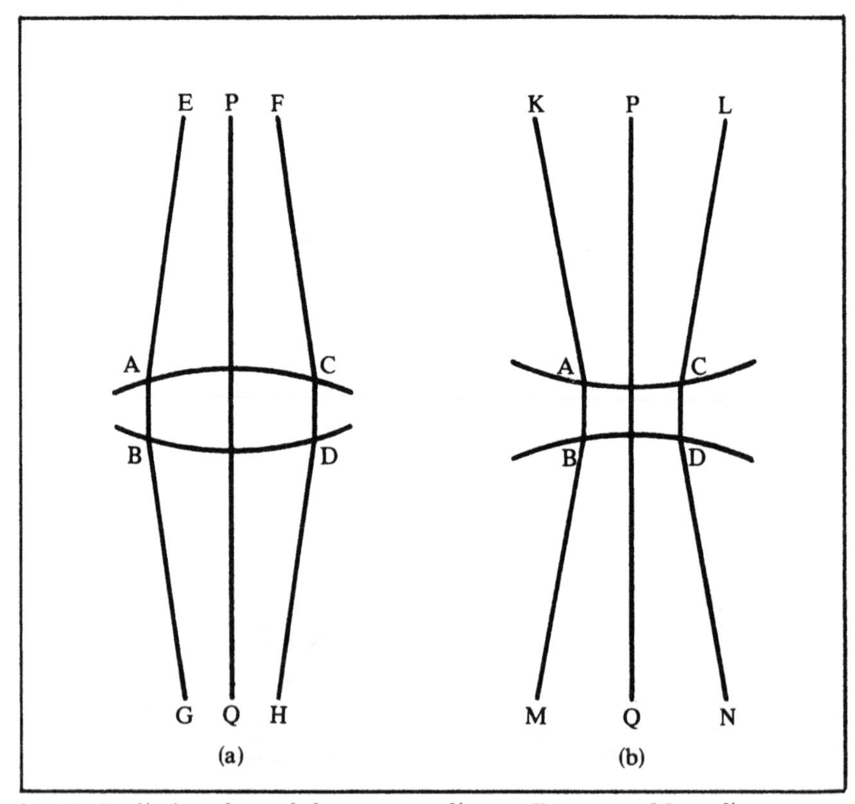

Fig. 30. Radiation through lenses according to Francesco Maurolico.

refract and transmit radiation according to the "law and covenant" of refraction.[12] But this means that Bacon and Pecham were wrong when they maintained that vision is produced only by rays perpendicularly incident on the crystalline humor:

> For if nature, not by chance but deliberately, gave the pupil [i.e., the crystalline humor] a lenticular form, certainly she made both the anterior and the posterior surface convex for the same reason. Therefore, whatever rule (*ratio*) holds for the entering rays must apply also to the emerging rays. Therefore, if the anterior surface of the pupil receives the visual rays perpendicularly, as Roger Bacon and John Pecham (*Petsan*) believe, then the posterior surface ought also to discharge them perpendicularly. But this cannot occur unless the pupil is spherical and all rays are along diameters, intersecting one another at the center. But this nature abhors, not only because of the unsuitability of the lenticular form [for the achievement of such an end], but also lest images appear inverted on account of the intersection of the rays. Therefore what Bacon and Pecham affirm, namely, that visual rays enter the pupil perpendicularly, is absurd.[13]

Apparently Maurolico was prepared to reject any theory of vision in which all rays do not enter and emerge from the crystalline humor symmetrically; he thus placed an exceedingly heavy evidential burden on the principle of symmetry. According to Maurolico, only the axis of the radiant pyramid passes through the crystalline humor without refraction, and it produces the "primary and most perfect" vision; all other rays are refracted toward the axis both upon entering and upon leaving the crystalline humor (as in fig. 30a) and, although they participate in vision, yield "less certitude."[14] The rays are convergent as they emerge from the crystalline humor, but before actually converging to an apex they reach the opening of the optic nerve, thus carrying the species of the visible object, properly arranged, to the nerve and ultimately to the brain.[15]

The causes of myopia and hypermetropia and the function of spectacles in correcting these defects are now apparent. Myopia occurs when the convex crystalline humor is excessively curved, so that the radiation passing through it converges too much or too soon, "and this excessive crowding together confuses the [interior] judgment and discrimination of sense."[16] In hypermetropia the crystalline humor is insufficiently curved, convergence is therefore delayed, and again vision is impaired. The purpose of spectacles, then, is to "correct the defects of nature"[17]—to hasten or delay convergence of the rays and hence to compensate for the insufficient or excessive curvature of the crystalline lens. For hypermetropia this requires a double convex lens, for myopia a double concave lens.[18]

This is an interesting and novel theory. Indeed, Maurolico's stress on the refracting properties of the crystalline humor and his proper understanding of the causes of myopia and hypermetropia and the function of spectacles have led some historians to see him as an important precursor of Kepler.[19] And it is true that Maurolico was the first to provide a theoretical analysis of spectacles or any other nonspherical lens. Moreover, he stressed more fully than any predecessor (except perhaps Leonardo) the refracting properties of the crystalline humor: at the beginning of his discussion of the eye he pointed out that "since the organ [of sight] is transparent, the problem is entirely one of transparent substances," and he concluded by noting that pupils (i.e., crystalline humors) "are the spectacles of nature, and glass spectacles the pupils of art."[20] Nevertheless, Maurolico regarded the crystalline humor not merely as a refracting device, but also as an instrument of perception. "Among those [ocular] structures that pertain to vision," he insisted, "the summit of dignity is occupied by the glacial or crystalline humor, . . . in which, as throne, the visual power resides."[21] The crystalline humor may refract and transmit visual species, but in the process it perceives them.

But Maurolico's position regarding the seat of visual power is not the only obstacle to seeing him as Kepler's precursor. Maurolico also failed even to perceive the problem raised by the multiplicity of rays emanating from all points in the visual field to all parts of the eye. He acknowledged that if distinct vision is to occur, every point in the visual field must be observed along a single line;[22] but his theory offered no means by which this could be achieved. The rays passing through the double convex lens in figure 30a (which can be conceived for present purposes as the crystalline lens of the eye) do not emanate from a single point in the visual field, but each ray comes from a different point. If we regard the rays in figure 30a as those by which vision is produced (as clearly we must), it is apparent that far from developing a mechanism by which the visual power can confine its attention to a single ray emanating from each point in the visual field (the solution of the medieval perspectivists) or by which rays from a given point are brought to a focus in the eye (the solution later proposed by Kepler), Maurolico has simply ignored the problem; he has restricted his attention to a single ray from each visible point and dismissed the rest without any attempt to justify such a step beyond a few remarks about the need for symmetry.[23] He has thus presented a theory of vision that effectively ignores the most fundamental premise of the geometrical theory of vision—that visible points radiate in all directions and that all rays entering the eye have the power (if not weakened or otherwise eliminated from consideration) of stimulating vision. Not only did Maurolico not anticipate Kepler's theory of vision; he appears not even to have understood (clearly, at any rate) the issues to which Kepler's theory of vision was addressed.

Nearly forty years after Maurolico completed his optical studies, the perspectivist cause was taken up by another Italian, this time a member of the Neapolitan nobility, Giovanni Battista Della Porta. Della Porta was born in

1535, probably in the family villa outside the town of Vico Equense.[24] Little is known of his education, though he was surely influenced by the learned circles in which his father moved. Clubb writes that Nardo Antonio Della Porta (Giovanni Battista's father) "so delighted in the company of philosophers, mathematicians, poets, and musicians that this house in Naples became a veritable academy."[25] Giovanni Battista appears also to have frequented the formal "academies" of Naples and to have come under the influence of Giovanni Antonio Pisano and perhaps others.[26] In adulthood he became a playwright and polymath. His *Magia naturalis* was first published in 1558, and a much expanded version of the same work appeared in 1589. He also published books on agriculture, physiognomy, pneumatics, military fortifications, and mathematics; and he translated the first book of Ptolemy's *Almagest* into Latin.[27] His major work on optics, *De refractione optices parte libri novem,* appeared in 1593. Della Porta died in 1615.

De refractione, as its title might suggest, is an attempt to deal with the problem of refraction in all of its ramifications. Book 1 treats refraction in general terms, and book 2 treats refraction through glass spheres. Books 3-7 are devoted to the visual process, and books 8 and 9 treat spectacles and meterological phenomena involving refraction. The theory of vision expressed in this work is in most respects traditional, and no medieval perspectivist would have been much surprised by the contents, though he might have been disappointed by the paucity and superficiality of mathematical argument and by Della Porta's failure to grasp certain crucial issues.

Vision, according to Della Porta, occurs through the intromission of forms or rays and their reception by the crystalline humor. He attempts to prove the truth of the intromission theory by reference to afterimages, and also by an original, but fallacious, geometrical argument.[28] For the sensitivity of the crystalline humor he presents a variety of traditional arguments drawn from the medical and perspectivist traditions.[29] On ocular anatomy Della Porta appears to follow Vesalius (whom he mentions by name), although he modifies the Vesalian scheme by placing the crystalline lens toward the front of the eye, so that its front surface will be concentric with the cornea, and the visual pyramid will pass into it without refraction and emerge before intersection (and hence inversion of the image) can occur.[30]

Della Porta seems to understand that vision is produced by perpendicular rays alone, for he argues that

> the visual pyramid extends unrefracted through the cornea and the pupil, through the aqueous humor to the crystalline humor, as follows. Let there be a quadrilateral visual pyramid *CBAD* [fig. 31] directed toward the center of the eye, *M*. It passes through the cornea *GH,* the pupil *EF,* and thus all the way to the crystalline humor *IL*. Therefore the whole pyramid passes through unrefracted and unimpaired, and represents the object as it [truly] is.[31]

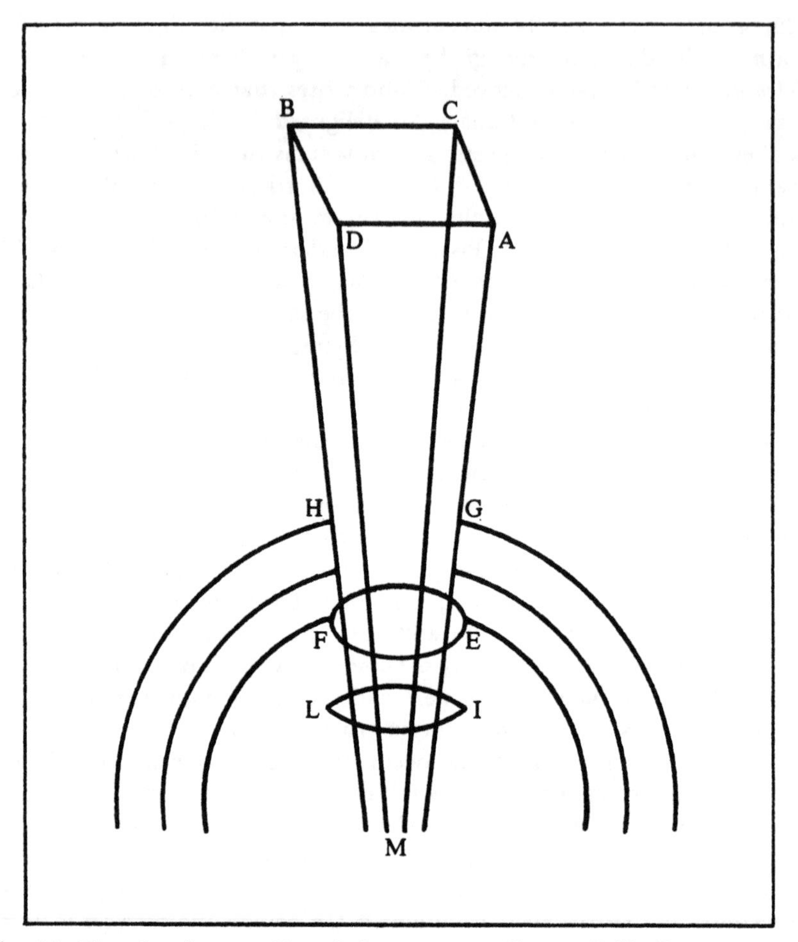

Fig. 31. The visual pyramid and the eye according to Della Porta.

However, he fails to explain how oblique rays can be ignored, and in certain other demonstrations he makes explicit reference to the influence of oblique rays on the visual power.[32] It may be that he was following the medieval perspectivists in first banishing and then restoring oblique rays, but I suspect that he was merely unclear on the matter.

In one significant respect Della Porta went beyond his perspectivist predecessors, including Maurolico. Della Porta, whose *Magia naturalis* helped to popularize the *camera obscura* as a spectacular toy, regarded the eye as a miniature *camera obscura.* He did not, however, employ this important insight as a point of departure for a new understanding of the visual process. On the contrary, instead of conceiving the crystalline humor as a lens inserted into the opening of a *camera* (it was Della Porta who, in the *Magia naturalis,* had pointed to the advantages of placing a glass lens in the opening of the artificial camera),[33] he maintained that the crystalline humor is the screen onto which the images are projected: "I say that just as light [passing] through the narrow opening of a window portrays bodies illuminated by the sun on paper placed opposite, so also does it, proceeding through the aperture of the pupil, portray the images of things seen on the crystalline [humor]."[34] The reason why this must be so, of course, is that the crystalline humor is the principal organ of sight.

The other leading perspectivist of the sixteenth century was Friedrich Risner, a German from Hesse, who spent most of his scholarly career at the University of Paris, where he collaborated closely with Peter Ramus.[35] Risner's most important contribution to the science of *perspectiva* was clearly his edition of the works of Alhazen and Witelo, prepared at the suggestion of Ramus and published in 1572. Risner did his editorial work well—transcribing the texts carefully, supplying section headings for Alhazen's treatise, and cross-referencing the two works—and through his edition the vast majority of late sixteenth-century and early seventeenth-century mathematicians and natural philosophers, including Kepler, became acquainted with the optics of Alhazen and Witelo.[36] If Risner's edition of Alhazen and Witelo leaves any doubt that he perfectly understood and accepted their theories of vision, that doubt is dispelled by a perusal of his own optical work (probably written in collaboration with Peter Ramus), which was posthumously published in 1606.[37] This treatise seems to have had little influence[38] and contains no novelties, and therefore it need not detain us. But it reveals that at least this one sixteenth-century perspectivist had a thorough grasp (surpassing that of Maurolico and Della Porta) of the visual theory of the medieval perspectivists.

It is curious that none of the works thus far mentioned in this chapter, except Risner's edition of Alhazen and Witelo and Della Porta's *Magia naturalis,* influenced Kepler. Maurolico's *Photismi* and Risner's *Optica* were posthumously published after Kepler had already completed his first and major optical work, and there is no reason to believe that he had access to them in manuscript. Kepler knew about Della Porta's *De refractione* (it was referred to by Della Porta in the 1589 edition of the *Magia naturalis,* which Kepler had read), but he complained of his inability to obtain a copy from the booksellers.[39] Thus when Kepler turned his attention to the science of *perspectiva,* he found it necessary to return to the original sources—to the writings of Euclid, Alhazen, Witelo, and Pecham.

KEPLER'S FIRST OPTICAL STUDIES: THE PROBLEM OF THE *Camera Obscura*

Johannes Kepler was born in the Swabian city of Weil in 1571, the son of Lutheran parents.[40] In 1587 he matriculated at the University of Tübingen and there came under the influence of the professor of astronomy, Michael Maestlin, who instructed him in both the Ptolemaic and the Copernican systems. Kepler entered the theological faculty at Tübingen in 1591, intending to become a Lutheran clergyman, but in 1594 he interrupted his theological studies to take a position as mathematician in the Lutheran academy in Graz. His duties there (which included the production of annual calendars and prognostications) granted him time to engage in serious astronomical endeavor and thus to enter upon those studies that would bring him ultimate fame. In 1600, he became Tycho Brahe's assistant at the court of Rudolf II in Prague, and when Tycho died in 1601 Kepler succeeded him as imperial mathematician. Twelve years later Kepler moved to a new post as district mathematician in Linz. He died in Regensburg in 1630.

Although Kepler undoubtedly encountered the science of optics while studying with Maestlin at Tübingen,[41] his serious interest in the subject appears to have originated in 1600, in connection with the solar eclipse of 10 July. Some years earlier Tycho had observed that the lunar diameter, measured with the use of a pinhole camera, was smaller during a solar eclipse than at other times, though the moon was no farther away; and as Straker has convincingly argued, Kepler concluded in July of 1600 that the cause of this "enigma" was to be found in optical theory, specifically in the theory of radiation through small (but finite) apertures.[42] He turned immediately to the works of the medieval perspectivists, especially Witelo and Pecham, but found their theory confused and useless:

> Three centuries ago, Witelo believed that [the roundness of solar radiation sent through an opening of any shape] occurred on account of some sort of parallelity of rays. . . . But the defect of his demonstration he himself does not conceal, saying, in Proposition 35, "perhaps a property of the rays cooperates greatly towards that very [result]." In these twisting equivocations, he showed that he himself did not understand the true cause, which was [actually, but without his recognizing it] brought forward in another part of his demonstration. Following this, Johannes Pisanus [Pecham] . . . rejects the statement of others, and among them that most true explanation . . . [and instead] retreats into the shadows of an arcane nature of light along with Witelo.[43]

The true cause, which Witelo and Pecham stumbled over but failed to recognize, was the intersection of rectilinear rays; their preferred solution (in Kepler's view) was to endow light with an "arcane nature"—the ability to withdraw from rectilinearity.[44]

Kepler discovered the proper solution to the problem by an experimental technique that perhaps drew its inspiration from Albrecht Dürer's *Underweysung der Messung* (Nuremberg, 1525)—namely, by stretching a thread through an aperture from a luminous source, or rather from a book meant to simulate a luminous source, to a surface on which the image was formed.[45]

Tracing out the image cast by each point of the book in this manner, Kepler could see the geometry of radiation in material, three-dimensional terms and was able to formulate a satisfactory theory of radiation through apertures, based firmly on the principle of rectilinear propagation. Kepler later wrote:

> A certain light drove me out of the shadows of Pisanus [Pecham] several years ago. For indeed, since I could not comprehend the obscure sense of [his] words from the diagram on the page, I had recourse to a personal observation in three dimensions. I placed a book on high to take the place of the shining body. Between it and the floor I set a table having a many-cornered aperture. Next, a thread was sent down from one corner of the book through the aperture and onto the floor; it fell on the floor in such a way that it grazed the edges of the aperture; I traced the path produced and by this method created a figure on the floor similar to the aperture. Likewise, by means of a thread attached to another, a third, a fourth corner of the book, and finally to an indefinite number [of points] along the edge, there resulted on the ground an indefinite number of traced figures [each having the shape] of the aperture, which together produced a great and four-cornered [figure having the] shape of the book.[46]

This is substantially identical to the theory formulated some eighty years earlier by Maurolico, though of course Kepler arrived at his own version of the theory independently and was the first to publish it.[47]

But Kepler did not stop with the theory of radiation through apertures—although his original intent was probably to treat this problem alone.[48] Having observed the power of optics, properly understood, to resolve Tycho's problem of the (seemingly) variable lunar diameter, Kepler proceeded to recognize that all astronomical observation depends for its validity on a proper understanding of visual theory:

> Since the diameters of the [celestial] luminaries and the quantities of solar eclipses are given a fundamental place by astronomers, [it must be recognized that] some deception of sight intrudes, arising partly from the art of observing [through small aperatures] . . . and partly from simple vision itself; and this deception, so long as it is not taken into account, creates great difficulty for investigators and detracts from the ability of the art to judge [properly]. And so the explanation of error in sight is to be sought in the structure and functioning of the eye itself. Had the opticians Alhazen and Witelo, or after them the anatomists, treated these matters clearly, distinctly, and without risk of uncertainty, they would have freed me from the labor of continuing the *Paralipomena ad Vitellionem* in this chapter.[49]

In short, if astronomical observations depend on the propagation of light and its perception by an observer, then a knowledge of visual theory is indispensable for astronomers. Indeed, in a letter of 1601 to his teacher Maestlin, Kepler noted that the eye itself possesses an aperture and therefore should be prone to the same errors that attend the observation of eclipses through an aperture.[50] Thus the man who had solved the problem of radiation through apertures resolved to turn his attention also to the general problem of vision. He would ultimately publish his solutions to both problems in the *Ad Vitellionem paralipomena* (*Additions to Witelo*) of 1604.

Kepler and the Perspectivists

Kepler began his serious study of the works of the medieval perspectivists in the course of his work on radiation through apertures; and from the understanding thus acquired came his conviction that something was seriously wrong with the perspectivist theory of vision. Kepler outlined its failings in chapter 5 of *Ad Vitellionem paralipomena*.

The perspectivist theory of vision was defective, in the first place, from an anatomical standpoint. According to Kepler, the perspectivists (or, more specifically, Witelo, to whom Kepler's objections were addressed) assigned the power of visual perception to the crystalline humor on the grounds that its surrounding web or capsule, the *aranea,* is connected directly to the retina and hence to the optic nerve. The crystalline humor thus "feels" (as a form of touch) the light and color entering it and communicates its feelings or perceptions to the brain through the connecting nervous tissue.[51] But this "opinion . . . is destroyed, since the crystalline humor is disconnected from the nerve and retina and connected to the uvea," as revealed by Felix Platter.[52] Moreover, the notion that sight is a form of touch is untenable, for light and color pass through the eye instantaneously, and "are far more subtle than to admit of perception by a corporeal tunic as a form of touch."[53]

Another anatomical detail destructive of the perspectivist theory of vision is the shape of the crystalline humor. According to Kepler, Witelo assigned the crystalline humor a flat posterior surface in order to insure that rays from the visible object would not intersect at the center of the eye and thus yield an inverted visual impression.[54] Rather, the rays would be refracted upon passing from the crystalline to the vitreous humor in such a way that intersection at the center of the eye would be averted and upright vision would be saved. But, unfortunately for the theory of Witelo, the posterior surface of the crystalline humor is not flat, but rounded or gibbous, as experience plainly teaches.[55] Kepler confesses that he too was much troubled by the problem of inversion and sought to discover a second intersection in the eye (the first, in his theory, occurring in the pupil) by which to reinvert the rays:

> And so I truly and dutifully tortured myself in order to show that the cones intersecting when they pass through the aperture of the uvea intersect again behind the crystalline humor in the middle of the vitreous humor, so that

> another inversion is produced. . . . And there was no end of this useless labor until I came upon Propositions 11 and 12 above, by which this opinion is plainly refuted.[56]

But even if he had obtained the desired result, Kepler adds, the problem would by no means have been solved. An upright image in the eye, each of its points directly opposite the imaged point in the visual field, would have the same defect as a mirror image—left and right would be interchanged.[57] The only remedy, Kepler argues, is for the image to be both reversed and inverted—that is, for points in the visual field and their images in the eye to be opposite one another in a spherical sense, with respect to a center of intersection.[58] This is perfectly achieved by his theory of the retinal image, as I will make clear below.

The visual theory of the perspectivists is thus contradicted by the facts of ocular anatomy; it is also discredited, and with equal ease, by mathematical and physical considerations. The weak point in the theory from a mathematical and physical standpoint is its unworkable and inconsistent scheme for eliminating superfluous radiation falling on the eye. Since every point within the eye receives radiation from every point in the visual field, there will be total confusion unless some of the radiation is ignored or the whole cone of radiation is somehow reorganized. (This difficulty is present no matter which ocular tunic or humor is conceived to be the seat of the visual power.) Witelo and the other perspectivists solved this problem by pointing out that only a single ray from each point in the visual field passes into the eye without refraction, that refracted rays can be ignored on account of their weakness (since bending weakens), and that the unrefracted rays which are thus made responsible for vision possess the same order within the eye as do the points in the visual field from which they issued. But the perspectivists then proceeded to contradict their own solution by readmitting nonperpendicular radiation into the kingdom of vision-producing rays. "Shown very clearly by experience that more than the hemisphere entering the eye perpendicularly is seen," they admitted that objects at the periphery of the visual field (outside the cone of perpendicular rays) are seen by means of rays that fall obliquely onto the eye.[59] This, of course, revives the confusion of rays within the eye so nicely eliminated by restricting attention to perpendicular rays; for these oblique rays from the edges of the visual field will mingle with perpendicular rays (coming from other points in the visual field) in the crystalline humor. Kepler comments on the outcome:

> I refute Witelo by this very confusion of rays. For, as he says, oblique radiation too is seen insofar as oblique rays intersect perpendicular rays [within the eye]; therefore the same point [of the eye] receives both oblique and perpendicular radiation. Consequently two things [the point sources, respectively, of the oblique and perpendicular radiation] will be judged to be situated in the same place.[60]

But perhaps a more basic flaw than the return of oblique radiation is the physical reasoning by which oblique radiation was originally eliminated from consideration. The position of the perspectivists was that all radiation falling obliquely on the eye could be ignored, however slight its obliquity, so that the pyramid of rays responsible for vision consists of perpendicular rays alone. This theory, if viewed solely from a mathematical standpoint, might have a certain plausibility, for perpendicular and oblique rays do indeed constitute mathematically distinct classes of rays; however, the medieval perspectivists did not view the matter solely from a mathematical standpoint, but defended their position with physical arguments about the strength of motion (and hence of radiation) along the perpendicular. Neither did Kepler view the matter strictly in mathematical terms. He perceived the absurdity of an abrupt physical discontinuity between perpendicular rays and rays that are nearly perpendicular, arguing that slightly oblique radiation is only slightly refracted and therefore should be only slightly weakened. How then would such radiation fail to stimulate the visual power?

> Now since Witelo forms his *simulacrum* by means of perpendiculars alone [Kepler wrote], it is a monstrous thing how subtly he distinguishes between the perpendiculars and the immediately adjacent rays. If light acts on sense and sense is affected by this action, then when the action is very strong sense will be severely affected. And yet the perpendiculars and the [oblique] rays adjacent to them differ scarcely at all in their power of illuminating because the adjacent rays are scarcely refracted. Therefore the reception or sensing of the perpendiculars and the rays adjacent to them [should be] almost equal. Thus perception has been thrown into disorder, and Witelo has labored in vain.[61]

This is the Achilles' heel of the perspectivists' visual theory, and here Kepler inflicted the mortal wound. It may be legitimate to attach less significance to refracted rays on account of their bending; however, since perpendicular and oblique rays scarcely differ in their ability to illuminate, and "the impression made on the visual power follows in manner and proportion the action of illumination," there are no grounds for ignoring refracted rays altogether.[62] This is a devastating argument, for if refracted rays cannot be ignored, the visual theory of the perspectivists is without justification.

Kepler on the Anatomy of the Eye

If the visual theory of the perspectivists is untenable, a new theory must be found to take its place. But Kepler recognizes that this new visual theory will have to contend with anatomical facts. Like Alhazen before him, Kepler acknowledges that he has not personally investigated ocular anatomy; he is a mathematician (in modern terms, a mathematical physicist), and he has necessarily relied on others for anatomical knowledge:

> It helps to generate confidence in the demonstration that I will set forth that I present not my own experiments on the eye, but those published by outstanding physicians. For what if somebody should charge me with bad faith, of attempting to establish my own opinion, or of incompetence in dissection, in which I have never participated either as spectator or as administrator? Therefore, let men of acknowledged authority speak for me concerning matters well known to them.[63]

Kepler relates that he consulted chiefly Felix Platter's *De corporis humani structura et usu* and the *Anatomia Pragensis* of his friend Johannes Jessen. Kepler's reliance on Jessen is easy to understand, in view of their friendship;[64] but how did he happen upon Platter's work, with its crucially important claim about the sensitivity of the retina? We have no answer to the question, although it is possible that Jessen called Platter's book to his attention—even though Jessen himself assigned sensitivity to the crystalline humor.

Whatever the method of their selection, Platter and Jessen were fairly dependable guides to the gross anatomy of the eye and optic nerves. The cerebrum, they taught, is surrounded by a pair of tunics. The outer one (which is thick) is continuous with (or gives rise to) the outer covering of the optic nerve and the outermost tunic of the eye; the interior cerebral tunic (which is thin) is continuous with the middle tunic of the eye; and finally, the substance of the cerebrum gives rise to the optic nerve, which becomes the innermost tunic of the eye, the retina.[65]

As for the eye itself, the outermost tunic is divided into two hemispheres. The posterior hemisphere, called the *sclerodis* (the sclera or sclerotic coat), is hard, thick, opaque, white, and almost cartilaginous; the anterior hemisphere of this outermost ocular tunic, the cornea, is transparent and slightly protruding, thus revealing itself to be a portion of a smaller sphere (or perhaps spheroid) than is the sclerotic coat.[66] The second tunic of the eye is also divided into hemispheres, the posterior being called the choroid, the anterior the uvea. These two hemispheres differ in thickness (the uvea being twice as thick); in addition, the choroid is joined to the sclerotic coat by small fibers at the point where the sclerotic coat becomes the cornea. The uvea is not attached to the cornea, but curves away from it and "swims" in the aqueous humor. The uvea is perforated by an aperture (the pupil) in front;[67] it is black on the inside, but seen through the cornea the outside is variegated in color and called the iris. Finally, the third and innermost tunic is the retina or retiform tunic, which arises from the spreading out of the optic nerve itself into a hollow hemisphere or funnel.[68]

Because of its significance for his visual theory, Kepler goes into additional detail concerning the structure and properties of the retina. It is neither dark nor dazzling white, Kepler notes, lest it alter the apparent colors of visible things; rather, it is reddish white. It must be hemispherical in order to be properly proportioned to the pictures formed on it; but in actuality it extends beyond a hemisphere (reaching as far as the ciliary processes) "so that when the globe [of the eye] is filled with vitreous humor, the retina is stretched out because the neck is narrower than the belly."[69] The retina cannot be bound to the tissue behind it because of the subtlety of the visual spirit contained within its channels, and it must therefore occupy more than a hemisphere lest it "become wrinkled and slip back to its junction with the [optic] nerve."[70]

To this point, Kepler's guides, Jessen and Platter, have been in substantial agreement. However, on the physical relationship between the retina and the crystalline humor they part company. Jessen follows Witelo in maintaining that the retina joins the crystalline humor at its equator. Platter, on the other hand, denies any connection between the two organs and argues that the crystalline humor is connected, through the ciliary processes (girdling the crystalline humor and resembling eyelashes), to the choroid and uvea. Kepler reveals that he "agrees more with Platter" on the grounds that Witelo (followed by Jessen) assigned the visual power to the crystalline humor and was therefore disposed to find a connection by which the visual power could be communicated from the optic nerve through the retina to the crystalline.[71] Platter, who recognized the sensitivity of the retina, needed no such physical connection.

The traditional three humors were held by Kepler to fill the eye. The largest of them is the vitreous humor, which is almost spherical and occupies the posterior region of the eye.[72] At the front of the eye is the aqueous humor, highly transparent and the most tenuous of the humors. Between the aqueous and vitreous humors are the crystalline humor (surrounded by the arachnoid membrane) and the ciliary processes. The anterior surface of the crystalline is either circular or "a portion of a lenticular spheroid"; the posterior surface is a "hyperbolic conoid" produced by rotating a hyperbola around its axis.[73] The crystalline humor has a greater density than either the aqueous or vitreous humor. Between the small hairlike structures that make up the ciliary processes, Kepler believes there are small fissures, and through these fissures pass small quantities of light to fall on the borders of the retina and thus extend vision beyond a hemisphere.[74] Finally, to illustrate his description of ocular anatomy, Kepler reproduces the figures from Platter's *De corporis humani structura et usu* (see fig. 28, above).

As his admitted dependence on Platter and Jessen implies, Kepler's anatomical descriptions contain nothing new. But neither do they contain major errors or misconceptions that would stand in the way of the theory of the retinal image. It is true that Kepler thought the entrance of the opitc nerve into the eye was on the axis of the eye, rather than to the nasal side of the axis; that the existence of the blind spot was as yet unrecognized; that the shape, and therefore the refractive power, of the crystalline was thought to be rigidly fixed; that the shape, proportions, and structure of the other ocular organs were not yet sufficiently known; and, indeed, that the minute anatomy of the eye had not been studied at all. But these were matters that would have no effect on the broad outlines of visual theory as established by Kepler. He had as much anatomical knowledge as he needed, and he was ready to turn from ocular anatomy to the geometry of sight.

The Theory of the Retinal Image

The foundation of the geometry of sight, for Kepler as for Alhazen and the Western perspectivists, was the punctiform analysis of the visible body. If luminous rays issue in all directions from every point in the visual field, then it is necessary to trace these rays to and through the eye and to establish an orderly one-to-one correspondence between the point sources of radiation and points stimulated within the eye. The principal achievement of the medieval perspectivists was to answer this need with a reasonable geometrical scheme; and here too Kepler would make his major contribution to visual theory.

Kepler lays down the principle of punctiform analysis in the second proposition of the *Paralipomena*: "lines [of light] infinite in number issue from every point" in the visual field.[75] It follows that each visible point opposite the eye completely bathes the cornea and fills the pupil with its rays, which thus form a cone with apex on the visible point and base within the eye. The rays are slightly refracted as they pass through the cornea and converge to a base on the surface of the crystalline lens:

> And since the cornea and the aqueous humor beneath it (which I take to be a single medium as far as density is concerned) are of greater density than air, rays from a point sent obliquely at the cornea are refracted toward the perpendicular. Therefore, rays that first diverged in the air converge upon entering the cornea to such an extent that although the largest circle described on the cornea by rays that [subsequently] descend to the edges of the pupil is larger than the pupil, nevertheless these rays are made to converge so greatly in the small depth of the aqueous [humor] before they reach the pupil that the extreme rays graze the edge of the pupil, and by their further descent they illuminate a portion of the surface of the crystalline smaller than the pupil.[76]

Each visible point produces a similar cone; all the cones intersect in the pupil, and at this point right and left, top and bottom, are interchanged.[77]

So much for radiation from the visible object to the surface of the crystalline humor. Kepler has conceptualized this radiation in terms of an infinity of cones, each with its apex on a different visible point and sharing a common base (or nearly so) in the eye, and no medieval perspectivist would have found any aspect of this scheme the least bit surprising.[78] It is when Kepler begins to investigate what happens to radiation after it reaches the crystalline humor that we first encounter significant innovations in the geometry of vision. The perspectivists, of course, had vastly simplified this portion of the problem by confining their attention to perpendicular rays; these were propagated in straight lines until they encountered the posterior surface of the crystalline humor, whereupon they were diverted from their convergence toward an apex and proceeded instead through the opening in the optic nerve and ultimately to the optic chiasma— all the while maintaining their proper arrangement. But Kepler, for the reasons outlined above,[79] could not ignore nonperpendicular rays and therefore had to concern himself not only with many more rays, but also with a multiplicity of refractions. He would, in short, have to come to an understanding of the focusing properties of lenses.

No satisfactory analysis of lenses could be found in the available literature. Lenses had been dealt with at length only by Maurolico and Della Porta, and Kepler did not have access to their works. In the writings of the medieval perspectivists, Kepler would have discovered only the most elementary analysis of the burning sphere, showing all the rays issuing from a luminous point converging to a single focus on the other side of the sphere.[80] However, if the perspectivist tradition did not provide Kepler with a theory of lenses, it furnished the principles by which such a theory could be constructed. The medieval perspectivists had dealt extensively and successfully with refraction at a single transparent interface, and although their analysis employed only a single ray from each visible point, it could easily be extended to more complicated cases involving additional rays and a second refracting interface.[81]

Kepler's own analysis, like that of his predecessors, is confined to transparent spheres.[82] He begins at an elementary level, arguing that the image of point *A* (fig. 32) observed by eyes *B* and *C* through sphere *EFHK* will be situated at *D*, where rays *AE*, *AI*, and *AG* converge to a focus.[83] Moving quickly to the case of parallel rays from a distant source, Kepler recognizes the phenomenon of spherical aberration—that rays passing close to the center of the sphere intersect its central axis farther from the sphere than do rays passing through the periphery. Thus ray *KB* (fig. 33), which passes close to central axis *DAF*, is refracted at *B* and *C* and intersects the axis at *F*; whereas ray *LG*, which passes through the periphery of the sphere, is refracted at *G* and *H* and intersects the central axis at *I*.[84] The final outcome of this ray geometry is the formation of a hyperbolic envelope—the caustic surface illustrated in figure 34.[85] Finally, Kepler moves closer to the actual circumstances of vision by placing an aperture before the transparent sphere and arguing that an inverted image of the scene opposite the aperture is produced on a sheet of paper placed behind the sphere. Let *A* (fig. 35) be the center of the sphere and *EF* the aperture placed before it.[86] Let *HI* be the visible object,

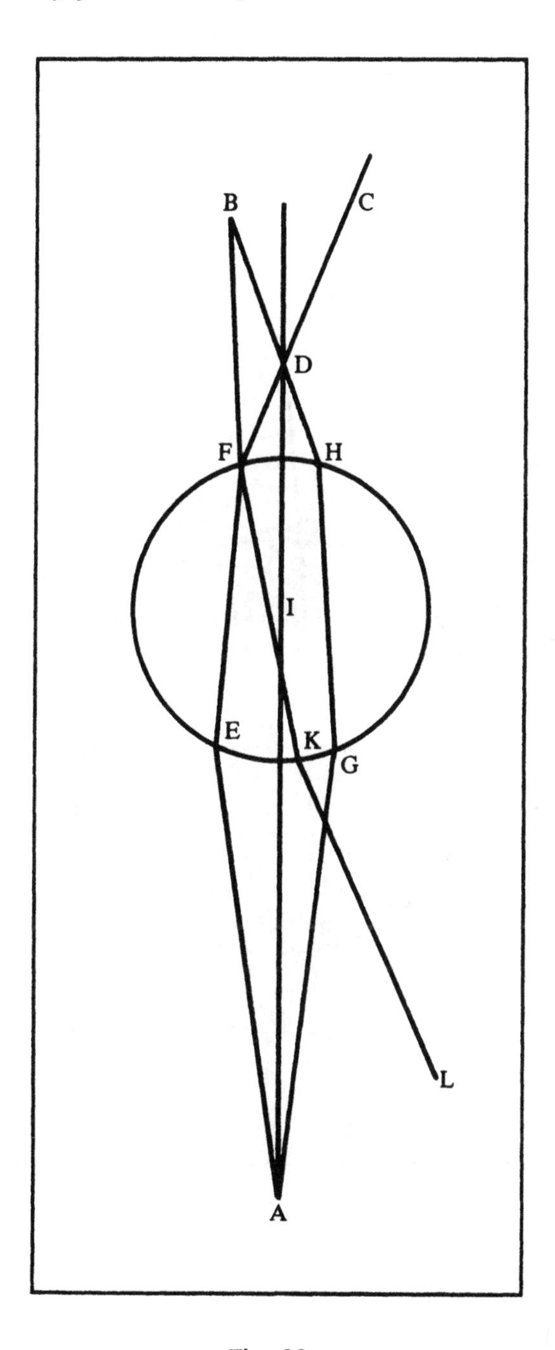

Fig. 32

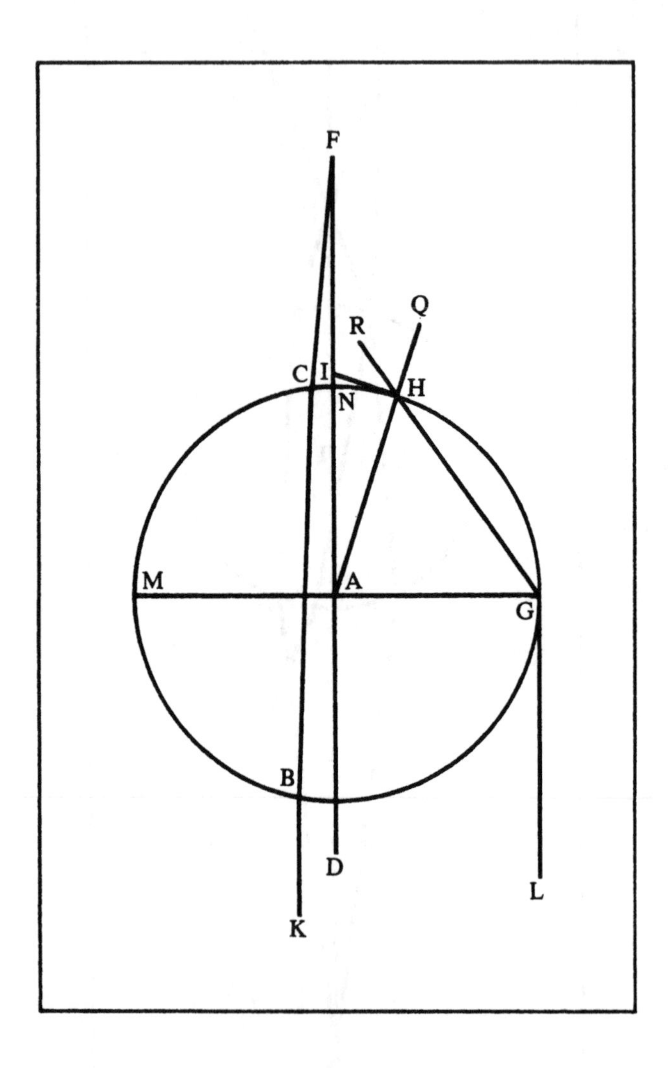

Fig. 33

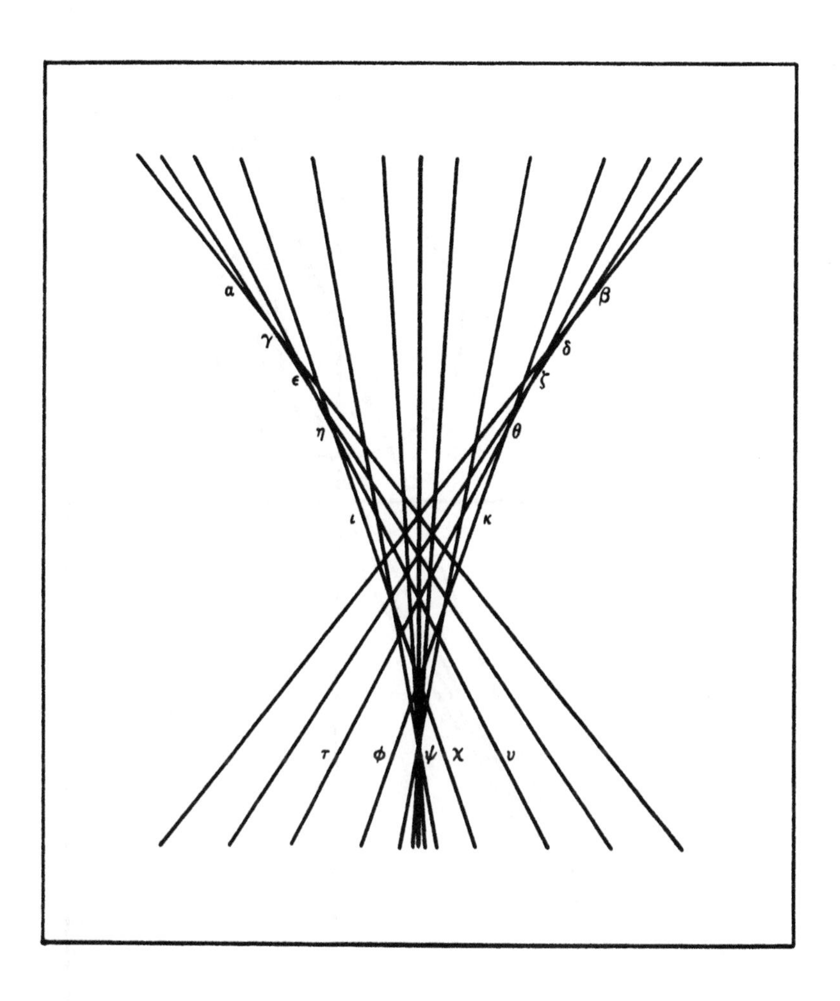

Fig. 34

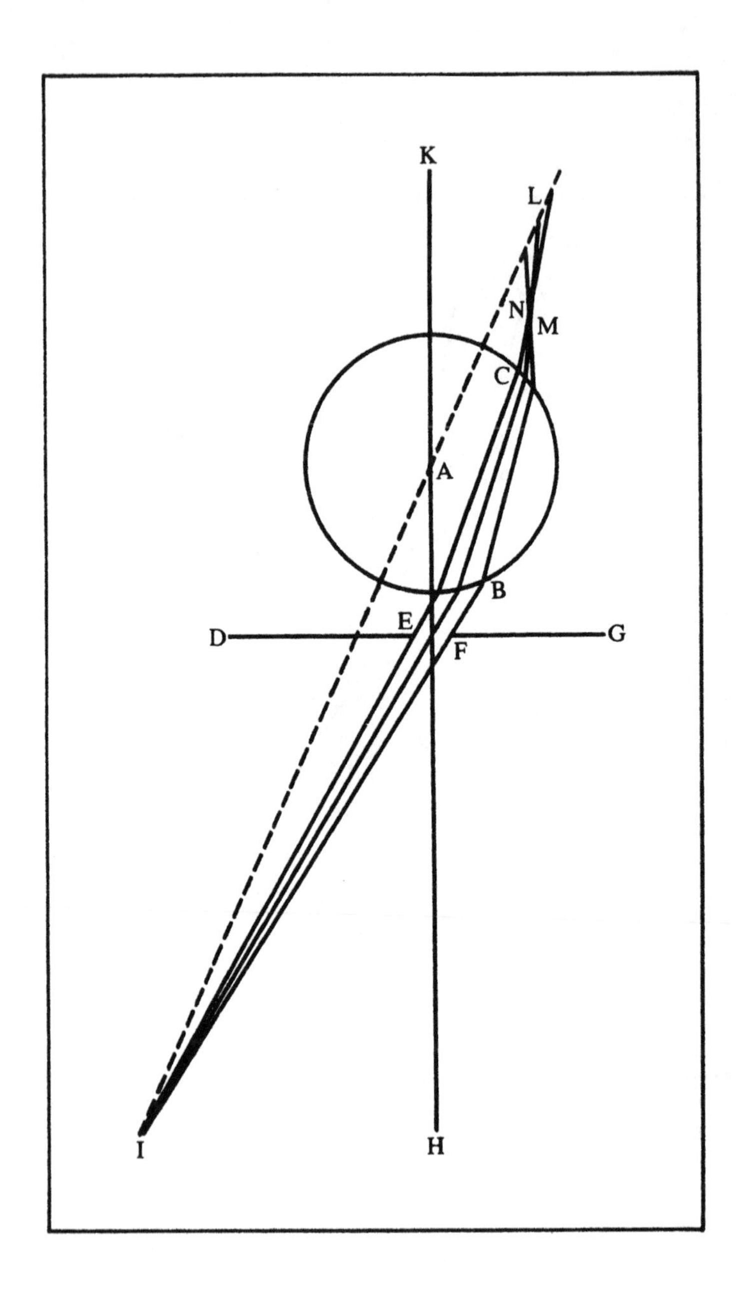

Fig. 35

and suppose the paper to be placed at *KL* (*K* and *L* being the cusps of the caustics formed by radiation issuing, respectively, from points *H* and *I*). Rays from *I* pass through the aperture and, after two refractions, come to an intersection in the region *NM*. Without the screen and aperture, the strongest image of point *I* would be at *L* rather than at *MN*; but since the screen prunes away the rays directed toward the center of the sphere, the peripheral rays that remain produce their sharpest image somewhat closer to the sphere, and at *L* there will be a certain blurring of the image. Rays from point *H*, on the other hand, pass only through the central region of the sphere and produce a relatively sharp image at *K*. It is clear from this demonstration that the paper *KL* bears an inverted image of the object at *HI* and that the image is most distinct near *K*.[87]

Kepler regards this analysis of radiation through transparent spheres as applicable, with minor modifications, to the dioptrical mechanism of the eye. It is true that the crystalline humor is lenticular rather than spherical, but that in itself is of little concern, since Kepler does not treat the crystalline in isolation from the humors that surround it. Rather, he argues that radiation is refracted by the aqueous and crystalline humors in combination—that radiation is refracted first when it encounters the outer surface of the cornea (the covering of the aqueous humor) and refracted again when it emerges through the posterior surface of the crystalline into the vitreous humor. Refraction at the interface between the aqueous and crystalline humors is insignificant, Kepler argues, because the radiation (at least if it comes from points in the visual field near the central axis of the eye) is usually close to the perpendicular as it crosses this surface:

> All those rays entering the anterior surface of the crystalline humor fall on it almost perpendicularly (if they originated from a point at a determined and proportionate distance, unique for each eye and clearly not the same for all) because of the similar convexity of the cornea and the crystalline humor. Therefore, at the anterior surface of the crystalline humor there is almost no refraction of rays coming from a point directly opposite the eye and at an appropriate distance, even though the crystalline humor is denser than the aqueous.[88]

Are the aqueous and crystalline humors spherical, then, in combination? Not quite, but close enough for Kepler's analysis of transparent spheres to be largely applicable to them. Kepler maintains that the anterior surface (the outer surface of the eye) is, in fact, spherical or spheroidal[89] and that the posterior surface (which is also the posterior surface of the crystalline humor) is hyperbolic.[90] Together these two surfaces refract the radiation in such a way as to avoid at least the most severe aberrational effects.

We are now prepared to return to the cones of radiation, which we earlier traced from points in the visual field to bases on the anterior surface of the crystalline lens. Kepler has already argued that each cone is first refracted at the cornea and converges slightly as it proceeds to the surface of the crystalline lens. Here the radiation is formed into a second, converging cone, having the same base as the first cone and an apex behind the eye.[91] This second cone proceeds until it encounters the hyperbolic posterior surface of the crystalline lens, whereupon each ray is refracted away from the perpendicular to this surface at its point of incidence, and a shorter and blunter cone is formed with its apex on the retina. "Therefore all the rays coming from one visible point," Kepler concludes, "finally converge at another point."[92]

It is perhaps unfortunate that Kepler presented no drawing to illustrate his overall scheme, but a figure from Descartes's *La dioptrique* (published thirty-three years later) will adequately reveal what Kepler had in mind.[93] Cones of radiation issue from points *V*, *X*, and *Y* (fig. 36) in the visual field to form a base on the cornea or front of the crystalline humor. Each base then gives rise to another cone converging toward an apex behind the eye; these cones are refracted as they pass from the crystalline to the vitreous humor and converge to apexes at *R*, *S*, and *T*. The cone pairs intersect in the pupil, and radiation originating from *V*, the point on the extreme left, will reach an apex at *R* on the right-hand side of the retina; radiation from *Y* on the right reaches an apex at *T* on the left, and radiation from the central point, *X*, is focused on the extremity of the optic nerve at *S*. We thus have a reversed and inverted picture of the visual field on the retina:

> Therefore vision occurs through a picture of the visible thing [being formed] on the white, concave surface of the retina. And that which is to the right on the outside is portrayed on the left side of the retina; that which is to the left is portrayed on the right; that which is above is portrayed below; and that which is below is portrayed above. . . . Therefore, if it were possible for that picture on the retina to remain after being taken outside into the light, by removing the anterior portions [of the eye], . . . and if there were a man whose vision was sufficiently sharp, he would perceive the very shape of the hemisphere [i.e., the visual field] on the extremely narrow [surface] of the retina.[94]

One thing more must be said about the geometry of image-formation on the retina. The final point of focus of rays coming from points near the edge of the visual field is not as deep within the eye as the point of focus of rays coming from the center of the visual field. Moreover, as a result of refraction in the vitreous humor, the cones in the eye are bent toward one another, so that radiation from a hemisphere of the visual field occupies less than a hemisphere in the eye. These facts suggest that the picture in the eye should be distorted. However, nature has compensated for these potential distortions by adjusting the shape and position of the retina:

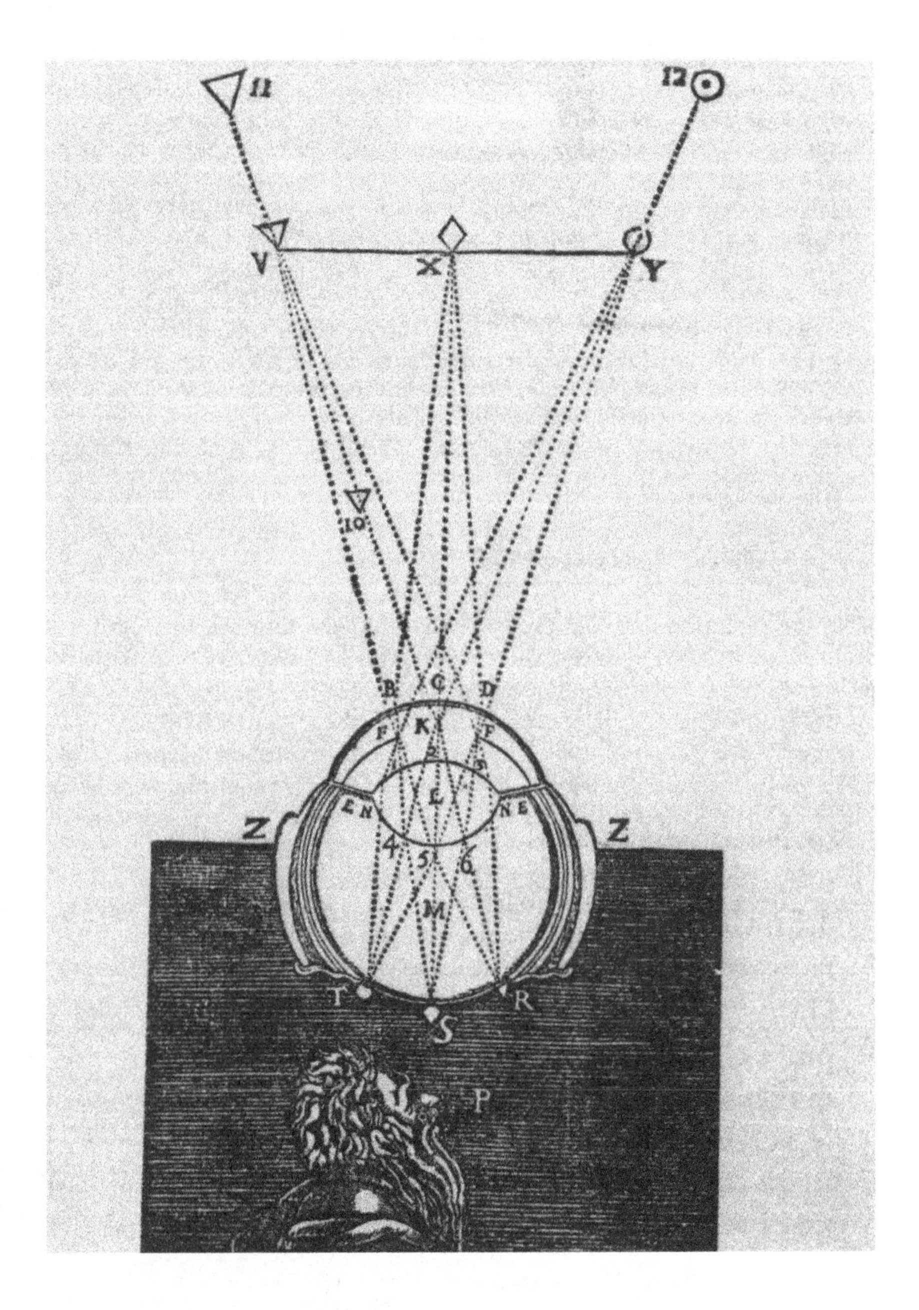

Fig. 36. Descartes's illustration of the theory of the retinal image.

> But in order that the proportions (*ratio*) of the hemisphere [of the visual field] should not be disturbed — [as they would be] if points outside in the air mutually opposite each other with respect to the center of the eye were deflected from opposition by a threefold refraction produced by the surfaces of the cornea and the crystalline, and if they were to descend at an angle into the depths of the eye and so converge on a portion of the retina smaller than a hemisphere — nature has devised an excellent plan, placing the center of the retina not at the intersection of the axes of the cones traversing the vitreous humor, but far inside [this intersection] and advancing the edge of the retina at the sides, so that the longer cones, which are more widely separated, intercept perpendicularly placed (and therefore smaller) parts of the retina, whereas the shorter cones, which are less widely separated at the sides of the retina, mark off at an acute angle large sections of the retina obliquely situated with respect to them. And the cones from opposite radiating points, though not opposite after refraction, are nevertheless incident on opposite points of the retina. And thus compensation is made.[95]

Thus Kepler concludes his defense of the theory of the retinal image. He has painstakingly demonstrated that all the radiation from a point in the visual field entering the eye must be returned to a point of focus on the retina. If all the radiation entering the eye must be taken into account (and who could gainsay that proposition after reflecting on Kepler's argument?),[96] and if the requirement of a one-to-one correspondence between the point sources of rays in the visual field and points in the eye stimulated by those rays is accepted, then Kepler's theory appears to be established beyond serious dispute. An inverted picture is painted on the retina, as on the back of the *camera obscura*, reproducing all the visual features of the scene before the eye. The fact that Kepler's geometrical scheme perfectly complemented Platter's teaching about the sensitivity of the retina surely helped to confirm this conclusion.[97] It is perhaps significant that Kepler employed the term *pictura* in discussing the inverted retinal image,[98] for this is the first genuine instance in the history of visual theory of a real optical image within the eye — a picture, having an existence independent of the observer, formed by the focusing of all available rays on a surface.[99]

But there is a severe difficulty posed by Kepler's theory, which, if not resolved, must place the entire theory in doubt: the picture on the retina is upside down and reversed from right to left. The attempt to deal with this difficulty forced Kepler to consider the nonmathematical (he calls them "physical") aspects of vision.

Kepler and the "Physics" of Vision

The inversion and reversal of the retinal image caused Kepler no little grief. In a passage quoted above, Kepler acknowledges that he "dutifully tortured" himself "in order to show that the cones intersecting when they pass through the aperture of the uvea intersect again behind the crystalline humor in the middle of the vitreous humor, so that another inversion is produced."[100] We have seen the lengths to which Leonardo went to escape the same awkward result; and Maurolico argued that the crystalline humor could not be spherical, lest an inverted image be sent to the visual power.[101] Kepler's friend, Johann Brengger, undoubtedly summarized the apprehensions of many when he wrote to Kepler after reading the *Paralipomena*:

> The means of vision you explain skillfully and elegantly, in which endeavor you surpass in diligence all those who have written on the matter before you. From what I had seen earlier on the use of the dark chamber (*camera tenebricosa*) by J. Bapt. Porta..., I had always convinced myself that vision was accomplished by the reception of the species of visible things on the retina. I was put in some doubt, however, since everything would be received there inverted, whereas vision occurs uprightly.[102]

Clearly Brengger and the rest were operating under the implicit assumption that the image or species in the eye is *seen* rather than interpreted.[103]

Kepler lacked the means to cope fully and successfully with the problem, and he could do little more than argue (as Straker has expressed it) that "geometrical laws leave no choice in the matter"[104] and that it is necessary anyway to have direct opposition (in a spherical sense) between the visible point and the recipient of its radiation, which is available only in the theory of the retinal image with its intersection of cones in the pupil.[105] But if Kepler could not satisfactorily solve the problem posed by inversion and reversal of the image, he could escape it by excluding it from optics—that is, by distinguishing between the optical and the nonoptical aspects of vision in such a way as to place this problem in the latter category. Optics, he therefore argues, ceases with the formation of the picture on the retina, and what happens after that is somebody else's business:

> I say that vision occurs when the image of the whole hemisiphere of the world that is before the eye ... is fixed on the reddish white concave surface of the retina. How the image or picture is composed by the visual spirits that reside in the retina and the [optic] nerve, and whether it is made to appear before the soul or the tribunal of the visual faculty by a spirit within the hollows of the brain, or whether the visual faculty, like a magistrate sent by the soul, goes forth from the administrative chamber of the brain into the optic nerve and the retina to meet this image, as though descending to a lower court—[all] this I leave to be disputed by the physicists. For the armament of opticians does not take them beyond this first opaque wall encountered within the eye.[106]

Witelo, Kepler notes, attempted to carry the optics of vision beyond the retina, arguing that "images of light" pass through the optic nerves to the optic chiasma and ultimately to the cerebral cavities, but he is

> not to be listened to. . . . For what can be said, according to the laws of optics, concerning this hidden journey, which, since it occurs through opaque and therefore dark places and is brought about by means of spirits which differ totally from humors and other transparent things, already exempts itself entirely from optical laws.[107]

Witelo also argued that the species from the two eyes must be united in the optic chiasma and therefore that "refraction must occur at the posterior surface of the vitreous humor, and spirit must be transparent."[108] But Kepler reverses the argument:

> Spirit is not an optical body, and the thin hollow nerve is not optically straight; and if it were, it would nevertheless immediately become crooked because of the twisting of the eye, and its opaque parts would become opposed to the tiny opening or entrance into the passageway. Therefore light neither passes through, nor is refracted at, the posterior surface of the vitreous humor, but impinges there.[109]

Therefore, if species are to unite in the optic chiasma, this must "occur according to the principles of physics, for it is surely certain that no optical image penetrates to that point."[110]

But while rejecting any obligation to discuss the post-retinal transmission (and disclaiming the ability to do so successfully), Kepler does concede some of the fundamental conceptions of Witelo and others in the past optical tradition regarding the "physics" of vision. He grants the existence of visual spirit and admits that it is influenced by the qualities of light and color; moreover, he seems to acknowledge that the impression made by light and color on the visual spirit constitutes the act of visual perception:

> This one optical conclusion from the first chapter can now be stated: the spirit is affected by colors and lights (*lumina*), and this effect is, so to speak, a certain coloring and illuminating. For the species of strong colors remain in sight after they have been looked at, and these are mixed with the colors impressed [on sight] from a new observation, and confusion of the colors occurs. This species, existing independently of the presence of the visible object, is not in the humors or tunics, as was proved above. Therefore vision occurs in the spirits and through this impression of species on the spirit. However, this impression is not optical but physical and mysterious.[111]

The phenomenon of the afterimage does not submit to interpretation solely in terms of pictures on the retina, and this drives Kepler to concede that visual spirit exists and participates in the visual process. Confirmation of the existence of visual spirit came from a story told Kepler by a certain Matthaeus Wagger:

> A certain man, who had lost one eye and happened to cover the other with the palm of his hand, observed that if a bright object were placed in any

> way beneath his nose, he could perceive and even to some extent differentiate the brightness. And so I am in doubt whether there is some opening in the hollow chamber of the eye, extended by that line upward into the head and to the extremities of the optic nerve (the seat of the spirits), or whether the same spirit is also diffused throughout the organ of smell. [However,] this example eminently confirms that the spirits of light and colors are plentiful.[112]

Finally, the existence of visual spirit and its participation in sight are attested by the acuity of central vision, for this is best explained by the concentration of visual spirit where the optic nerve meets the retina. "From that point," Kepler argues, "it is spread out over the sphere of the retina; and as it departs from its source it becomes weaker."[113]

In the final analysis, Kepler's position on the character of visual perception is not very different from Alhazen's. Both recognized the existence of optical and nonoptical (or quasi-optical) aspects of the visual process.[114] Optics strictly defined ends when radiation reaches the seat of visual power within the eye (the crystalline humor for Alhazen, the retina for Kepler), whereupon there ensues a transmission involving visual spirit (which in some measure violates the laws of optics while still belonging to the science of vision) and an act of visual perception.[115] It is true that Kepler adopted a much firmer position against admitting this second stage of the process into the optical realm, and he acknowledged more fully than Alhazen that he did not understand what goes on in this second stage, but both were wrestling with the same difficulty and employing basically the same distinction in the attempt to resolve it.

Conclusion

How then are we to view Kepler's achievement? It has often been argued that Kepler's theory of the retinal image was the natural outcome of comparing the eye to a *camera obscura* and applying to the eye the knowledge of image-formation acquired in solving the problem of the *camera.*[116] This interpretation, if properly qualified, conveys important truth. Kepler's initial interest in visual theory surely grew out of his recognition that the aperture of the eye must introduce the same errors into vision as the aperture of a *camera obscura* introduces into its images. It is clear, moreover, that Kepler recognized the analogy between the eye and the *camera*: he was familiar with Della Porta's attempt to construe the eye as a *camera obscura,*[117] and in the *Paralipomena* he asserted quite plainly that "very nearly the same thing occurs [in the eye] as . . . in a closed chamber."[118] We do not know exactly how much the analogy between the eye and the *camera* contributed to Kepler's understanding of vision, but it might have suggested that a picture is formed on the retina (although Platter's book and purely geometrical considerations could have brought Kepler to the same idea), and it might have taught Kepler that visual images are inverted and reversed in the eye (though it is hard to see how one could have been certain of this until the dioptrical mechanism of the eye was fully understood).

But although the analogy between the eye and the *camera* may have contributed to Kepler's visual theory, I do not believe that it provided the key to the riddle of vision. The crucial questions facing Kepler in his search for an adequate and viable visual theory—the course of radiation through the manifold transparent humors of the eye, and how one ought to deal with the superfluity of rays in the eye in order to establish a one-to-one correspondence between points in the visual field and points in the chamber of the eye—had no analogue at all in the theory of the *camera*. Nor could the principal feature of Kepler's analysis of the *camera*—an infinity of overlapping cones all bearing the shape of the aperture, each with its apex at a point in the visual field and its base at the back of the *camera*—find easy and direct application to vision and the eye. In short, the critical features of the respective theories were largely independent of one another, and the fact that Kepler had already solved the problem of the *camera obscura* did not remove any of the major obstacles confronting his attempt to devise a satisfactory theory of vision.[119]

Moreover, Kepler himself made no attempt to construe his theory of vision as an outgrowth or an instance of the theory of the *camera*. On only one occasion did he explicitly compare the eye to a *camera obscura,* and then merely to explain the overlapping of the bases of cones on the anterior surface of the crystalline lens;[120] and given the purpose of the analogy on this occasion, he of course compared the screen at the back of the *camera* not to the retina, but to the front of the crystalline lens. We must therefore recognize that the functioning of the eye raised its own unique problems, and that in Kepler's view the theory of the retinal image had its own justification independent of analogies with the *camera obscura*.

A second interpretation of Kepler's achievement, surely the most frequent and serious alternative to the foregoing, contends that the theory of the retinal image was simply the inevitable outcome of skillful ray-geometry industriously pursued.[121] If it is understood that rays issue from a point before the eye, and if the principles of refraction are known and properly applied, one will inevitably discover that the rays issuing from that point in the visual field converge to a focus on the retina; and by tracing many such points, one will perceive that an inverted picture of the visual field is traced on the surface of the retina. Again there is undoubtedly an element (perhaps even a large element) of truth in the interpretation. Surely ray-geometry carried out with Kepler's characteristic rigor and dedication was a determining factor in the theoretical outcome of his labors. But there is something wrong with an analysis that considers Kepler in a historical vacuum—that makes him an isolated mathematical genius without paying due attention to his membership in a historical tradition. Kepler labored within a conceptual framework given him by the medieval perspectivists, and this furnished him with a variety of theoretical commitments (both general and specific), factual information, unresolved problems, methodological rules, criteria of success, and the like; and if these are not taken into account, there can be no full understanding of his achievement. We must, in short, see Kepler's theory of the retinal image against the background of medieval theories of vision.

The crucial requirement of medieval theories of vision from Alhazen onward was that they establish a one-to-one correspondence between points in the visual field and points in the eye. The perspectivists achieved this goal by ignoring refracted rays and assigning efficacy in the visual process only (or chiefly) to those rays making up the visual cone, with its base on the object of vision and its apex at the center of curvature of the cornea. When Kepler destroyed this theory by pointing to the illegitimacy of ignoring rays that were only slightly refracted and therefore only slightly weakened, he posed for himself the problem of finding an alternative method of achieving the required one-to-one correspondence. Operating entirely within the framework given him by the medieval perspectivists, with the sole exception of his unwillingness to ignore refracted rays,[122] Kepler had few alternatives. Indeed, he had only one with any practical viability: if confusion of rays within the eye could not be eliminated by ignoring most of the rays, it could be eliminated only by drawing all those rays issuing from a single point in the visual field back to a single point of focus within the eye. To one who possessed even an elementary understanding of the anatomy of the eye and the focusing properties of lenses and transparent spheres (not to mention an awareness of Platter's claim about the location of the visual power) it would be obvious that the point of focus must be on the retina.

To be explicit, I am arguing that Kepler was the culminating figure in the perspectivist tradition, and I must strenuously object to Crombie's and Straker's attempt to view him as a revolutionary figure who transformed visual theory by "mechanizing" it.[123] That his theory of vision had revolutionary implications, which would be unfolded in the course of the seventeenth century, must not be allowed to obscure the fact that Kepler himself remained firmly within the medieval framework. The theory of the retinal image constituted an alteration in the superstructure of visual theory; at bottom, it remained solidly upon a medieval foundation. Kepler attacked the problem of vision with greater skill than had theretofore been applied to it, but he did so without departing from the basic aims and criteria of visual theory established by Alhazen in the eleventh century.

Thus neither extreme of the continuity-discontinuity spectrum will suffice to describe Kepler's achievement: his theory of vision was not anticipated by medieval scholars; nor did he formulate his theory out of reaction to, or as a repudiation of, the medieval achievement. Rather, Kepler presented a new solution (but not a new kind of solution) to a medieval problem, defined some six hundred years earlier by Alhazen. By taking the medieval tradition seriously, by accepting its most basic assumptions but insisting upon more rigor and consistency than the medieval perspectivists themselves had been able to achieve, he was able to perfect it.

APPENDIX
The Translation of Optical Works from Greek and Arabic into Latin

In this appendix I shall summarize what I take to have been established or to be establishable regarding the translation of optical works (or more general works containing a significant portion of optical material). For convenience, I divide the treatises into three categories: (1) works of a purely optical nature, usually mathematical or physical in their approach; (2) works on ophthalmology, including comprehensive medical texts containing a significant section on the eye and its diseases; and (3) more general works on natural philosophy, especially meteorology and psychology. Within each category I alphabetize the treatises according to author's name. I do not claim to be exhaustive, especially for the second and third categories.[1]

Purely Optical Works

Abhomadi Malfegeyr (Abu 'Abd Allah Muhammad ibn Mu'adh),
De crepusculis
Translated by Gerard of Cremona (d. 1187). Authorship of this work was long assigned to Alhazen.[2]

Alhazen, *De aspectibus* or *Perspectiva*
As this was easily the most significant of all the translated works, it has been a source of some frustration to historians that they have been unable to identify its translator. Actually, three candidates have been put forward—Gerard of Cremona (d. 1187), Witelo (fl. 1250-70), and Friedrich Risner (d. 1580 or 1581).[3] The last two can be immediately dismissed, for the Latin version of *De aspectibus* began to circulate in the first half of the thirteenth century, the earliest known citation being in the *Liber de triangulis* of Jordanus de Nemore, composed between 1220 and the early 1230s; it must therefore have been translated late in the twelfth century or early in the thirteenth, too early for Witelo or Risner to have had anything to do with it.[4] Early attribution of the translation to Gerard seems to have been based solely on the fact that *De aspectibus* was printed in 1572 along with a work that Gerard did translate, the *De crepusculis* of Abhomadi Malfegeyr. That Alhazen was believed, at the time, to be the author of *De crepusculis* of course strengthened the case; the high quality of the translation may also have been taken into account. The absence of *De aspectibus* from the list of Gerard's translations drawn up by associates shortly after his death did not seem crucial, since we know this list to be incomplete.[5] However, this argument is no longer convincing: we

know that Alhazen did not write *De crepusculis*, and in any case combination of the two works in a Renaissance edition is of no significance; moreover, I have examined all known extant manuscripts of Alhazen's *De aspectibus*, and none of them assigns the translation to Gerard.[6] It may be significant that Abhomadi's *De crepusculis* and Alhazen's *De aspectibus* appear together in seven manuscripts (the earliest dating from 1269 and four more from the fourteenth century). However, this hardly establishes Gerard as translator of *De aspectibus*, since there are sixteen additional copies of *De crepusculis* and twelve additional copies of *De aspectibus* (including several of each from the thirteenth century), which circulated independently of one another. Moreover, it is clear that the two treatises could have been brought together either by the translator or by a later copyist. Thus we have virtually no evidence on which to base a judgment. The only further clue is the spelling of Alhazen's name in the rubric of manuscripts of *De aspectibus*; there the name (which we transliterate al-Hasan) is universally spelled with a "c," that is, "Hacen" or "Alhacen" or "Alhaycen." The use of the letter "c" with the value of our voiceless "s" (as in "so") strongly suggests a Spanish origin for the translation. What then can we conclude from these scraps of data? Clearly there is not enough evidence to permit more than a guess at the translator's identity. However, judging by the date of translation, the probable provenance, the exceedingly high quality of the translation, and the absence of any other candidates, I would hazard the *speculation* that the translation emanates from Gerard's school—possibly from Gerard himself.

Alhazen, *De speculis comburentibus*
The translation of this work has customarily been attributed to Gerard of Cremona, primarily on the grounds that it circulated with other works known to be his.[7]

Euclid, *Catoptrica* (Theon of Alexandria's recension?)
Murdoch and Ito have argued on internal grounds that the translator, though anonymous, is identical with the translator of Euclid's *Data* and *Optica* and Proclus's *Elementatio physica*.[8] Translation was directly from Greek to Latin, and the translated work circulated under the title *De speculis*. As far as I am aware, this version has never been printed, the version commonly appearing in editions of Euclid's works being the Renaissance translation by Bartolomeo Zamberti.

Euclid, *Optica*
There are at least three distinct translations of the *Optica*.[9] The most widely circulated of these, entitled *De visu*, was rendered directly from the Greek and begins with the incipit: "Ponatur ab oculo eductas rectas lineas . . ." or some variant of this. I have identified seven distinct recensions of *De visu*,[10] but all seem to be based on a single translation. The date of translation appears to be the twelfth century, and if Murdoch and Ito are correct[11] it was translated by the same anonymous translator who rendered the *Catoptrica*. The other two translations of the *Optica* are both from Arabic to Latin. The first, entitled *De aspectibus*, has the incipit: "Radius egreditur ab oculo super

lineas equales rectas . . ." and is probably by Gerard of Cremona.[12] The second, entitled *De radiis visualibus* or *Liber de fallacia visus*, begins with the incipit: "Visualis radius ab oculo super procedit lineas . . ." There have been no speculations on the identity of its translator.

There is, in addition, a version entitled *De aspectuum diversitate*, extant in a single fifteenth-century manuscript and presumably of late date. Theisen takes this to be Theon of Alexandria's recension of the *Optica*, even though it lacks the preface by which Theon's recension is usually identified.[13]

Pseudo-Euclid, *De speculis*
An Arabic compilation of theorems drawn primarily from Hero's *Catoptrica* and Euclid's *Optica* and *Catoptrica*. Probably translated by Gerard of Cremona.[14]

Hero of Alexandria, *Catoptrica*
Translation from Greek to Latin by William of Moerbeke, completed 31 December 1269.[15] During the Middle Ages this work circulated as Ptolemy's *De speculis*.

Al-Kindi, *De aspectibus*
Translated by Gerard of Cremona.[16]

Ptolemy, *Optica* or *De aspectibus*
Translated from Arabic to Latin by Admiral Eugene of Sicily in the mid-twelfth century, perhaps between 1156 and 1160.[17]

Tideus, *De speculis* or *De speculis comburentibus* or *De aspectibus*
Tideus was probably a late Greek author; his work was rendered from Arabic to Latin by Gerard of Cremona.[18]

Ophthalmology

Alcoati, *Congregatio sive liber de oculis*
It has been debated whether this work was originally composed in Arabic or Latin, but since there is a Catalan version, rendered from the Arabic by Johannes Jacobi (Joan Jacme, d. 1384), there can be little doubt that it was composed originally in Arabic and translated into both Catalan and Latin.[19]

'Ali ibn al-'Abbas al-Majusi (Haly Abbas), *Liber regalis,* or *Pantegni*
A medical encyclopedia, containing a discussion of the eye, which was twice translated—first by Constantinus Africanus in the eleventh century, later by Stephen of Antioch (completed about 1127).[20]

'Ali ibn 'Isa (Jesu Haly), *Epistola de cognitione infirmitatum oculorum*, or *De oculis*
There were two translations of this work, one by Magister Dominicus Marrochini about 1270 (according to the explicits of the Bern and Prague manuscripts), the other by an unknown translator, possibly through a Hebrew intermediary.[21]

Averroes, *Colliget*
A comprehensive medical encyclopedia, translated into Latin by the Jew Bonacosa in the second half of the thirteenth century.[22]

Avicenna, *Liber canonis*
A medical encyclopedia of enormous influence in the West, translated by Gerard of Cremona. The section on the eye and its diseases sometimes circulated independently of the remainder of the work.[23]

Galen, *De usu partium*, or *De utilitate particularum*
The first translation to include bk. 10 on the eye was a Greco-Latin version by Peter of Abano near the end of the thirteenth century. It was retranslated (again from the Greek) by Niccolò da Reggio in 1317. Other works of Galen treating vision, such as *De placitis Hippocratis et Platonis*, were not translated until the Renaissance.[24]

Hunain ibn Ishaq, *De oculis*
A rendering of Hunain's *Ten Treatises on the Eye* by Constantinus Africanus (d. 1087). Two versions have been discovered, but both appear to be based on a single translation.[25] Latin manuscripts attribute the treatise either to Constantinus or to Galen, and it was not recognized as Hunain's until modern times.

Al-Razi, *Liber ad Almansorem*
A medical encyclopedia, containing a section on the eye, which also had an independent circulation, translated by Gerard of Cremona.[26] During the Renaissance, bk. 9 of this work, within which the optical material falls, was frequently published separately.

Yuhanna ibn Masawaih (Johannes Mesue), *De infirmitatibus oculorum* (Summa 5 of the *Grabadin*).
Translator and date of translation unknown. The optical section sometimes circulated independently.[27]

Yuhanna ibn Sarabiyun (Johannes Serapion), *Practica*, or *Breviarium*
An encyclopedic work, containing a section on diseases of the eye (which also circulated independently), translated by Gerard of Cremona. There was also a later Renaissance translation by Andreas Alpago.[28]

Works on Natural Philosophy

Alexander of Aphrodisias, *Commentary on Aristotle's Meteorologica*
Translated by William of Moerbeke, completed 24 April 1260.[29]

Aristotle, *De anima*
First translated from Greek to Latin about 1150 by Jacob of Venice; revised ca. 1260-70 by William of Moerbeke.[30] *De anima* was also translated from the Arabic, perhaps by Michael Scot, who died in the mid-1230s.[31]

Aristotle, *De sensu*
This work, along with the other *Parva naturalia*, was rendered from Greek to

Latin about the end of the twelfth century and revised by William of Moerbeke in the second half of the thirteenth century.[32]

Aristotle, *Meterologica*
The first three books (at least) were translated from Arabic to Latin in the twelfth century by Gerard of Cremona; the fourth book is also available in an Arabo-Latin translation, probably by Michael Scot. There was an early Greco-Latin translation of the fourth book by Henricus Aristippus (d. ca. 1162). William of Moerbeke translated the remaining three books from Greek to Latin and, according to the standard view, revised Aristippus's translation of bk. 4; however Minio-Paluello has argued that William's version of bk. 4 was an independent translation.[33]

Averroes, *Commentarium magnum in de anima libros.*
Translated perhaps by Michael Scot, in the first half of the thirteenth century.[34]

Averroes, *Epitome of the parva naturalia*
Manuscripts assign the translation both to Michael Scot and to a certain Gerard (of Cremona?), but recent scholarship has tended to favor Scot. *De senso et sensato*, included within this *Epitome*, was sometimes attributed to al-Farabi in the West.[35]

Avicenna, *De anima* (pt. 4, bk. 6 of the *Shifa* or *Sufficientia*).
Translated in the second half of the twelfth century (from Arabic to Castilian to Latin) by Avendauth and Dominicus Gundissalinus.[36]

Nemesius of Emesa, *De natura hominis*
Important because of its early date, this work was rendered first by Alphanus, bishop of Salerno (d. 1085), later by Burgundio of Pisa (dedication dated 1159).[37]

Plato, *Timaeus*
There was an early translation by Cicero (of portions, at least), but only the first third is extant. A later and far more influential translation by Chalcidius, probably made near the beginning of the fourth century, included the first half of the work.[38]

Themistius, *Commentary on Aristotle's De anima*
Translated by William of Moerbeke, ca. 1268.[39]

NOTES

Chapter 1 The Background: Ancient Theories of Vision

1. On the history of these earliest optical artifacts and investigations, see Bruno Schweig, "Mirrors," *Antiquity* 15 (1941): 257-68; H. C. Beck, "Early Magnifying Glasses," *Antiquaries Journal* 8 (1928): 327-30; Max Meyerhof, "Eye Diseases in Ancient Egypt"; Arlington C. Krause, "Ancient Egyptian Ophthalmology," *Bulletin of the Institute of the History of Medicine* 1 (1933): 258-76; and Joseph Needham, *Science and Civilisation in China*, vol. 4, pt. 1 (Cambridge, 1962), pp. 78-101.

2. On early attitudes toward vision as compared with the other senses, see David C. Lindberg and Nicholas H. Steneck, "The Sense of Vision and the Origins of Modern Science," pp. 33-40.

3. The role of scenography in the early development of optics is apparent from remarks of Geminus, quoted below, chap. 1, n. 72. See also Willem van Hoorn, *As Images Unwind*, p. 49.

4. For recent studies of ancient visual theory, broad in scope, see van Hoorn, *As Images Unwind*, pp. 42-107; David Hahm, "Early Hellenistic Theories of Vision and the Perception of Color"; and Bernard Saint-Pierre, "La physique de la vision dans l'antiquité." Useful older studies are Hugo Magnus, *Die Augenheilkunde der Alten*; Arthur Erich Hass, "Antike Lichttheorien"; Julius Hirschberg, "Die Seh-Theorien der griechischen Philosophen in ihren Beziehungen zur Augenheilkunde"; John I. Beare, *Greek Theories of Elementary Cognition from Alcmaeon to Aristotle*; Walter Jablonski, "Die Theorie des Sehens im griechischen Altertume bis auf Aristoteles"; and Vasco Ronchi, *Storia della luce* (translated by J. Taton as *Histoire de la lumière* and by V. Barocas as *The Nature of Light*), chap. 1. For the most part, however, one must rely on specialized studies.

5. See Cyril Bailey, *The Greek Atomists and Epicurus*, pp. 103-4, 165-70, 406-13; Kurt von Fritz, "Democritus' Theory of Vision."

6. Leucippus, frag. A. 29-30 (Hermann Diels, *Die Fragmente der Vorsokratiker*, ed. Walther Kranz, 6th ed., 2: 78-79); translated by G. S. Kirk and J. E. Raven, *The Presocratic Philosophers*, pp. 421-22.

7. Democritus, frag. A. 135 (Diels-Kranz, 2: 114); translated by Kirk and Raven, *Presocratic Philosophers*, p. 421; cf. *Theophrastus and the Greek Physiological Psychology before Aristotle*, trans. George M. Stratton, pp. 109-11. The dangers of using Theophrastus as a source for the opinions of Pre-Socratic philosophers have been emphasized by John B. McDiarmid, "Theophrastus on the Presocratic Causes," *Harvard Studies in Classical Philology* 61 (1953): 85-87, echoing Harold Cherniss's similar position with regard to Aristotle (*Aristotle's Criticism of Presocratic Philosophy*, especially p. 347); however, see the misgivings of W. K. C. Guthrie, "Aristotle as a Historian of Philosophy: Some Preliminaries," *Journal of Hellenic Studies* 77 (1957): 35-41.

8. Von Fritz speculates that Democritus's theory of imprints in the air was meant to explain how the effluences from visible objects retain their shape in transit to the eye so as to give accurate knowledge of the shape of the object ("Democritus' Theory of Vision," pp. 94-95). Rudolph E. Siegel traces the theory to the observation of mirages and their interpretation as impressions in the air ("Theories of Vision and Color Perception of Empedocles and Democritus," p. 146). See also van Hoorn, *As Images Unwind*, pp. 53-55.

9. Epicurus, "Letter to Herodotus," in Diogenes Laertius *Lives of Eminent Philosophers* 10. 48-49, trans. R. D. Hicks, 2: 577-79.

10. *De rerum natura* 4. 54-61, trans. W. H. D. Rouse, pp. 251-53. We probably should resist the temptation (encouraged by several of Lucretius's comparisons) of supposing that the constituent atoms of an individual *simulacrum* or *eidolon* must be physically hooked. As Edward Lee has recently emphasized, *eidola* may be no more than "convoys of atoms," which maintain a fixed configuration as they travel outward from their source; see Lee's excellent study, "The Sense of an Object."

11. Rudolph E. Siegel, *Galen on Sense Perception*, pp. 17-23. It is Siegel's position that the statements attributed to Democritus convey "the ideal that the visual impression did not depend on the transmission of minute particles but on an alteration of the space between eye and object," and that "Epicurus only said that the illuminated object induced a definite three-dimensional [but not corpuscular] pattern in the space between object and eye." Cf. Siegel, "Did the Greek Atomists Consider a Non-Corpuscular Visual Transmission?"

12. Aristotle censures Democritus for reducing all sensation to touch (*De sensu* 4. $442^{a}30$-$442^{b}1$).

13. See this volume, pp. 40, 49-50, 55, 163-64. On Democritus's theory of vision, see also Saint-Pierre, "La physique de la vision," pp. 89-97.

14. Aristotle *De sensu* 2. $438^{a}5$-12, trans. J. I. Beare, in *The Works of Aristotle*, ed. W. D. Ross, vol. 3. Theophrastus (who was heavily dependent on Aristotle) gives a similar account; see Stratton, *Greek Physiological Psychology*, pp. 109-13.

15. Von Fritz, "Democritus' Theory of Vision," p. 93.

16. Ibid. This conception of reflection was still common during the Middle Ages; see David C. Lindberg, *John Pecham and the Science of Optics*, p. 235.

17. Our accounts of Democritus's theory are not specific. The image actually appears to an observer to be located a short distance within the eye.

18. See Beare, *Theories of Elementary Cognition*, pp. 11-13. On Alcmaeon's dates, see Kirk and Raven, *Presocratic Philosophers*, pp. 232-33.

19. Alcmaeon, frag. A. 5 (Diels-Kranz, 1: 212); translated by Stratton, *Greek Physiological Psychology*, p. 89.

20. Ibid.

21. Beare, *Theories of Elementary Cognition*, p. 12; W. J. Verdenius, "Empedocles' Doctrine of Sight," pp. 161-62.

22. See this volume, p. 2.

23. Empedocles, frag. B. 84 (Diels-Kranz, 1: 341-42); translated by W. K. C. Guthrie, *A History of Greek Philosophy*, vol. 2 (Cambridge, 1965), p. 235. For another translation, see Clara E. Millerd, *On the Interpretation of Empedocles*, pp. 82-83.

24. *De sensu* 2. $437^{b}24$-25, trans. W. S. Hett, p. 223.

25. Ibid., $438^{a}4$-5. Cf. Beare, *Theories of Elementary Cognition*, p. 14. Jean Bollack, *Empédocle*, vol. 1 (Paris, 1965), pp. 237-39; Millerd, *Interpretation of Empedocles*, pp. 82-83.

26. *Aristotle's Criticism of Presocratic Philosophy*, p. 318, n. 106.

27. *Interpretation of Empedocles*, p. 84: "There seems no ... sufficient reason for doubting Aristotle's implication that the two modes of vision were not related." Guthrie, *History of Greek Philosophy*, 2: 237, shares Millerd's view.

28. "Empedocles' Doctrine of Sight," p. 162: "When we are seeing anything, we are not only aware of receiving an impression but also suppose this impression to correspond to reality. Our being conscious of a correspondence takes the form of a projection by which we return our impressions to the object. Thus seeing is imagined to be at once something passive (to receive impressions) and active (to be directed towards the object)." See also D. O'Brien, "The Effect of a Simile: Empedocles' Theories of Seeing and Breathing," which came to my attention after this section had been completed.

29. A. A. Long, "Thinking and Sense-Perception in Empedocles," p. 264. For yet another analysis, see Charles Mugler, "Sur quelques fragments d'Empédocle," *Revue de philologie de littérature et d'histoire anciennes*, ser. 3, 25 (1951): 33-65.

30. Stratton, *Greek Physiological Psychology*, pp. 69-71. Theophrastus's misunderstanding has been passed on by many recent historians, including Beare, *Theories of Elementary Cognition*, p. 49; and A. E. Taylor, *A Commentary on Plato's Timaeus*, pp. 277-78.

31. *Timaeus* 45b-d, trans. Francis M. Cornford, *Plato's Cosmology*, pp. 152-53. The necessity of sunlight is also stressed in the *Republic* 6. 507d-508c.

32. Taylor is surely wrong (*Commentary on Timaeus*, p. 278), when he interprets vision of a mountain to be the result of the coalescence of visual rays with light reflected from the mountain, whereafter the homogeneous body thus formed transmits "the sensation due to its contact with the mountain." Rather, the homogeneous body is formed by coalescence of visual rays and daylight, whereafter motions can be transmitted from the mountain to the eye to produce the sensation of sight.

33. *Theaetetus* 156d-e, trans. Francis M. Cornford, *Plato's Theory of Knowledge*, p. 47.

34. 76d, in Plato, *Laches, Protagoras, Meno, Euthydemus*, trans. W. R. M. Lamb (*Plato, with an English Translation*, vol. 4) (London, 1924), p. 285.

35. See Cornford, *Plato's Cosmology*, p. 153, n. 1.

36. 67c-d, trans. Cornford, ibid., p. 276.

37. On Aristotle's theory of vision, see Beare, *Theories of Elementary Cognition*, pp. 56-92; Cherniss, *Aristotle's Criticism of Presocratic Philosophy*, pp. 313-20; Thomas J. Slakey, "Aristotle on Sense Perception"; Charles H. Kahn, "Sensation and Consciousness in Aristotle's Psychology"; van Hoorn, *As Images Unwind*, pp. 74-107; and Saint-Pierre, "La physique de la vision," pp. 159-83. On the development of Aristotle's thought on psychology, see François Nuyens, *L'évolution de la psychologie d'Aristote* (Louvain, 1948); and Irving Block, "The Order of Aristotle's Psychological Writings," *American Journal of Philology* 82 (1961): 50-77. (Nuyens and Block reach opposite conclusions on the chronology of Aristotle's psychological works); see also Kahn's evaluation of the dispute between Nuyens and Block, "Sensation and Consciousness," pp. 64-69.

38. *De anima* 2. 7. 418^{b}14-16, trans. W. S. Hett, p. 105. The principal fallacy of the atomistic theory, according to Aristotle, is its attempt to reduce sensation to touch; see Cherniss, *Aristotle's Criticism of Presocratic Philosophy*, pp. 313-16.

39. *De sensu* 2. 438^{a}26-438^{b}2, trans. Hett, p. 225. A different view appears in certain other Aristotelian works. Aristotle adopts a more moderate position toward the extramission theory in *De generatione animalium*, where he writes: "Animals with prominent eyes do not see well from a distance, but those with sunken eyes placed in a hollowed recess are able to see things at a distance, because the movement does not get scattered into space but follows a straight course. It makes no difference to this which of the two theories of sight we adopt. Thus, if we say, as some people do, that seeing is effected 'by the sight issuing forth,' then on this theory, unless there is something projecting in front of the eyes, the 'sight' of necessity gets scattered and so less of it strikes the object. . . . If we say that seeing is effected 'by a movement derived from the visible object,' then on this theory, the clarity with which the sight sees will of necessity vary directly as the clarity of the movement" (5. 1. 780^{b}36-781^{a}8, trans. A. L. Peck, rev. ed. [London, 1953], p. 505). It is true that Aristotle makes no real concession to the extramission theory in this passage, but his failure to dismiss it as sheer absurdity represents a different position from that exhibited in *De anima* and *De sensu*.

In the *Meteorologica*, on the other hand, Aristotle assumes the theory of visual rays throughout. In discussing why the sea appears to flash when struck by a stick at night, he concludes that "the water seems to flash when struck because our line of vision is reflected from it to some bright object" (2. 9. 370^{a}18-19, trans. H. D. P. Lee, p. 231). Aristotle argues that the rainbow is produced by reflection of our sight from droplets of moisture to the sun (373^{a}35-374^{a}3), and explains the origin of dark colors in the rainbow by noting that "dark colour is a kind of negation of vision, the appearance of darkness being due to the failure of our sight; hence objects seen at a distance appear darker because our sight fails to reach them" (3. 4. 374^{b}11-15, trans. Lee, p. 259). Still more compelling is his account of a man of weak sight who was able to see his own image walking before him, an account to be often repeated in medieval optical treatises: "Air reflects when it is condensed; but even when not condensed it can produce a reflection when the sight is weak. An example of this is what used to happen to a man whose sight was weak and unclear: he always used to see an image going before him as he walked, and facing toward him. And the reason why this used to happen to him was that his vision was reflected back to him; for its enfeebled state made it so weak and faint that even the neighbouring air became a mirror and it was unable to thrust it aside" (3. 4. 373^{b}2-10, trans. Lee, p. 253).

The explanation of these inconsistencies undoubtedly lies in the development of Aristotle's thought, though some have preferred to question the authenticity of the *Meteorologica*; on the authenticity and dating of the *Meteorologica*, see Lee's introduction to his translation, pp. xii-xxv. My own inclination would be to assign the *Meteorologica* to an early period in Aristotle's career, when he was still under Plato's influence. In any case, the universal teaching of Aristotle's psychological works is that sight does not occur through the extramission of an ocular ray.

40. *De sensu* 6. 446^{b}10-13, trans. Beare, in *Works of Aristotle*, ed. Ross, vol. 3.

41. *De anima* 2. 7. 419^{a}12-22, trans. Hett, p. 107.

42. Ibid., 418^{b}5-9, p. 105. Cf. *De sensu* 3. 439^{a}22-25.

43. *De anima* 2. 7. 418^{b}9-13, trans. J. A. Smith, in *Works of Aristotle*, ed. Ross, vol. 3.

44. *De sensu* 6. 447^{a}2-4, trans. Beare. The question arose among Aristotle's commentators whether the state of the transparent medium associated with light is the result of an actual modification of the medium or merely of a certain relation. Alexander of Aphrodisias clearly defended the latter position: "Qualitative change is motion and occurs in time and by a gradual transition. But this is not the way in which a transparent medium receives light and colour. It does not undergo a change but rather the situation is similar to someone becoming a right-hand neighbour without any motion or action on his part. Such is the turn which a transparent medium takes with regard to light and colours. . . . And just as the right-hand neighbour ceases to be on the right when the man on his left leaves his place, so does light disappear when the illuminating source is removed" (quoted by S. Sambursky, *The Physical World of Late Antiquity*, p. 112); cf. Saint-Pierre, "La physique de la vision," pp. 238-56. On theories of light among Peripatetic commentators of late antiquity, see ibid., pp. 207-79; Sambursky, "Philoponus' Interpretation of Aristotle's Theory of Light"; and Walter Böhm, *Johannes Philoponos, Grammatikos von Alexandrien* (Munich, 1967), pp. 174-207.

45. As in *De generatione animalium* 5. 1. 780^{b}33-36.

46. *De anima* 2. 7. 418^{a}32-418^{b}3, trans. Hett, p. 103.

47. Ibid., 419^{a}14-15, p. 107.

48. *De sensu* 3. 439^{b}11-12, trans. Hett, p. 231; cf. 439^{a}27-30.

49. Above, n. 47.

50. *De sensu* 2. 438^{b}8-12, trans. Hett, p. 227.

51. Ibid. 438^{b}19-20.

52. Cherniss, *Aristotle's Criticism of Presocratic Philosophy*, p. 320; cf. Aristotle *De anima* 2. 5. 417^{b}28-418^{a}6.

53. Aristotle, like Theophrastus, oversimplifies Plato's theory of vision.

54. Cornford, *Plato's Cosmology*, p. 153, n. 1.

55. On the Stoic pneuma, see G. Verbeke, *L'évolution de la doctrine du pneuma de stoicisme à S. Augustin*; S. Sambursky, *Physics of the Stoics*, especially pp. 1-11; Owsei Temkin, "On Galen's Pneumatology." On the Stoic theory of vision, see Hahm, "Early Hellenistic Theories of Vision"; Saint-Pierre, "La physique de la vision," pp. 107-14; Sambursky, *Physics of the Stoics*, pp. 23-29; Harold Cherniss, "Galen and Posidonius' Theory of Vision"; Siegel, *Galen on Sense Perception*, pp. 37-40; F. Ogereau, *Essai sur le système philosophique des Stoïciens* (Paris, 1885), pp. 91-94.

56. Quoted by Sambursky, *Physics of the Stoics*, p. 28. Apparently not all Stoics regarded the air itself as percipient; Diogenes Laertius reports that according to Chrysippus and Apollodorus, "the apex of the cone [of stressed air] . . . is at the eye, the base at the object seen. Thus the thing

seen is reported to us by the medium of the air stretching out towards it, as if by a stick" (*Lives of Eminent Philosophers* 7. 157, trans. Hicks, 2: 261).

57. Quoted by Sambursky, *Physics of the Stoics*, p. 124. Cf. the quotation from Diogenes Laertius in the preceding note. The analogy of the walking stick was criticized by Galen *De placitis Hippocratis et Platonis* 2.5 and 2.7, in *Opera Omnia*, ed. C. G. Kühn, 5: 627, 642; and by Tideus, *De speculis*, in Axel Anthon Björnbo and Sebastian Vogl, "Alkindi, Tideus und Pseudo-Euklid. Drei optische Werke," p. 74.

58. I do not mean to gloss over very real differences among Aristotle, Plato, and the Stoics. For example, it is clear that Aristotle did not regard the transparent medium as percipient, as an extension of the observer, in the way that certain Stoics regarded air activated by pneuma and light.

59. *De placitis* 7. 5, trans. Philip De Lacy (forthcoming in the *Corpus Graecorum medicorum*). I am grateful to Professor De Lacy for making available to me his text and translation of *De placitis* and for giving me permission to quote from them. Cf. *Opera omnia*, ed. Kühn, 5: 618. On Galen's theory of vision, see Saint-Pierre, "La Physique de la vision," pp. 126-57; Siegel, *Galen on Sense Perception*, pp. 40-78; Siegel, "Principles and Contradictions of Galen's Doctrine of Vision"; and Cherniss, "Galen and Posidonius' Theory of Vision." For a late-antique defense of Galen's theory of vision, see Tideus, *De speculis*, in Björnbo and Vogl, "Drei optische Werke," pp. 73-94 (esp. pp. 74-75); as Björnbo points out (ibid., pp. 150-51), there is no justification for identifying Tideus with the Greek geometer Diocles.

60. *De placitis* 7. 5, *Opera omnia*, ed. Kühn, 5: 618. The quoted words are from De Lacy's translation.

61. *De placitis* 7. 5, trans De Lacy; cf. *Opera omnia*, ed. Kühn, 5: 619.

62. *De placitis* 7. 4, *Opera omnia*, ed. Kühn, 5: 612-13. See below, chap. 3, n. 18.

63. *De placitis* 7. 4, trans. De Lacy; cf. *Opera Omnia*, ed. Kühn, 5: 617.

64. *De placitis* 7. 5, trans De Lacy; cf. *Opera omnia*, ed. Kühn, 5: 619. See also ibid., p. 638.

65. *De placitis* 7. 5, trans. De Lacy; cf. *Opera omnia*, ed. Kühn, 5: 625.

66. *De placitis* 7. 7, trans. De Lacy; cf. *Opera omnia*, ed. Kühn, 5: 642. Note that Galen requires the action of sunlight as well before the air is transformed into a visual instrument; in this respect, he is very close to Plato.

67. *De placitis* 7. 5, trans. De Lacy; cf. *Opera omnia*, ed. Kühn, 5: 625. On the walking-stick analogy see also ibid., p. 627. Galen has yet another argument by which to demonstrate that the air itself becomes sentient: the air is for us the same kind of instrument for discerning visible things as the nerve is for tangible things, and in the latter case it is evident that sensation is in the individual parts rather than in the ruling part of the soul, because "pain would not be felt in the member that is cut or crushed or burned unless the power of sensation is also in the members" (*De placitis* 7. 7, trans. De Lacy; cf. *Opera omnia*, ed. Kühn, 5: 641-42).

68. I do not mean to imply that Galen was the first to study the anatomy and physiology of the eye; on his predecessors, see Saint-Pierre, "La physique de la vision," pp. 19-20.

69. Galen *On the Usefulness of the Parts of the Body* 10. 1, trans. Margaret T. May, pp. 463-64. On cataracts, see also *De placitis* 7. 6, in *Opera Omnia*, ed. Kühn, 5: 635. On the compatibility of this stress on the crystalline lens with other aspects of Galen's theory of vision, see chap. 3.

70. See Galen *Usefulness of Parts* 10. 1-7, trans. May, pp. 463-82; *On Anatomical Procedures: The Later Books* 10. 2-4, trans. W. L. H. Duckworth, pp. 33-50. The most detailed discussion of these matters is to be found in Siegel's *Galen on Sense Perception*, pp. 40-45, 51-70. See also the remarks on Hunain's anatomy and physiology of the eye in chap. 3 of this volume.

71. *Physics* 2. 2. 194a10-11, trans. R. P. Hardie and R. K. Gaye, in *Works of Aristotle*, ed. Ross, vol. 2.

72. Geminus (first century B.C.) reflects the same narrow conception of optics in his division of it into optics proper, catoptrics, and scenography: "Again, optics and canonics are derived from geometry and arithmetic, respectively. The science of optics makes use of lines as visual rays and makes use also of the angles formed by these lines. The divisions of optics are: (*a*) the study which is

properly called optics and accounts for illusions in the perception of objects at a distance, for example, the apparent convergence of parallel lines or the appearance of square objects at a distance as circular; (*b*) catoptrics, a subject which deals, in its entirety, with every kind of reflection of light and embraces the theory of images; (*c*) scenography (scene-painting), as it is called, which shows how objects at various distances and of various heights may so be represented in drawings that they will not appear out of proportion and distorted in shape" (quoted by Proclus and translated by Morris R. Cohen and I. E. Drabkin, *A Source Book in Greek Science*, p. 4). Sambursky has pointed out (*Late Antiquity*, pp. 111-14) that geometrical conceptions penetrated Peripatetic optics in late antiquity.

73. *Usefulness of Parts* 10. 12, trans. May, pp. 490-91; cf. Siegel, *Galen on Sense Perception*, pp. 10-12, 86-87.

74. See Siegel's discussion, ibid., pp. 46-47, 86-94. For a medieval discussion of the distinction between the mathematical and "natural" approaches to optics, see "Le 'Discours de la lumière' d'Ibn al-Haytham (Alhazen)," trans. Roshdi Rashed, p. 205; or "Abhandlung über das Licht von Ibn al-Haitam," trans. J. Baarmann, p. 197.

75. On the authenticity of the *Optica* as a Euclidean work, see J. L. Heiberg, *Litterargeschichtliche Studien über Euklid*, pp. 129-33; cf. H. Weissenborn, "Zur Optik des Eukleides." As Heiberg has demonstrated (pp. 148-53), there is, in addition to the genuine version, a recension of the *Optica* prepared by Theon of Alexandria. The *Catoptrica* attributed to Euclid is probably a later recension of a genuine work, perhaps also prepared by Theon. See also Euclid, *L'Optique et la Catoptrique*, trans. Paul Ver Eecke, pp. xxviii-xxix.

76. Albert Lejeune, *Euclide et Ptolémée*, p. 172. I am much indebted to Lejeune's perceptive analysis of Euclid's optics.

77. More literally, "in the eye"; the Greek text reads ἐν τῷ ὄμματι. See Euclid, *L'Optique et la Catoptrique*, trans. Ver Eecke, p. 57, n. 1.

78. Cohen and Drabkin, *Source Book*, pp. 257-58. The full text of the *Optica* (Greek and Latin) appears in *Euclidis opera omnia*, ed. J. L. Heiberg and H. Menge, vol. 7. Heiberg's Greek text has been translated into French by Ver Eecke, and into English by H. E. Burton, "The Optics of Euclid." For a critical edition and English translation of various medieval Latin versions of the *Optica*, see "The Medieval Tradition of Euclid's *Optics*," ed. and trans. Wilfred R. Theisen.

79. Whereas Euclid assumed the rectilinear propagation of light, others attempted to demonstrate it metaphysically (Hero of Alexandria), teleologically (Damianos), or experimentally (Ptolemy, Alhazen, and Witelo). Galen mentions that geometers take rectilinear propagation to be self-evident, though some have attempted to prove it. See Lejeune, *Euclid et Ptolémée*, pp. 37-41; Eilhard Wiedemann, "Zu Ibn al-Haitams Optik," pp. 24-25, 40-41; Witelo, *Perspectiva*, bk. 2, theor. 1, in *Opticae thesaurus Alhazeni Arabis libri septem . . . . Item Vitellonis . . . libri X*, ed. Friedrich Risner, pp. 61-63; Galen *De placitis* 7. 5, in *Opera omnia*, ed. Kühn, 5: 626-27.

80. See, for example, Ver Eecke's introduction to Euclid, *L'Optique et la Catoptrique*, p. xxiv; cf. ibid., p. 50, n. 1.

81. Hirschberg and Mach have claimed that Euclid took no position regarding the direction in which the rays proceed; see Hirschberg, "Optik der alten Griechen," p. 327; Ernst Mach, *The Principles of Physical Optics: An Historical and Philosophical Treatment*, trans. J. S. Anderson and A. F. A. Young (London, 1926), pp. 8-9.

82. See Lejeune, *Euclide et Ptolémée*, pp. 28-29, 62-66, 72; Damianos, *Schrift uber Optik*, ed. and trans. Richard Schöne, p. 5.

83. Among Euclid's Greek followers were Hipparchus, Ptolemy, Damianos, and Theon of Alexandria. Lejeune's *Euclide et Ptolémée* is by far the best source on this Euclidean tradition; see also the same author's *Recherches sur la catoptrique grecque*.

84. Cohen and Drabkin, *Source Book*, pp. 261-62. The entire *Catoptrica*, in William of Moerbeke's Latin translation (accompanied by a modern German translation of Moerbeke's Latin text), is available in Hero of Alexandria, *Opera quae supersunt omnia*, ed. L. Nix and W. Schmidt, vol. 2, fasc. 1, pp. 301-65.

85. See Carl B. Boyer, "Aristotelian References to the Law of Reflection."

86. Cohen and Drabkin, *Source Book*, p. 263.

87. Ibid., pp. 263-64. Hero takes a similar position regarding the corpuscularity of external luminous rays in his *Pneumatica*, trans. Bennet Woodcroft, p. 9.

88. Lejeune, *Euclide et Ptolémée*, pp. 22-23, 32, 57-58, 83-84.

89. See the recent critical edition by Albert Lejeune, *L'Optique de Claude Ptolémée*.

90. Besides Lejeune's edition of Ptolemy's *Optica*, cited in the previous note, see his *Euclide et Ptolémée*. Other studies of Ptolemy's optics include G. G. A. Caussin de Perceval, "Mémoire sur L'Optique de Ptolémée"; J. B. J. Delambre, "Sur l'Optique de Ptolémée comparée à celle qui porte le nom d'Euclide at à celles d'Alhazen et de Vitellion"; Baldassarre boncompagni, "Intorno ad una traduzione latina dell'Ottica di Tolomeo," *Bullettino di bibliografia e di storia delle scienze matematiche e fisiche* 4 (1871): 470-92; 6 (1873): 159-70; Lejeune, *Recherches sur la catoptrique*.

91. *Euclide et Ptolémée*, pp. 62-84.

92. Quoted in ibid., p. 65. Seth's terminology suggests that Ptolemy had Stoic leanings. Lejeune, however, would prefer to explain away Seth's reference to "pneuma" by assuming that he had learned of Ptolemy's ideas through encyclopedic works where those ideas were expressed in Stoic terms.

93. *De anima* 2. 7. $418^{b}8$-9.

94. *Euclide et Ptolémée*, pp. 67-71.

95. Ibid., pp. 68-69. Lejeune also insists (p. 69) on the energetic character of external light: "A travers ces considérations perce la conviction de la quasi-identité des deux radiations. Sans doute l'energie des rayons visuels se traduit concrètement par une capacité plus ou moins parfaite de percevoir l'objet, l'énergie des rayons lumineux par une aptitude plus ou moins grande à le rendre visible. Mais elle est conçue de part et d'autre comme une propriété mécanique identique."

96. *Optica* 2. 50, ed. Lejeune, p. 37. The term "ray," in the quotation, does not denote a single line or pencil of radiation, but the entire cone of light; such usage was common throughout the Middle Ages. On the continuity of the radiation, see also Lejeune, *Euclide et Ptolémée*, pp. 81-83; cf. al-Kindi, this volume, chap. 2.

97. Lejeune, *Euclide et Ptolémée*, p. 68.

98. Ibid., pp. 23-28. Cf. Saint-Pierre, "La physique de la vision," pp. 48-61.

99. So far as the extant portions of the *Optica* reveal, Ptolemy did not deal explicitly with the nature of external light except to assign it to the same genus as visual radiation. However, Lejeune argues that Ptolemy conceived the two radiations to be basically identical; see above, n. 95.

100. *Optica* 2. 16, ed. Lejeune, p. 18; cf. Lejeune, *Euclide et Ptolémée*, pp. 29-31.

101. *Optica* 2. 20, ed. Lejeune, p. 20; cf. Lejeune, *Euclide et Ptolémée*, pp. 35-37, 70-71.

102. On the importance of this conception in the Middle Ages, see this volume, pp. 26-28, 85.

103. See Albert Lejeune, "La dioptre d'Archimède."

104. See Lejeune, *Euclide et Ptolémée*, pp. 51-57. On the later history of the idea, see this volume, pp. 74, 80-81, 109.

Chapter 2 Al-Kindi's Critique of Euclid's Theory of Vision

1. This chapter is a revised version of a previously published article, "Alkindi's Critique of Euclid's Theory of Vision," which appeared in *Isis*; however, the article and the chapter are not identical, and each treats certain matters omitted by the other. A brief excerpt from this chapter also appears in my article "The Intromission-Extramission Controversy in Islamic Visual Theory: Alkindi versus Avicenna," forthcoming in *Perception: Philosophical and Scientific Themes and Variations*, edited by Peter Machamer and Robert G. Turnbull (Columbus, Ohio: Ohio State University Press).

The date traditionally given for al-Kindi's death is 873, but recent sentiment seems to favor 866. See George N. Atiyeh, *Al-Kindi: The Philosopher of the Arabs* (Rawalpindi, 1966), p. 8; Majid Fakhry, *A History of Islamic Philosophy*, p. 83. On Al-Kindi's life and writings, see G. Flügel, "Al-Kindi, gennant 'der Philosoph der Araber.' Ein Vorbild seiner Zeit und seines Volkes," *Abhandlungen der deutschen morgenländischen Gesellschaft*, 1, pt. 2 (1857): 1-54; Lucien Leclerc, *Histoire de la médecine arabe* 1: 160-68; Albino Nagy, "Sulle opere di Ja'qub ben Ishaq al-Kindi," *Rendiconti della Reale Accademia dei Lincei, classe di scienze morali, storiche e filologiche*, ser. 5, 4 (1895): 157-70; Francis J. Carmody, *Arabic Astronomical and Astrological Sciences in Latin Translation: A Critical Bibliography* (Berkeley, 1956), pp. 78-85; Richard Walzer, "New Studies on al-Kindi," *Oriens* 10 (1957): 203-32 (reprinted in Walzer, *Greek into Arabic: Essays on Islamic Philosophy* [Oxford, 1962]); Nicholas Rescher, *Al-Kindi: An Annotated Bibliography* (Pittsburgh, 1964); *The Fihrist of al-Nadim: A Tenth-Century Survey of Muslim Culture*, trans. Bayard Dodge (New York, 1970), 2: 615-26.

2. "New Studies," p. 203; cf. Walzer's "Arabic Transmission of Greek Thought to Medieval Europe," *Bulletin of the John Rylands Library* 29 (1945-46): 176-78. For a useful summary of Mu'tazilitism, see F. E. Peters, *Aristotle and the Arabs: The Aristotelian Tradition in Islam* (New York, 1968), pp. 136-46; Fakhry, *Islamic Philosophy*, pp. 58-80. It has sometimes been claimed that al-Kindi was himself a translator, but it seems now to be generally conceded that he was only a patron of the art; see Fakhry, p. 82.

3. Quoted by Walzer, "Arabic Transmission," pp. 172-73.

4. Ibid., p. 175. This aspect of al-Kindi's work has also been discussed by Franz Rosenthal, "Al-Kindi and Ptolemy," in *Studi Orientalistici in onore di Giorgio Levi Della Vida* (Rome, 1956), 2: 455-56.

5. *The Fihrist*, trans. Dodge, 2: 615-26, attributes 265 works to al-Kindi. On the published works, see Rescher's bibliography.

6. No copy of the Arabic text has so far been discovered. The Latin text has been published in Axel Anthon Björnbo and Sebastian Vogl, "Alkindi, Tideus und Pseudo-Euklid. Drei optische Werke," pp. 3-41; see also the review of Björnbo and Vogl's edition by Alexander Birkenmajer, *Bibliotheca Mathematica*, ser. 3, 13 (1912-13): 273-80. Björnbo has demonstrated (pp. 148-50) that the translator of *De aspectibus* was Gerard of Cremona. Latin MSS are relatively numerous and have been described by Björnbo and Vogl; Carmody, *Astronomical and Astrological Sciences*, p. 79, adds an additional MS to the list.

The *Fihrist* attributes to al-Kindi a number of other optical works, several of which are extant. There are three Arabic MSS of an extract of al-Kindi's recension of Euclid's *Optica*: Paris, Bibl. Nationale, MS Arab. 2467(2), fols. 56a-58a; Berlin, Staatsbibl., MS Mf. 258, fols. 163b-165a; and Berlin, Staatsbibl., MS Mq. 559, fols. 117a-119a. On this treatise, see Eilhard Wiedemann, "Aus al Kindis Optik," pp. 247-48 (I owe the precise manuscript references to Emilie Smith). Arabic MSS of al-Kindi's treatise on the burning mirror (probably no. 112 in Flügel's list) have recently been discovered (and lost again?); a photocopy of one of them has been published as *Propagations of Ray: The Oldest Arabic Manuscript about Optics [Burning-Mirror], from Ya'kub ibn Ishaq al-Kindi*, ed. Mohamed Yahia Haschmi. Rescher, *Al-Kindi*, p. 47, lists two other works of marginal optical interest, both available in Arabic editions: "On the Reason for the Azure Color Seen in the Air towards the Sky" and "On the Body That Bears Color Naturally from among the Four Elements"; the former of these has been translated into English by Otto Spies, "Al-Kindi's Treatise on the Cause of the Blue Colour of the Sky," *Journal of the Bombay Branch of the Royal Asiatic Society*, n.s., 13 (1937): 7-19.

7. *De Aspectibus*, ed. Björnbo and Vogl, p. 3. Subsequent citations of *De aspectibus* will be to this edition.

8. Graziella Federici Vescovini, *Studi sulla prospettiva medievale*, pp. 44-47; on the same treatise, see Lynn Thorndike, *A History of Magic and Experimental Science*, 1: 642-46.

9. The Latin text is quoted by Vescovini, *Studi sulla prospettiva*, p. 46.

10. Thorndike, *History of Magic*, 1: 645, summarizes al-Kindi's view of the power of words as

follows: "Thus by words motion is started, accelerated, or impeded; animal life is generated or destroyed; images are made to appear in mirrors; flames and lightnings are produced; and other feats and illusions are performed."

11. The dependence of the doctrine of multiplication of species on al-Kindi's *De radiis* has been claimed (rightly, I believe) by Thorndike, *History of Magic*, 1: 646, and Vescovini, *Studi sulla prospettiva*, p. 40. The large number of Latin MSS (16) of *De radiis* listed by Carmody, *Astronomical and Astrological Sciences*, p. 82, gives some indication of its Western influence. I do not mean to suggest, however, that al-Kindi was the only or the principal source of the doctrine; nor am I able to demonstrate that Grosseteste or Bacon made firsthand use of *De radiis*. I. R. F. Calder has argued that al-Kindi's *De radiis* was also, in all probability, one of the sources of John Dee's emanation doctrine ("John Dee Studied as an English Neoplatonist," doctoral dissertation [University of London, 1952], 1: 514-15).

12. Vescovini, *Studi sulla prospettiva*, p. 48, n. 58; Richard Lemay, *Abu Ma'shar and Latin Aristotelianism in the Twelfth Century* (Beirut, 1962), pp. 47-48. A similar view was later expressed by al-Farabi, *Fusul al-Madani, Aphorisms of the Statesman*, ed. and trans. D. M. Dunlop (Cambridge, 1961), p. 73.

13. It is astonishing how little attention *De aspectibus* has received from historians of optics. In 1907 Eilhard Wiedemann discussed the first few propositions in his brief article, "Aus al Kindis Optik." Five years later a critical edition of the Latin text prepared by A. A. Björnbo, was published along with a free paraphrase by Sebastian Vogl in their "Drei optische Werke," pp. 3-70. The only discussion of any significance to appear in the meantime is in chap. 3 of Vescovini's *Studi sulla prospettiva*, but (as in the rest of her book) Vescovini is there chiefly concerned with epistemological developments rather than with optical theory. As I have already pointed out (n. 6, above), al-Kindi did prepare a recension of Euclid's *Optica*, but *De aspectibus* is not it.

14. See above, chap. 1.

15. The influence of Theon's preface is obvious elsewhere in al-Kindi's *De aspectibus* as well. Indeed, there is no way to demonstrate that al-Kindi had access to Euclid's *Optica* except in Theon's recension (although it is demonstrable that both versions were available in Arabic), but this in no way affects my position. It remains that al-Kindi's argument is directed toward Euclid, and his reliance on others who had also elaborated or criticized the Euclidean theory, though interesting, is immaterial to the claim that *De aspectibus* is essentially a critique of Euclid's theory of vision.

16. The identity of luminous and visual radiation seems to have been taken for granted throughout antiquity. It was specifically defended by Hero, Damianos, Theon, and apparently Ptolemy; see chap. 1, n. 82.

17. Nothing in al-Kindi's account suggests that he recognized the existence of a penumbral shadow or the difficulty of determining precisely where the penumbra ends and the umbra begins. This might suggest that there was nothing experimental about al-Kindi's procedure—that he was arguing from the presumed behavior of light as represented in his geometrical diagrams rather than from actual observations—in which case his argument is circular.

18. See above, chap 1.

19. *De aspectibus*, prop. 7, p. 9.

20. *De aspectibus*, prop. 8, p. 10. Cf. Aristotle, *Meteorologica* 3. 4. 373b3-10. The same argument would later be repeated by Albert the Great, *De homine*, appendix to qu. 22, in *Opera omnia*, ed. A. Borgnet, 35: 215.

21. *De aspectibus*, prop. 10, p. 12. Cf. Theon's preface to his recension of Euclid's *Optica*, in Euclid, *L'Optique et la Catoptrique*, trans. Paul Ver Eecke, pp. 55-56.

22. *De aspectibus*, prop. 9, pp. 11-12.

23. Ibid., prop. 7, p. 9. This argument appeared first in Theon's preface to Euclid's *Optica*, trans. Ver Eecke, p. 56; it would be repeated by Albert the Great, *De homine*, appendix to qu. 22, in *Opera omnia*, ed. A. Borgnet, 35: 215.

24. And, of course, al-Kindi says nothing about the corporeality of the forms. On the coherence of forms in the Epicurean theory, see above, chap. 1; I remind the reader that when I refer to the coherence of Epicurean *eidola*, I do not mean to imply that the atoms comprising an *eidolon* were necessarily thought to be physically hooked (see, chap. 1, n. 10). I wish to maintain only that the atoms maintain a fixed configuration as they leave the body; this might be called mathematical, as opposed to physical, coherence. Alexander of Aphrodisias, not himself an atomist, rejected any attempt to consider atomistic forms as incoherent entities; see Saint-Pierre, "La phyisque de la vision," pp. 224-28, 290-91.

25. Except the negative justification that Plato and Aristotle issued no denial of the theory of coherent forms that al-Kindi wishes to attribute to them; indeed, since they presented no alternative theory to explain how shapes are perceived, al-Kindi might well suppose that he has accurately captured their view. On the perception of shape in the Platonic and Aristotelian theories, see Robert Turnbull, "The Role of the 'Special Sensibles' in the Perception Theories of Plato and Aristotle."

26. Al-Kindi gives no detail regarding the process of radiation, but I think one may surmise that the form, as it is propagated toward the eye, would maintain the same "edgewise" orientation with respect to the eye possessed by the circle of which it is the form. Once inside the eye the form is not perceived according to the laws of perspective because the visual power entirely surrounds it. The principle that seems to be implicit in this analysis—that the circle placed before the eye has a "real" or "proper" shape, which must be perceived if the form enters the observer's eye—has an obvious Platonic flavor. In the *Sophist* (235e-236c), Plato condemns sculptors who fail to represent things as they "really are" when they distort the proportions of a very tall image in such a way as to make it appear properly proportioned to an observer stationed at ground level; cf. E. H. Gombrich, *Art and Illusion: A Study in the Psychology of Pictorial Representation*, rev. ed. (Princeton, 1961), pp. 191, 253.

27. Al-Kindi himself (in a different context) stated the principle of punctiform analysis on which Alhazen would build a successful intromission theory; see below.

28. See above, chap. 1.

29. Al-Kindi was surely dependent, for his conception of a cone of continuous radiation, on Ptolemy, whose *Optica* circulated widely in Islam. On the Islamic tradition of the *Optica*, see *L'Optique de Claude Ptolémée*, ed. Albert Lejeune, pp. 28*-30*. On Ptolemy's theory of the visual cone, see above, chap. 1. Stoic influence also appears likely: on the Stoic theory of vision, see above, chap. 1; on al-Kindi's familiarity with Stoic ideas, see Vescovini, *Studi sulla prospettiva*, pp. 33-43.

30. *De aspectibus*, prop. 11, pp. 12-13.

31. This seems to be merely a semantic point about the proper use of the term "ray" (*radius* in the Latin text), unless we assume that in al-Kindi's view luminous rays and the entities (whatever we call them) issuing from the eye share the same fundamental nature. In the latter case, the three-dimensionality of luminous rays (which al-Kindi defends not only on the basis of their origin in three-dimensional bodies, but also on the basis of their perceptibility) speaks not merely against using the term "ray" to denote the geometrical line issuing from the eye, but also against the very conception of such a one-dimensional entity.

32. This point had been clearly stated by Aristotle in *De sensu* 7. 449^{a}21-22.

33. *De aspectibus*, prop. 11, pp. 13-14.

34. The Latin term "*visibilis*," here translated, can mean "visual," i.e., having the capacity to perceive by visual means, as well as "visible."

35. Ibid., p. 14.

36. Ibid.

37. Ibid.

38. Cf. Avicenna's similar argument, chap. 3, p. 46.

39. *De aspectibus*, prop. 11, p. 15.

40. Al-Kindi thus reminds us once again of his reverence for the philosophers of antiquity. The full passage is worth quoting: "However, it is essential that we not judge evil of the opinion of this

man, nor compare it to error, because we know the excellence of his rank in this art and because he himself is one of the causes of our knowledge of good things; but we would think good things of him and convert his opinion to good purpose, since this [possibility] lies open to us" (ibid.).

41. Ibid.

42. Perhaps Damianos; see the following note.

43. Damianos argued that the visual cone consists of discrete rays and also that vision through the axis is clearest; see Richard Schöne (trans.), *Damianos Schrift über Optik*, pp. 3, 9, 11. Although the position ascribed by al-Kindi to his unnamed opponent is not identical to this, I cannot find anybody else to whom al-Kindi's accusations would apply as well, and I would thus suggest that Damianos was a possible (perhaps even the probable) object of al-Kindi's attack.

44. *De aspectibus*, prop. 11, p. 17.

45. Ibid., prop. 14, p. 23.

46. This is true, of course, only if the visible object lies in a plane perpendicular to the axis of the cone.

47. This argument occurs in proposition 12.

48. Al-Kindi limits himself to the following remark: "Therefore if it is known *per se* by them [i.e., certain of Euclid's followers] and by us and by all learned men what the effect of sight is, that it converts that which is opposite . . ." (*De aspectibus*, prop. 12, p. 19). However, later in *De aspectibus*, in connection with a problem of reflection, al-Kindi touches once more on the transformation of the medium, referring to it as a *resolution*: "A difficulty arises from the claim that air is affected by sight suddenly, for if sight suddenly resolves (*resolvit*) all of the air opposite the eye, why is that which is reflected [also] resolved, since it is not opposite the eye, but only opposite the mirror?" (Ibid., prop. 20, p. 34). The Latin terms "*resolutio*" and "*resolvere*" denote a relaxation, dilation, dispersion, or separation into component parts; but unfortunately neither al-Kindi nor any subsequent medieval commentator reveals how we are to apply this conception to the process of vision. Avicenna confesses his ignorance of the effect of sight on air, according to the extramission theory, when he writes: "If only the effect of sight on air were known!" (*Avicenna Latinus: Liber de anima seu sextus de naturalibus, I-II-III*, ed. Simone Van Riet, p. 224).

Nor does an examination of ancient theories of vision shed much light on al-Kindi's intention. The extant portions of Ptolemy's *Optica*, for example, contain no conception of visual or luminous radiation that could be construed as a relaxation or dispersion. Tideus describes the effect of sight on air simply as an alteration of the latter to the nature of the former; see Tideus, *De speculis*, in Björnbo and Vogl, "Drei optische Werke," pp. 74-75. Vogl asserts in a note to his paraphrase of *De aspectibus* (p. 65, n. 2), that according to Plato light dilates the air, resolving it into its component parts; but in fact the effect of visual light on air in Plato's theory is one of fusion rather than of resolution (although in his theory of color perception Plato does admit that visual light can be dilated or dispersed by smaller particles coming from external objects—see *Timaeus* 67d). The same can be said for the Stoic theory, in which a tension is produced in the air by the efflux of visual pneuma—though again Stoic theories of color perception acknowledge a process of dilation or dispersion (see David Hahm, "Early Hellenistic Theories of Vision and the Perception of Color").

49. Literally, "mark."

50. *De aspectibus*, prop. 14, pp. 24-25.

51. On Euclid and Damianos, see Euclid, *L'Optique et la Catoptrique*, trans. Ver Eecke, p. 57, n. 1; Lejeune, *Euclide et Ptolémée*, pp. 55-56. On Ptolemy, see the latter, pp. 51-55; and above, chap. 1.

52. Ptolemy clearly held the latter view: because visual rays issue from the center of curvature of the eye, they emerge without refraction and proceed to the object of sight, thereby producing a one-to-one correspondence between points in the visual field and points on the surface of the eye; moreover, this correspondence is a matter not merely of mathematics, but of physical process, for it is the ray or energy issuing from a given point on the surface of the eye that perceives directly opposed points in the visual field. See Lejeune, *Euclide et Ptolémée*, pp. 54-55.

53. The following figures may help to clarify my point. Figure 37*a* illustrates the visual cone of traditional (i.e., Euclidean and Ptolemaic) optics; it consists of rays issuing from the apex of the cone within the eye (not necessarily at the center in Euclid's theory) to points on the visible object. (We need not be concerned here with the question of the continuity of radiation within the cone.) These are the rays, if we take this argument of al-Kindi seriously, that cease to represent the physical process of sight; rather, according to al-Kindi, the physical process of sight is carried out through rays emanating from every point on the surface of the cornea, as in figure 37*b*.

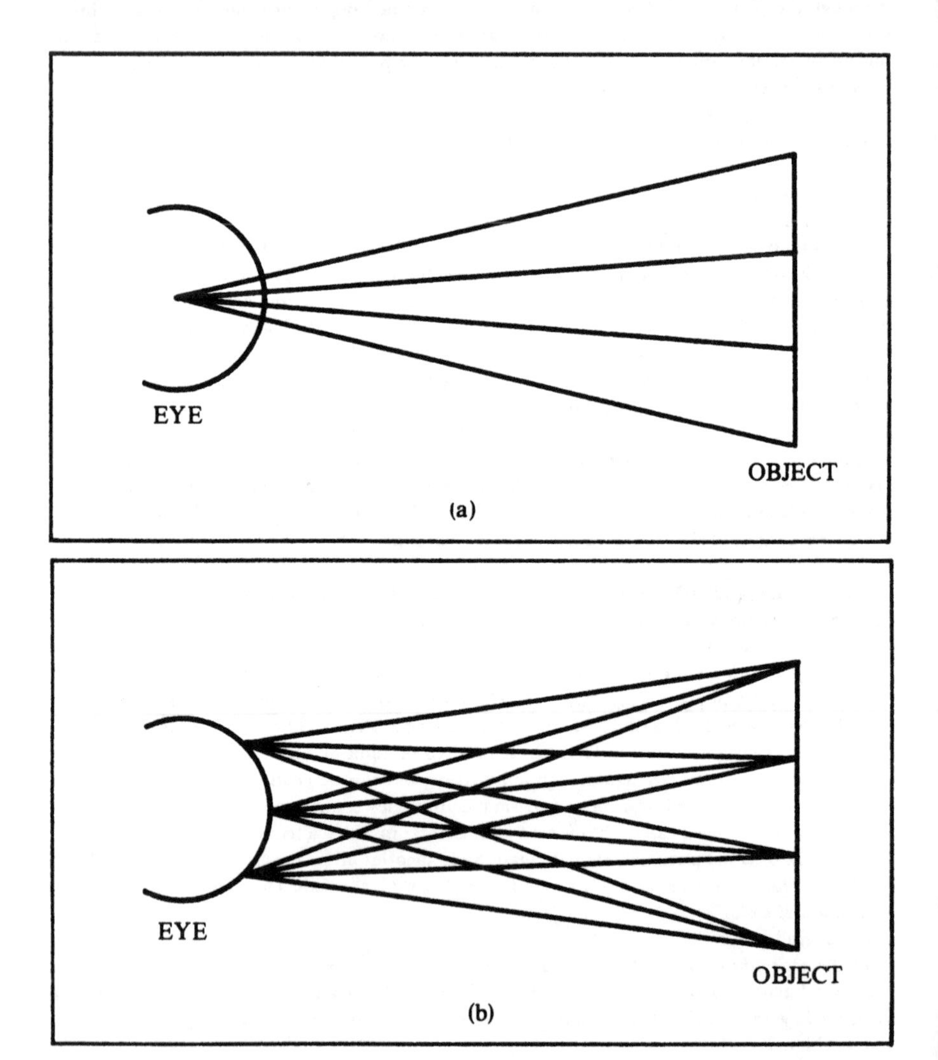

Fig. 37

54. *De aspectibus*, prop. 13, p. 23.

55. I believe that one can perceive faint foreshadowings or implicit applications of this principle in a number of ancient sources; see, for example, Pseudo-Euclid, *Catoptrica*, prop. 30,

in Euclid, *L'Optique et la Catoptrique*, trans. Ver Eecke, pp. 122-23; Saint-Pierre, "La physique de la vision," pp. 226-27. But it was al-Kindi who first gave the principle clear, explicit, and extended articulation. Moreover, many of those who stated this principle during the next seven centuries followed al-Kindi's exposition closely and also reproduced his geometrical diagram; see my article, "Alkindi's Critique," p. 488.

56. See chap. 4.

57. See above, chap. 2, nn. 30 and 35.

58. See chap. 2, n. 48.

59. Ibid.

60. On the Stoic and Galenic theories, see above, chap. 1. It is a pity that we lack the first book of Ptolemy's *Optica* and hence any firm knowledge of his theory of the nature of visual radiation. However, by Lejeune's reconstruction (*Euclide et Ptolémée*, pp. 62-71), Ptolemy was not Galenic or Stoic in this regard.

61. *The Book of the Ten Treatises of the Eye Ascribed to Hunain ibn Is-haq*, ed. and trans. Max Meyerhof, p. 33; Hunain's theory is discussed more fully in chap. 3. For Galen's similar view, see chap. 1, above.

62. See chap. 1, pp. 10-11; chap. 3, pp. 38-39. Al-Kindi maintains that the visual cone is sensitive throughout in the sense that an object intercepting a ray anywhere within the cone is perceived, but he does not indicate whether perception occurs at the surface of the perceived object or in the observer's eye.

63. See Eilhard Wiedemann, "Ueber das Leben von Ibn al Haitam und al Kindi," *Jahrbuch für Photographie und Reproduktionstechnik* 25 (1911): 6-7; Max Meyerhof, "Die Optik der Araber," p. 20.

64. On 'Ubaid Allah, see Max Meyerhof, "An Arabic Compendium of Medico-Philosophical Definitions," *Isis* 10 (1928): 347-48; on the others, see the same author's "Optik der Araber," pp. 21-22, 24-27, 52-54, 89, 90. Several of these authors presented differing views on the theory of sight in different works. Al-Farabi, although accepting the theory of visual rays in his *Catalogue of the Sciences*, defended the Aristotelian theory of vision in his *The Model State*; see this volume, chap. 3. Similarly, Nasir al-Din al-Tusi, although adopting the extramission theory in his answer to al-Qazwini regarding Avicenna's theory of the influence of heat and cold on the colors of dry and moist bodies, admits in his recension of Euclid's *Optica* that in fact sight occurs through intromission; see Meyerhof, "Optik der Araber," p. 53; George Sarton, *Introduction to the History of Science*, 2: 1009; Eilhard Wiedemann, "Ueber die Entstehung der Farben nach Nasir al Din al Tusi," *Jahrbuch für Photographie und Reproduktionstechnik* 22 (1908): 88; Wiedemann, "Ueber die Reflexion und Umbiegung des Lichtes von Nasir al Din al Tusi," ibid., 21 (1907): 39. I omit Galenists from this list, although they too may exhibit al-Kindi's influence.

65. See the German translation of Salah al-Din's work in *Die arabischen Augenärzte*, trans. J. Hirschberg, J. Lippert, and E. Mittwoch, 2: 207-30.

66. There were many "questions" written in the Latin West during the Middle Ages devoted to the discussion of visual rays. See, for example, Albert the Great, *De homine*, appendix to qu. 22 ("Utrum visus est per emissionem radiorum") (in which al-Kindi is frequently cited), ed. Borgnet, pp. 215-28; Pseudo-Grosseteste, *Summa philosophie*, bk. 19, qu. 14 ("Disputatio subtilis utrum videamus intussuscipientes, an extramittentes, an utroque modo"), qu. 15 ("De positione radiorum visualium"), qu. 16 ("Quod secundum Platonem et alios visus fiat per emissionem radiorum visualium"), in *Die philosophischen Werke des Robert Grosseteste*, ed. Ludwig Baur, pp. 499-504; Rudolph of Erfurt, *Questiones de anima*, bk. 2, qu. 18 ("Utrum visio fiat per emissionem radiorum"), Cracow, Bibl., Jagiellońska, MS 753, fols. 32v-34r, and MS 635, pp. 306-9; Anonymous, "An visio fiat per extramissionem," Erfurt, Wissenschaftliche Bibl., MS Amploniana Q.369, fol. 4r (14th c.). The extramission theory continued to be debated as late as the seventeenth century; see the *Supplementum* of Hugo Cavellus to Duns Scotus's *Questiones in libros de anima*, disputatio 2, sectio 9, dubium 5 ("Quod est organum visus, et an videat per extramissionem?"), in Duns Scotus, *Opera omnia*, ed. Lucas Wadding (Paris, 1639), 2: 619-20. That light issues from the eyes of nocturnal animals was still being admitted in the sixteenth and

seventeenth centuries; see, for example, Hieronymus Fabricius (or Girolamo Fabrici) of Aquapendente, *Tractatus anatomicus triplex*, p. 28.

67. For analysis of the combined intromission-extramission theory in the West, see this volume, pp. 100-101, 114-16, 118.

68. Crombie, *Grosseteste*, p. 117, n. 2.

69. *De iride*, in Edward Grant, *A Source Book in Medieval Science*, p. 389.

70. Roger Bacon, *The Opus Majus of Roger Bacon*, ed. John H. Bridges, 2: 49; Lindberg, *Pecham*, pp. 35, 127-29; also, this volume, chap. 6. For a slightly fuller discussion of al-Kindi's influence in the West, see Lindberg, "Alkindi's Critique," pp. 487-89.

Chapter 3 Galenists and Aristotelians in Islam

1. The best general sources on Islamic ophthalmology are Julius Hirschberg, *Geschichte der Augenheilkunde im Mittelalter und in der Neuzeit*; and J. Hirschberg, J. Lippert, and E. Mittwoch, *Die arabischen Lehrbücher der Augenheilkunde*. See also Pierre Pansier, *Collectio ophtalmologica veterum auctorum*, 7 fascs.; Thomas H. Shastid, "History of Ophthalmology," pp. 8691-8722; Stephen Polyak, *The Vertebrate Visual System*, pp. 14-21; Polyak, *The Retina*, pp. 104-23: and the many works of Max Meyerhof (see Bibliography). In 1905, Hirschberg claimed to know of more than thirty texts on ophthalmology ("Arabian Ophthalmology," p. 1127), and a number have been discovered since then.

2. See C. Prüfer and Max Meyerhof (trans.), "Die Augenheilkunde des Juhanna b. Masawaih," pp. 217-56; J. C. Sournia and G. Troupeau, "Médecine Arabe: Biographies critiques de Jean Mésué (VIII° siècle) et du prétendu 'Mésué le Jeune' (X° siècle)," *Clio Medica* 3 (1968): 109-17. The medieval Latin translation of Mesue's work on ophthalmology, *De egritudinibus oculorum*, is available in *Mesue cum expositione Mondini super canones universales* . . . (Venice, 1497), and other early editions.

3. The standard works on Hunain are Gotthelf Bergsträsser, *Hunain ibn Ishak und seine Schule* (Leiden, 1913); and Bergsträsser, *Hunain ibn Ishaq über die syrischen und arabischen Galen-Übersetzungen* (Leipzig, 1925). See also Hunain, *Ten Treatises*, ed. and trans. Max Meyerhof, pp. xvi-xxxiii; Meyerhof, "New Light on Hunain ibn Ishaq and His Period," pp. 685-724 (which summarizes and interprets some of Bergsträsser's findings); Giuseppe Gabrieli, "Hunáyn ibn Isháq," *Isis* 6 (1924): 282-92; Lufti M. Sa'di, "A Bio-Bibliographical Study of Hunayn ibn Ishaq al-Ibadi (Johannitius) (809-877 A.D.)," *Bulletin of the Institute of the History of Medicine* 2 (1934): 409-46; and G. Strohmaier, "Hunayn b. Ishak al-'Ibadi," *The Encyclopaedia of Islam*, new ed., 3: 578-81.

4. The *Ten Treatises* is available in Meyerhof's Arabic text and English translation. For editions of the medieval Latin version, see chap. 3, n. 58. Portions of this work have also been translated into German in Max Meyerhof and C. Prüfer, "Die Augenanatomie des Hunain b. Ishaq," pp. 163-90; and "Die Lehre vom Sehen bei Hunain b. Ishaq," pp. 21-33. The *Book of Questions on the Eye* is available in an Arabic text and French translation in P. Sbath and Max Meyerhof (trans.), *Le livre des questions sur l'oeil de Honain ibn Ishaq*; see also the brief study of this work by P. Sbath, "Le livre des questions sur l'oeil de Hunain ben Ishaq," pp. 129-38. Yet another extant work by Hunain, *On Light and Its True Nature*, has been translated into German by C. Prüfer and Max Meyerhof in "Die aristotelische Lehre vom Licht bei Hunain b. Ishaq."

5. Hunain personally translated many Galenic works, including *De usu partium* and *De placitis Hippocratis et Platonis*, from which he drew most of his material on vision; see Meyerhof, "New Light on Hunain," p. 695.

6. On the translation of Hunain's *Ten Treatises*, see chap. 3, n. 58, below. Galen's *De placitis* was not translated into Latin until the sixteenth century; on the several Renaissance translations, see Richard J. Durling, "A Chronological Census of Renaissance Editions and Translations of Galen," *Journal of the Warburg and Courtauld Institutes* 24 (1961): 285 (under *De Hippocratis et Platonis decretis*). *De usu partium* was translated into Latin in separate stages: the translation

by Burgundio of Pisa in the second half of the twelfth century (which circulated as *De iuvamentis membrorum*) included only the first nine books, and it was not until the translations of Peter of Abano and Niccolò da Reggio in the late thirteenth and early fourteenth centuries that book 10 was available in Latin; see Lynn Thorndike, "Translations of Works of Galen by Peter of Abano," *Isis* 33 (1941-42): 649; Thorndike, "Translations of Works of Galen by Niccolò da Reggio (c. 1308-1345)" *Byzantina Metabyzantina* 1 (1946-49): 214, 232; Emilie Savage Smith, "Galen on Nerves, Veins and Arteries: A Critical Edition and Translation from the Arabic, with Notes, Glossary and an Introductory Essay," doctoral dissertation (University of Wisconsin, 1969), pp. 29-36. Certain aspects of the Galenic theory of vision were also defended in the *De speculis* of Tideus, translated from Greek to Arabic and subsequently from Arabic to Latin by Gerard of Cremona; for Tideus's theory of vision, see Björnbo and Vogl, "Drei optische Werke," pp. 74-75; on the translation of *De speculis* from Arabic to Latin, see ibid., pp. 151-53.

7. Apparently Hunain conceived the lens to be not simply concentric with the anterior tunics of the eye, as was generally maintained in the West during the Middle Ages, but truly central to the eye; see fig. 5. When Hunain claims that the lens is luminous, he is alluding not to its function as a source of visual rays, but to its ability to be influenced by light and color; indeed, Hunain's remarks undoubtedly reflect a passage from Galen's *De placitis*, where it is claimed that "the organ of sight, . . . since it must discriminate colors, is luminous, for only such bodies are by nature capable of being altered by colors" (7. 7, trans. Phillip De Lacy [forthcoming in the *Corpus Graecorum medicorum*]; cf. *Claudii Galeni Opera omnia*, ed. C. G. Kühn, 5: 637). The same idea appears in Galen's *Usefulness of Parts*, trans. May, p. 464.

8. *Ten Treatises*, trans. Meyerhof, p. 4; cf. *Livre des questions*, trans. Sbath and Meyerhof, p. 83. Hunain here distorts and confuses Galen's correct claim (*Usefulness of Parts*, trans. May, p. 503) that the crystalline humor, if flattened, will have more of its parts exposed to the impressions of visible things than if it is perfectly spherical. The medieval perspectivists, who believed that vision is caused only by perpendicular rays, would later point out that a completely flat ocular surface is disadvantageous because of the smallness of the visual field that it could perceive; see Lindberg, *Pecham*, p. 111. On the shape of the crystalline humor, see this volume, pp. 51, 67, 192, 273-74 (n. 141).

9. *Ten Treatises*, trans. Meyerhof, p. 4; cf. *Livre des questions*, trans. Sbath and Meyerhof, p. 83.

10. *Ten Treatises*, trans. Meyerhof, p. 6. Here and in other quotations from the *Ten Treatises*, I have converted Meyerhof's parentheses to brackets, since it is apparent that these enclose his own explanatory insertions.

11. Ibid., pp. 7-8. Cf. Galen, *Usefulness of Parts*, trans. May, p. 465.

12. *Ten Treatises*, trans. Meyerhof, p. 8. Cf. Galen, *Usefulness of Parts*, trans. May, pp. 465-66, 468-69; Galen, *On Anatomical Procedures: The Later Books*, trans. W. L. H. Duckworth, pp. 33 ff.

13. On the anterior tunics and humor, see *Ten Treatises*, trans. Meyerhof, pp. 8-10; cf. *Livre des questions*, trans. Sbath and Meyerhof, pp. 87-91; Galen, *Usefulness of Parts*, trans. May, pp. 469-70, 475-78.

14. On this manuscript, see *Ten Treatises*, trans. Meyerhof, pp. xlviii-xlix. I have copied the drawing, with certain modifications, from ibid., p. 5, and from Polyak, *Vertebrate Visual System*, p. 17; cf. Polyak, *Retina*, fig. 7, and Meyerhof and Prüfer, "Augenanatomie des Hunain b. Ishaq," p. 172. Meyerhof, so far as I can judge, has misdrawn the figure, and Polyak has mislabeled it; cf. the photographic reproduction of the original in Polyak, *Vertebrate Visual System*, p. 16. My reconstruction agrees with that of Meyerhof and Prüfer.

15. *Ten Treatises*, trans. Meyerhof, pp. 10-11. I have discussed a small fraction of the medieval literature on this subject in *Pecham*, pp. 249-50. Benvenutus Grassus, in his discussion of the number of ocular tunics, refers specifically to Hunain; see Benvenutus Grassus of Jerusalem, *De oculis eorumque egritudinibus et curis*, trans. Casey A. Wood, p. 28.

16. Actually the anterior capsule of the lens; see chap. 3, nn. 141-42, below.

17. *Ten Treatises*, trans. Meyerhof, p. 23. Cf. Galen *Usefulness of Parts*, trans. May, p. 491.

18. *Ten Treatises*, trans. Meyerhof, p. 21. Cf. Galen, *Usefulness of Parts*, trans. May, p. 491; *De placitis* 7. 4, *Opera omnia*, ed. Kühn, 5: 612-13; above chap. 1. On the theory of nerves and nerve function, including the doctrine of hollow nerves, see Friedrich Solmsen, "Greek Philosophy and the Discovery of the Nerves," *Museum Helveticum* 18 (1961): 150-67, 169-97; Edwin Clarke, "The Doctrine of the Hollow Nerve in the Seventeenth and Eighteenth Centuries," in *Medicine, Science, and Culture: Historical Essays in Honor of Owsei Temkin*, ed. Lloyd G. Stevenson and Robert P. Multhauf (Baltimore, 1968), pp. 123-41.

19. *Ten Treatises*, trans. Meyerhof, p. 25; I have slightly emended Meyerhof's translation. Cf. Galen, *Usefulness of Parts*, trans. May, p. 501.

20. Cf. Galen, *Usefulness of Parts*, trans. May, p. 476; *De placitis* 7. 4, *Opera omnia*, ed. Kühn, 5: 614; Pseudo-Aristotle *Problemata* 31. 4. 957^{b}18-20; Tideus, *De speculis*, in Björnbo and Vogl, "Drei optische Werke," p. 77.

21. *Ten Treatises*, trans. Meyerhof, p. 26; cf. Galen, *Usefulness of Parts*, trans. May, pp. 498-99. This conception is flawed by a serious difficulty, namely that there is no way for rays or visual spirit to proceed in straight lines through two adjacent branches of an X-shaped junction and fall on the same point. Later scholars were to remove the difficulty by permitting rays to deviate from rectilinearity once inside the vitreous humor and optic nerve; see chap. 4, pp. 82-83.

22. Cf. Pseudo-Aristotle *Problemata* 31. 7. 958^{a}25-26; Galen, *Usefulness of Parts*, trans. May, p. 494.

23. *Ten Treatises*, trans. Meyerhof, pp. 31-32; I have made minor improvements in the wording of Meyerhof's translation. Galen described only two alternatives; see above, chap. 1.

24. *Ten Treatises*, trans. Meyerhof, p. 32. For Galen's similar argument, see above, chap. 1.

25. *Ten Treatises*, trans. Meyerhof, p. 32; cf. Galen, *De placitis* 7.5, *Opera omnia*, ed. Kühn, 5: 618. This argument was to be frequently employed in subsequent centuries; see this volume pp. 45, 64. In the Euclidean and Ptolemaic theory, there was, of course, no encirclement of the object by visual rays.

26. I maintain this despite the fact that Hunain appears to classify Aristotle under the first theory and thus to dismiss him along with the atomists. Galen would not have objected to this comparison between his theory and Aristotle's, for he speaks favorably of the Aristotelian theory of vision in *De Placitis* 7. 7. *Opera omnia*, ed. Kühn, 5: 638-43.

27. Hunain points out that this transformation is not permanent, but must occur continuously, "for if it were the case that a permanent transformation took place, then its [i.e., the sun's] light would remain in the air for a time after the light-giving body had been removed" (*Ten Treatises*, trans. Meyerhof, p. 33). Cf. Galen *De placitis* 7. 5, *Opera omnia*, ed. Kühn, 5: 620.

28. *Ten Treatises*, trans. Meyerhof, p. 33. Cf. Galen's view, this volume, chap. 1.

29. On the air as a percipient organ, see just below.

30. Hunain says at one point that it "leaves the eye" (*Ten Treatises*, trans. Meyerhof, p. 31) and at another that it "penetrates into the air" (ibid., p. 28), from which it is evident that the spirit passes beyond the interface separating the eye from the air.

31. Galen said explicitly that the pneuma emerges from the eye "but does not itself extend out very far"; see this volume, p. 10. Later Islamic ophthalmologists made statements that can be construed to mean that the visual spirit itself proceeds to the visible object, and it is not clear whether this is the result of faulty exposition of the Galenic theory or a positive departure from it; see 'Ali ibn 'Isa, *Errinnerungsbuch für Augenärzte aus arabischen Handschriften*, in *Die arabischen Augenärzte*, trans. Julius Hirschberg, J. Lippert, and E. Mittwoch, 1: 28; Pansier, *Collectio ophtalmologica*, fasc. 3, p. 212.

32. *Ten Treatises*, trans. Meyerhof, p. 29. I have slightly altered Meyerhof's punctuation for the sake of clarity.

33. Ibid.; cf. above, p. 36.

34. Ibid., p. 37. Cf. Galen, above, chap 1.

35. *Ten Treatises*, trans. Meyerhof, p. 37. I have made minor modifications in Meyerhof's translation.

36. Ibid., p. 36.

37. Ibid. The analogy of the walking stick was mentioned in antiquity by Galen, Diogenes Laertius, and Alexander of Aphrodisias; see this volume, pp. 9-11.

38. Descartes employed the analogy of the walking stick to eludicate the nature of light in *La dioptrique* (see *Discourse on Method, Optics, Geometry, and Meteorology*, trans. Paul J. Olscamp, pp. 67-68), maintaining that pressure is propagated mechanically through a walking stick to a blind man's hand; analogously in the case of light, there is an inward propagation of mechanical pressure, rather than an outward propagation of sensitivity. Galen had rejected the walking-stick analogy precisely because of its mechanistic connotations (see above, chap. 1), and it is puzzling that Hunain, whose theory of vision does not differ from Galen's, should revive it.

39. With regard to the sensitivity of the eye, Hunain argues that "the instrument [i.e., the eye] is changed into the nature of the nerve that is conjoined to it and thereby becomes sensitive. It is therefore again evident that flesh perceives because of the power of vision existing in it, which is bestowed on it by the nerve" (Pansier, *Collectio ophtalmologica*, fasc. 7, p. 176). I quote here from the medieval Latin translation of the *Ten Treatises*, which is somewhat clearer than Meyerhof's direct translation from the Arabic; cf. the latter, p. 29.

40. *Ten Treatises*, trans. Meyerhof, p. 36.

41. Ibid., p. 35. For Galen's similar view, see above, chap. 1.

42. It is true, however, that the perception of size and position was frequently the object of mathematical analysis. Hunain briefly discusses the elements of this mathematical analysis, namely, the visual cone and axis, in his remarks on binocular vision; see *Ten Treatises*, trans. Meyerhof, pp. 25-26; cf. Galen, *Usefulness of Parts*, trans. May, pp. 494-95.

43. However, note the reflections of the pre-Socratic philosophers on this question; see J. I. Beare, *Greek Theories of Elementary Cognition from Alcmaeon to Aristotle*, pp. 9-42.

44. *Ten Treatises*, trans. Meyerhof, pp. 35-36.

45. Ibid., p. 38.

46. Ibid. Galen employs the same example in *De placitis* 7. 7, *Opera omnia*, ed. Kühn, 5: 637-38.

47. *Ten Treatises*, trans. Meyerhof, p. 4.

48. *Usefulness of Parts*, trans. May, pp. 464-65.

49. Hunain confirms the existence of this second phase of transmission when he writes that perception is transmitted "by the nerves, until it reaches the brain and is perceived by the allotted part of the soul, and the individual is informed [of the sensation]" (*Ten Treatises*, trans. Meyerhof, p. 29). This passage is missing from the Latin text; see Pansier, *Collectio Ophtalmologica*, fasc. 7, p. 176; *Liber de oculis* (Lyons, 1515), fol. 173va.

50. This would become an important issue in the sixteenth century (see chap. 8, pp. 171-72)—thus the need to understand Hunain's position on the matter. I employ the term "neurophysiological" only to denote a nervous process taking place through the tissue of the nerves; I do not intend any of the other modern connotations of the term.

51. Galen, *Usefulness of Parts*, trans. May, p. 467; cf. Galen, *De symptomatum causis*, ed. Kühn, 7 [Leipzig, 1824]: 103 (quoted by Emilie S. Smith, "Galen's Account of the Cranial Nerves and the Autonomic Nervous System," *Clio Medica* 6 [1971]: 183); Hunain, *Ten Treatises*, trans. Meyerhof, p. 8.

52. Galen, *Usefulness of Parts*, trans. May, pp. 464-65; Hunain, *Ten Treatises*, trans. Meyerhof, p. 6.

53. *Ten Treatises*, trans. Meyerhof, p. 4. Cf. Galen, *Usefulness of Parts*, trans. May, pp. 463-64; *De placitis* 7. 6, *Opera omnia*, ed. Kühn, 5: 635. This argument was to face many vicissitudes during the course of the Middle Ages; on its history, see Lindberg, *Pecham*, pp. 15-16, 250; Lindberg, "Lines of Influence in Thirteenth-Century Optics," pp. 77-80.

54. *Usefulness of Parts*, trans. May, p. 465.

55. The alternative view, that the post-crystalline transmission is optical, might be defended on the grounds that both Galen and Hunain maintained that a straight line could be drawn through the pupil and optic nerve to the optic chiasma, where the two optic nerves unite, and (as Galen made clear) beyond the chiasma to the brain (Hunain, *Ten Treatises*, trans. Meyerhof, pp. 25-27; Galen, *Usefulness of Parts*, trans. May, pp. 492-99). To be sure, at this point in their argument Galen and Hunain reverted to the extramission theory of Euclid, developing the theory of the visual cone and the geometrical perception of space and remarking that rectilinear visual rays originate in the optic chiasma; however, it could be explained that this utilization of the Euclidean theory simply reflected its historic association with the mathematical approach to vision, and that the rectilinearity of the optical path between the pupil and the optic chiasma or brain should be viewed as evidence that, in truth, Galen and Hunain believed visual impressions to be radiated from the crystalline humor to the brain. Surely this is a cogent argument, though I find the arguments for the neurophysiological interpretation more persuasive. For the views of a later Galenist, see this volume, pp. 171-72.

As for the other alternatives, it is surely possible that Galen and Hunain viewed the post-crystalline transmission as simultaneously optical and neurophysiological. For example, impressions might be thought to radiate through the transparent humors filling the visual pathway, all the while being somehow perceived by the surrounding tissues of the retina and optic nerve. Rudolph Siegel attributes to Galen two distinct and fully articulated theories on the subject: a mathematical one, which assumes the post-crystalline transmission to be optical in nature, and a physiological one, which postulates a post-crystalline transmission through the lateral attachments of the crystalline lens to the retina and optic nerve. Unfortunately, Siegel presents no textual evidence to support his position; see *Galen on Sense Perception*, pp. 53-54, 98-101; "Principles and Contradictions of Galen's Doctrine of Vision," pp. 265-69.

56. See Hirschberg, *Augenheilkunde im Mittelalter*, pp. 35-37. Hirschberg et al., *Die arabischen Lehrbücher*, pp. 19-20; Max Meyerhof, "Eine unbekannte arabische Augenheilkunde des 11. Jahrhunderts n. Chr.," pp. 66-67.

57. See *Memorandum Book of a Tenth-Century Oculist for the Use of Modern Ophthalmologists*, trans. Casey A. Wood, p. 4; 'Ali ibn 'Isa, *Erinnerungsbuch*, p. 5; Alcoati, *Liber de oculis*, in Pansier, *Collectio ophtalmologica*, fasc. 2, p. 87.

58. On Constantine's translation of the *Ten Treatises*, see Hirschberg, *Augenheilkunde im Mittelalter*, pp. 34-36; Moritz Steinschneider, *Die europäischen Übersetzungen aus dem arabischen bis Mitte des 17. Jahrhunderts*, pp. 9-10; *Ten Treatises*, trans. Meyerhof, pp. xxxiii-xxxiv. For a general study of Constantine's medical translations, see Steinschneider, "Constantinus Africanus und seine arabischen Quellen," *Virchow's Archiv für pathologische Anatomie und Physiologie und klinische Medizin* 37 (1866): 351-410; R. Kreutz, "Der Arzt Constantinus Africanus von Montekassino," *Studien und Mitteilungen zur Geschichte des Benediktiner-Ordens* 47 (1929): 1-44. The Latin text of the *Ten Treatises* is available in Pansier, *Collectio ophtalmologica*, fasc. 7; *Liber Constantini de oculis*, in *Omnia opera Ysaac* [*Israeli*] (Lyons, 1515), fols. 172r-178r; and Galen, *De oculis a Demetrio translatus*, in the nine Juntine editions of Galen's works, 1541-1625 (vol. 5, fols. 178r-187r in the edition of 1541). Hirschberg has argued that there were two distinct Latin translations, one by Constantine and the other by the otherwise unidentified Demetrius; see "Über das ältesten arabischen Lehrbuch der Augenheilkunde," pp. 1087-91; *Augenheilkunde im Mittelalter*, pp. 34-36. However, Pansier has been quite insistent that the two versions are identical; see *Collectio ophtalmologica*, fasc. 2, pp. 46-48; fasc. 7, pp. 165-66. After studying the manuscripts, my student Philip Teigen has concluded that there were two versions, based on a single Latin translation; see my *Catalogue of Optical Manuscripts*, pp. 98-100.

59. Benvenutus Grassus, *De oculis*, trans. Wood, pp. 28, 31; Bartholomaeus Anglicus, *De rerum proprietatibus*, pp. 64, 128 ff.; Vincent of Beauvais, *Speculum naturale*, in *Speculum quadruplex sive speculum maius*, vol. 1, col. 2024; Roger Bacon, *Opus maius*, ed. Bridges, 2: 13-17. The inclusion of the *Ten Treatises* among Galen's works in the printed editions of the

sixteenth century was particularly effective in spreading Hunain's influence in the later period.

60. The meaning of the term "tunic" was discussed as early as Galen, *On Anatomical Procedures*, trans. Duckworth, p. 42.

61. On the works of al-Razi, see Heinrich Ferdinand Wüstenfeld, *Geschichte der arabischen Aerzte und Naturforscher* (Göttingen, 1840), pp. 40-49; Leclerc, *Histoire de la médecine arabe*, 1: 350; Hirschberg, *Augenheilkunde im Mittelalter*, p. 106; George S. A. Ranking, "The Life and Works of Rhazes (Abu Bakr Muhammad ibn Zakariya ar-Razi)," *Seventeenth International Congress of Medicine*, section 23 (London, 1914), pp. 237-68; Julius Ruska, "Al-Biruni als Quelle für das Leben und die Schriften al-Razi's," *Isis*, 5 (1924): 26-50.

62. *Trois traités d'anatomie arabes par Muhammed ibn Zakariyya al-Razi, 'Ali ibn al-'Abbas et 'Ali ibn Sina*, trans. P. De Koning, p. 53.

63. On al-Farabi's life and works, see Moritz Steinschneider, *Al-Farabi (Alpharabius), des arabischen Philosophen Leben und Schriften* (Mémoires de l'Académie impériale des sciences de St.-Pétersbourg, ser. 7, vol. 13, no. 4 [St. Petersburg, 1869]); Nicholas Rescher, *Al-Farabi: An Annotated Bibliography* (Pittsburg, 1962); Richard Walzer, "Al Farabi," *The Encyclopaedia of Islam*, new ed., 2: 778-81.

64. I quote from Marshall Clagett's unpublished translation of the *Catalogue of the Sciences*, rendered from Arabic manuscripts; cf. Alfarabi, *Catálogo de las ciencias*, ed. and trans. Angel González Palencia, pp. 44-45, 100, 149-50.

65. In translating this passage into English, I have used both the German translation by Friedrich Dieterici, *Der Musterstaat von Alfarabi*, pp. 70-71, and the French translation by R. P. Jaussen, Youssef Karam, and J. Chlala, in Al-Farabi, *Idées des habitants de la cité vertueuse*, p. 66.

66. Dieterici, *Musterstaat*, p. 71; Jaussen et al., *Idées des habitants*, p. 66.

67. There have been attempts to attribute to al-Farabi a more developed Aristotelian theory of vision than that found in *The Model State* (see Eilhard Wiedemann, "Zur Geschichte der Lehre vom Sehen," pp. 471-72). However, such attempts rest on the false attribution of Averroes's *De sensu et sensato* to al-Farabi. The error seems to have originated with Albert the Great, who summarized Averroes's *De sensu et sensato* in his own *De sensu* and *De homine*, but identified al-Farabi as the author (see Albert the Great, *Opera omnia*, ed. A. Borgnet, 9: 11, 17; 35: 210-14). Albert's error was repeated by Vincent of Beauvais in his *Speculum naturale* (*Speculum quadruplex*, vol. 1, cols. 1800, 1803), by the author of the Pseudo-Grossetestian *Summa philosophie* (*Die philosophischen Werke des Robert Grosseteste*, pp. 504-6), and by a number of unsuspecting modern authors. However, Averroes's authorship of *De sensu et sensato* was recognized by Roger Bacon, *Opus Maius*, ed. Bridges, 2: 50. That the work summarized by Albert, Vincent, and the author of the *Summa philosophie* is Averroes's *De sensu et sensato* was demonstrated in modern times by R. de Vaux, "La première entrée d'Averroës chez les Latins," *Revue des sciences philosophiques et théologiques* 22 (1933): 238-40.

68. The better portion of this section also appears, with some variations and reorganization, in my paper "The Intromission-Extramission Controversy in Islamic Visual Theory: Alkindi versus Avicenna," forthcoming in *Perception: Philosophical and Scientific Themes and Variations*, edited by Peter Machamer and Robert G. Turnbull (Columbus, Ohio: Ohio State University Press).

69. Avicenna's autobiography (with additions made after his death by his friend al-Juzjani) is most readily available in a translation with commentary by A. J. Arberry, "Avicenna: His Life and Times," in *Avicenna: Scientist and Philosopher*, ed. G. M. Wickens (London, 1952), pp. 9-28. The secondary literature on Avicenna is immense, but particularly useful summaries (in addition to the articles assembled by Wickens) are Carra De Vaux, *Avicenne* (Paris, 1900); Soheil M. Afnan, *Avicenna: His Life and Works* (London, 1958); F. Rahman, "Ibn Sina," in *A History of Muslim Philosophy*, ed. M. M. Sharif, vol. 1 (Wiesbaden, 1963), pp. 480-506; A.-M. Goichon, "Ibn Sina," *The Encyclopaedia of Islam*, new ed., 3: 941-47; and Majid Fakhry, *A History of Islamic Philosophy*, pp. 147-83.

70. The best bibliography (in a western language) of Avicenna's extant works is that of M.-M. Anawati, O.P., "La tradition manuscrite orientale de l'oeuvre d'Avicenne," *Revue thomiste* 51 (1951): 407-40. On the translation and western dissemination of Avicenna's works, see M.-T. d'Alverny, "Les traductions d'Avicenne (Moyen Age et Renaissance)," in *Avicenna nella storia della cultura medioevale* (Accademia Nazionale dei Lincei, Problemi attuali di scienza e di cultura, Quaderno 40 [Rome, 1957]), pp. 71-87; d'Alverny, "Notes sur les traductions médiévales d'Avicenne," *Archives d'histoire doctrinale et littéraire du moyen âge* 19 (1952): 337-58; and d'Alverny, "Avendauth?" in *Homenaje á Millás-Vallicrosa*, vol. 1 (Barcelona, 1954), pp. 19-43.

Avicenna's fullest treatment of vision is in pt. 4, bk. 6 (the psychological section) of the *Kitab al-Shifa*. This section was translated into Latin in the second half of the twelfth century by Avendauth (who rendered it orally into Castilian) and Dominicus Gundissalinus (who put it into Latin) and circulated as *De anima* or *Liber sextus naturalium*; for a description of the Latin manuscripts, see M.-Th. d'Alverny, "Avicenna Latinus," *Archives d'histoire doctrinale et littéraire du moyen âge* 28 (1961): 281-316, and subsequent volumes; see also Simone Van Riet, "La traduction latine du 'De anima' d'Avicenne. Préliminaires à une édition critique," *Revue philosophique de Louvain* 61 (1963): 583-626. The Latin text is available in the edition of Venice, 1508, and a modern critical edition by Simone Van Riet, *Avicenna Latinus: Liber de anima seu sextus de naturalibus, I-II-III*. A critical edition and French translation of the Arabic text is available in *Psychologie d'Ibn Sina [Avicenne] d'après son oeuvre as-Sifa*, ed. Ján Bakoš. Both the Latin and French translations are frequently incoherent, but a careful comparison of the two removes nearly all ambiguities about Avicenna's intent. A brief synopsis of *De anima* is available in Martin Winter, *Über Avicennas Opus agregium de anima*; on visual perception, see pp. 42-53. For editions and translations of the other works in which Avicenna treats vision, see below.

Avicenna's theory of vision has been almost entirely neglected by historians of optics. The principal exceptions are Eilhard Wiedemann, "Ibn Sina's Anschauung vom Sehvorgang," pp. 239-41; and Vescovini, *Studi sulla prospettiva*, chap. 5; however, neither makes full use of the available sources. On nonvisual aspects of Avicenna's optics, see A. Süheyl Ünver, "Avicenna Explains Why Stars Are Visible at Night and Not during the Day," *Journal of the History of Medicine and Allied Sciences* 1 (1946): 330-34; M. Horten, "Avicennas Lehre vom Regenbogen, nach seinem Werk al Schifa," *Meteorologische Zeitschrift* 30 (1913): 533-44; and Carl B. Boyer, *The Rainbow*, pp. 77-79.

71."Die Psychologie des Ibn Sina," trans. S. Landauer, pp. 336-39.

72. Arberry, "Avicenna," pp. 20-22; Afnan, *Avicenna*, pp. 65-68.

73. Both the Latin translation of Avendauth and Gundissalinus and the French translation of Bakoš are obscure at this point, but Avicenna's intent is clear.

74. I quote here from the medieval Latin translation of *De anima* (pt. 4, chap. 6 of the *Shifa*), ed. Van Riet, p. 214. Cf. Bakoš's French translation, p. 82. I would like to express my gratitude to Mlle. Van Riet for permitting me to see her text while it was still in typescript.

75. One example will suffice to illustrate the kind of schema devised by Avicenna. In the *Najat*, he argues that what emanates from the eye is either (A) material or (B) immaterial. If it is material, then either (A1) it remains a coherent body, in contact with the eye and reaching to the fixed stars, or (A2) it is dispersed into discrete rays, or (A3) it is united with the air and the heavens to form a coherent and sentient instrument of vision. If it is immaterial, it must be a quality that transforms the air and renders it either (B1) a medium of transmission or (B2) sentient in itself. In the discussion below, I have largely followed the arrangement employed by Avicenna in the *Shifa*.

76. Avicenna does not refer to either Euclid or Galen by name, but it is clearly to their theories (broadly defined) that he directs his refutation. For example, theories A1 and A2 (see the preceding note) are fundamentally Euclidean, while A3, B1, and B2 are Galenic. The

extramission theory of vision would have been available to Avicenna not only in the works of Euclid and Galen, but also in those of al-Kindi and Hunain and perhaps of Ptolemy and Tideus.

77. The following account goes into considerable detail. Avicenna's discussion of vision merits such close attention because of its profound influence on those who shaped visual theory in the West; see this volume, pp. 101, 105-6, 112, 114, 116, 117, 273-74 (n. 141).

78. F. Rahman (trans.), *Avicenna's Psychology*, p. 27.

79. *De anima*, ed. Van Riet, p. 214; cf. Bakoš, p. 82. Avicenna also notes that the radiating substance assumes a conical shape and that vision occurs best through its central axis (Van Riet, pp. 212-13; Bakoš, p. 81); he thus reveals that he has in mind not simply the theory of Euclid in its original form, but the later elaborations of Ptolemy, Damianos, and al-Kindi.

80. At one point Avicenna maintains, of this fourth alternative, that the substance also loses contact with the eye, but later he appears to admit that in fact contact is maintained (*De anima*, ed. Van Riet, pp. 226, 234; Bakoš, pp. 87, 91). Avicenna's refutation will prove to be equally effective in either case.

81. Rahman, *Avicenna's Psychology*, p. 28. Cf. *De anima*, ed. Van Riet, pp. 226-27 (Bakoš, p. 87); Avicenna, *Le livre de science* [the *Danishnama*], trans. Mohammad Achena and Henri Massé, 2: 58. This objection of Avicenna to the extramission theory had been anticipated by Aristotle and Alexander of Aphrodisias; on the former, see *De sensu* 2. 438a26-27; on the latter, see Saint-Pierre, "La physique de la vision," p. 215.

82. Rahman, *Avicenna's Psychology*, p. 28.

83. *De anima*, ed. Van Riet, pp. 226-27; Bakoš, p. 87-88. *Livre de science*, 2: 59, 61. See also this volume, pp. 49-50.

84. There is no evidence that in fact Euclid, al-Kindi, or any other Euclidean ever inquired where (at the base of the cone or at its apex) perception occurs.

85. See this volume, pp. 13, 23. The position of the mathematicians was that one judges the size of an object by the angle between the visual rays that proceed from the eye to the extremities of the object. The apex of the visual cone thus serves as a center of perspective. Though not of the mathematical school, Galen agreed that only on an extramission theory can magnitude be discerned; see *De placitis* 7. 5, *Opera omnia*, ed. Kühn, 5: 618. A similar view was expressed by Tideus and Hunain; see Tideus, *De speculis*, in Björnbo and Vogl, "Drei optische Werke," p. 75; Hunain, *Ten Treatises*, trans. Meyerhof, p. 32. Even Alhazen, who rejected the extramission theory, admitted grudgingly that it is useful for a mathematical understanding of vision; see this volume, p. 66. And finally, Averroes acknowledged that the mathematical achievements of the Euclideans were evidence for the extramission theory of vision; see his *Epitome of Parva naturalia*, trans. Harry Blumberg, p. 15.

86. *De anima*, ed. Van Riet, pp. 227-28, 234; Bakoš, pp. 88, 91.

87. *De anima*, ed. Van Riet, p. 234; cf. Bakoš, p. 91.

88. Rahman, *Avicenna's Psychology*, p. 28. See also *De anima*, ed. Van Riet, p. 228 (Bakoš, p. 88); *Livre de science*, 2: 59. Cf. al-Kindi, above, p. 25.

89. *De anima*, ed. Van Riet, p. 228; cf. Bakoš, p. 88. With this supposition, the third version of the Euclidean theory becomes very like the Galenic theory of vision.

90. *De anima*, ed. Van Riet, p. 228; cf. Bakoš, p. 88

91. *De anima*, ed. Van Riet, p. 229; Bakoš, pp. 88-89.

92. The more appropriate question would seem to be: How is it possible to see the earth at all in the absence of void, or to see it continuously even if vacuous pores should exist?

93. *De anima*, ed. Van Riet, p. 230; cf. Bakoš, p. 89.

94. When he first describes the Galenic theory in the *Shifa*, Avicenna makes a statement that could be construed to mean that the ray itself meets the visible object (along with the air to which it is conjoined): "Another opinion is that of him who holds that a ray issues from the pupil; but out of the pupil comes that which touches a hemisphere of the heaven only by a dispersion [of the ray], from which comes the enlarging of vision [in the form of a cone]." However, not only is this

remark highly ambiguous, but it is followed immediately by another that seems to assert the traditional Galenic view that the ray is not itself conveyed to the visible object: "But when it [the ray] issues forth and is conjoined to illuminated air, the air becomes its instrument, and it apprehends through the air" (*De anima*, ed. Van Riet, p. 213; cf. Bakoš, p. 81). Moreover, a little later in the argument, Avicenna points out that the Galenists attack the Euclideans on the grounds that if a body actually traversed the space between the eye and the fixed stars, a perceptible time would be required; the obvious implication is that since no time is perceived, the substance issuing from the eye cannot be conveyed to the fixed stars (see *De anima*, ed. Van Riet, pp. 215-16; Bakoš, p. 82).

When Avicenna gets down to a systematic refutation of the Galenic theory, he describes two principal versions of it. The visual ray, which transforms the transparent medium into an instrument of vision, can be either a material substance or an immaterial quality. However, he refutes both views with the same argument, and I see no point in making any more of the distinction than he made. What is essential to the Galenic theory is that the air is transformed into an instrument of vision, and it is to this point alone that Avicenna directs his argument (see *De anima*, ed. Van Riet, pp. 219-20; Bakoš, p. 84; cf. Rahman, *Avicenna's Psychology*, pp. 28-29).

95. To be precise, Avicenna mentions yet a third possibility: the air can be converted into a visual organ capable of returning impressions to the eye (rather than percipient in itself). However, the distinction between this view and the claim that the air may become a *medium* capable of transmitting visual impressions to the eye is a very fine one indeed, and Avicenna makes little of it in *De anima* and overlooks it altogether in the *Najat*. In either case, the air returns visual impressions to the eye, and it is not relevant to a refutation of the theory whether one calls the transformed air a transmitting medium or a visual organ. See *De anima*, ed. Van Riet, p. 222; Bakoš, p. 85; cf. Rahman, *Avicenna's Psychology*, p. 29.

96. In this case the visual power.

97. *De anima*, ed. Van Riet, p. 222; cf. Bakoš, p. 85. See also Rahman, *Avicenna's Psychology*, pp. 28-29; *Livre de science*, 2: 58.

98. This recalls a similar distinction made by Tideus, *De speculis*, in Bjørnbo and Vogl, "Drei optische Werke," pp. 74-75; cf. Galen on the walking-stick analogy, above, chap. 1.

99. For the similar argument of Alexander of Aphrodisias, see Saint-Pierre, "La physique de la vision," p. 216.

100. Rahman, *Avicenna's Psychology*, p. 29. On the latter argument, see above, p. 45.

101. *De anima*, ed. Van Riet, pp. 222-23; cf. Bakoš, pp. 85-86.

102. In his own theory of vision, Avicenna can argue that light is a form or quality, rather than a body, and thus escape this difficulty.

103. *De anima*, ed. Van Riet, p. 223; Bakoš, p. 86.

104. *De anima*, ed. Van Riet, p. 224; Bakoš, p. 86.

105. Cf. Aristotle *De sensu* 2. 437^{b}17-24.

106. *De anima*, ed. Van Riet, p. 225; cf. Bakoš, p. 87.

107. I have translated this from the French version of Achena and Massé, *Livre de science*, 2: 59.

108. *De anima*, ed. Van Riet, pp. 213-14; cf. Bakoš, p. 81.

109. Rahman, *Avicenna's Psychology*, p. 21, with altered paragraphing. See also *Livre de science*, 2: 59-61; and Avicenna, *A Compendium on the Soul, by Abu-'Aly al-Husayn Ibn 'Abdallah Ibn Sina*, trans. Edward A. Van Dyck, pp. 51-52. In the latter two works, Avicenna specifically identifies Aristotle as the object of his loyalty.

110. Aristotle *De anima* 2. 7. 419^{a}13-19, trans. W. S. Hett, p. 107.

111. *Livre de science*, 2: 60; cf. *Compendium on the Soul*, p. 51. In *De sensu* Aristotle wrote that "Democritus is right when he says that the eye is water, but wrong when he supposes vision to be mere mirroring" (2. 438^{a}5-7, trans. Hett, p. 223). Of course, Aristotle only denies in this

passage that vision is *mere* mirroring, but the thrust of his position is to dismiss the Democritean theory that associated vision with the formation of a mirror image in the eye. On the theory of mirroring, see also this volume, pp. 3, 55, 163-64.

112. *Livre de science*, 2: 60; *De anima*, ed. Van Riet, p. 227; Bakoš, p. 87. The location of the visual power in the hollow nerve is specified in the *Najat*; see Rahman, *Avicenna's Psychology*, p. 26.

113. *Livre de science*, 2: 61.

114. Ibid.

115. See above, p. 45. The position adopted here by Avicenna, namely, the association of the visual pyramid and the intromission theory of vision, would be greatly extended by Avicenna's contemporary, Alhazen; see chap. 4.

116. The *Liber canonis* was translated into Latin in the twelfth century by Gerard of Cremona and again in the sixteenth century by Andreas Alpago. It has appeared in some thirty Latin editions. For a German translation of the optical portions (from the Arabic text), see *Die Augenheilkunde des Ibn Sina*, trans. J. Hirschberg and J. Lippert.

117. *Liber canonis* (Venice, 1507), bk. 3, fen. 3, chap. 1, fol. 203va; cf. *Die Augenheilkunde*, trans. Hirschberg and Lippert, pp. 11-12. On the function of the flattening of the glacial humor according to Galen and Hunain, see above, p. 34. For further discussion of Avicenna's ideas on the shape of the glacial humor, as well as the interpretation of his ideas in the sixteenth century, see chap. 8, n. 141, below.

118. Contrary to the view of Crombie, "Mechanistic Hypothesis," p. 16.

119. *Ten Treatises*, trans. Meyerhof, p. 4; Pansier, *Collectio ophtalmologica*, fasc. 3, p. 204 (cf. Wood, *Memorandum Book*, pp. 14-15).

120. See chap. 1, above.

121. See above, pp. 15-16, 24-26.

122. If somebody should argue that the ray is not a material substance, Avicenna would reply that a medium is then required, which brings us to the equally impossible Galenic theory.

123. The classic work on Averroes is Ernest Renan, *Averroès et l'averroïsme*, 3d ed. (Paris, 1869). See Gauthier, *Ibn Rochd*; R. Arnaldez, "Ibn Rushd," *The Encyclopaedia of Islam*, new ed., 3: 909-20; and Fakhry, *History of Islamic Philosophy*, pp. 302-25.

124. Averroes's commentaries are listed by Harry A. Wolfson, "Plan for the Publication of a *Corpus Commentariorum Averrois in Aristotelem* submitted to the Mediaeval Academy of America," *Speculum* 6 (1931): 415-16.

125. By concentrating on the *Epitome of the Parva naturalia* and the *Colliget* one does not ignore any important innovation. In addition to these two works, I have used Averroes's *Commentarium magnum in Aristotelis de anima libros*, ed. F. Stuart Crawford. On Averroes's theory of vision, see Gauthier, *Ibn Rochd*, chap. 7; Bruce S. Eastwood, "Averroes' View of the Retina—A Reappraisal," pp. 77-82; and David C. Lindberg, "Did Averroes Discover Retinal Sensitivity?"

126. Averroes, *Epitome of Parva naturalia*, trans. Blumberg, pp. 6, 18. "Behind the retina" is probably intended to denote the anterior part of the brain; see Lindberg, "Did Averroes Discover Retinal Sensitivity?" n. 23. On the location of the common sense, see also Aristotle *De juventute et senectute* 3. 469a10-19. For discussions of Aristotle's doctrine of the common sense, see Irving Block, "The Order of Aristotle's Psychological Writings," *American Journal of Philology* 82 (1961): 62-75; and Charles H. Kahn, "Sensation and Consciousness in Aristotle's Psychology," pp. 50-70. The classic study of the medieval theory of the common sense is Harry Austryn Wolfson, "The Internal Senses in Latin, Arabic, and Hebrew Philosophic Texts."

127. Trans. Blumberg, p. 18.

128. Ibid., p. 15.

129. Ibid.

130. Ibid., p. 16.

131. Ibid., p. 17. For Averroes's account of arguments put forward in defense of the extramission theory, see ibid., p. 15.

132. Cf. Aristotle *De sensu* 2. 437b18-22.

133. *Epitome of Parva naturalia*, trans. Blumberg, pp. 17-18. Cf. Avicenna on the need for a material bearer of the visual power, above, p. 44.

134. Vincenz Fukala, "Historischer Beitrag zur Augenheilkunde," pp. 203-6; Huldrych M. Koelbing, "Averroes' Concepts of Ocular Function—Another View," pp. 207-13; see also Polyak, *Retina*, pp. 134-35. Against Fukala and Koelbing, see Crombie, "Mechanistic Hypothesis," p. 49; Eastwood, "Averroes' View of the Retina"; and Lindberg, "Did Averroes Discover Retinal Sensitivity?" The remainder of this section reproduces portions of the latter article.

135. Trans. Blumberg, p. 9. It is this passage that Koelbing considers crucial.

136. I render this from the Latin editions of A. L. Shields and H. Blumberg, *Compendia librorum Aristoteleis qui Parva naturalia vocantur*, p. 38. Blumberg's English translation (p. 19) conveys the same view, though it appears somewhat defective at this point.

137. *Colliget* (Venice, 1562), bk. 3, chap. 38, fol. 53ra. Where I have translated "anterior" the text reads "inferior," but emendation is clearly required.

138. Trans. Blumberg, pp. 18-19.

139. *Colliget*, bk. 3, chap. 38, fol. 53 ra. On the comparison of vision to image-formation in a mirror, see this volume, pp. 3, 49-50.

140. *Colliget*, bk. 2, chap. 15, fol. 25r-v.

141. See this volume, p. 36; Galen, *On Anatomical Procedures*, trans. Duckworth, pp. 40-41; Hunain, *Ten Treatises*, trans. Meyerhof, p. 10, n. 3. For a later medieval view, see Lindberg, *Pecham*, pp. 115-17; as I failed to understand when preparing my notes to the latter passage, Pecham uses the term *aranea* in two senses—to designate the anterior capsule of the crystalline humor (the arachnoid membrane) and also to designate the combined anterior capsule and retina.

142. So far as I have been able to discover, Averroes was the first to suggest that the principal organ of sight might be the anterior capsule of the crystalline humor, the *aranea* or arachnoid membrane, rather than the crystalline humor itself. He defended this possibility on the grounds that the *aranea* is of the highest transparency and lucidity, and as Galen had indicated, these are the properties that must be possessed by the sensitive organ; see *Colliget*, bk. 2, chap. 15, fol. 25vb; and this volume, chap. 3, n. 7. As a result of Averroes's influence, it was much discussed in the sixteenth century whether sensitivity resides in the crystalline humor or the *aranea*; see this volume, p. 171.

143. I.e., without contact through arteries and veins; cf. Galen, *Usefulness of Parts*, trans. May, pp. 464-65.

144. *Colliget*, bk. 2, chap. 15, fol. 25va-vb.

145. *Usefulness of Parts*, trans. May, pp. 463, 465.

146. Cf. Hunain, above, chap. 3.

147. *Usefulness of Parts*, trans. May, pp. 463-64.

148. Vescovini, *Studi sulla prospettiva*, p. 104, claims credit for 'Ali ibn 'Isa as "one of the first Arabic scientists to have understood the importance of the nervous sensibility of the retina," but in fact 'Ali (like Averroes) was being strictly traditional.

149. Averroes states that the retina receives the light "from the humors of the eye, just as the humors receive the light from the air" (above, chap. 3, n. 135).

150. Rectilinear propagation of light is mentioned in the Pseudo-Aristotelian *Problemata* 15. 6, 911b5-20, and it is implicit in Aristotle's *Meteorologica*, especially 3. 5. 375b16 ff. Galen touches briefly upon rectilinear propagation in *Usefulness of Parts*, trans. May, pp. 492-93; and *De placitis* 7. 5, *Opera omnia*, ed. Kühn, 5: 626-27. Cf. Hunain, *Ten Treatises*, trans. Meyerhof, p. 25; Avicenna, *De anima*, ed. Van Riet, p. 212; Averroes, *Epitome of Parva naturalia*, trans. Blumberg, p. 15.

Chapter 4 Alhazen and the New Intromission Theory of Vision

1. On atomistic theories of vision, see above, chap. 1.

2. Aristotle's theory of vision is discussed in chapter 1.

3. *De anima* 2. 5. 418a3-6, trans. W. S. Hett, p. 101. Cf. Harold Cherniss, *Aristotle's Criticism of Presocratic Philosophy*, p. 320.

4. See above, p. 9.

5. The quoted phrase is Cherniss's, *Aristotle's Criticism of Presocratic Philosophy*, p. 320. The theory of light as a relation, presented by Alexander of Aphrodisias, may partially escape this difficulty.

6. *Ten Treatises*, trans. Max Meyerhof, p. 32. On Hunain's theory of vision, see above, chap. 3.

7. See above, pp. 23-24.

8. I have argued above, p. 30, that al-Kindi was the first to analyze a luminous body into point sources; this did not, however, have any significant effect on his theory of vision. In antiquity, Alexander of Aphrodisias had forcefully rejected any attempt to consider Epicurean forms or images in an incoherent or punctiform manner within the context of visual theory; see Saint-Pierre, "La physique de la vision," pp. 224-28, 290-91. If Alexander was replying to a position actually assumed by certain Epicureans, then it must be admitted that Alhazen's starting point had been anticipated in antiquity; in any case, no ancient philosopher successfully developed the idea.

9. See this volume, pp. 57, 85.

10. The identity of Alhazen and Ibn al-Haytham was first noted by Giambernardo De Rossi, *Dizionario storico degli autori arabi più celebri e delle principali loro opere* (Parma, 1807), pp. 98-99; and by A. L. M. M. B. Jourdain, "Alhazen," in *Biographie universelle ancienne et moderne*, vol. 1 (Paris, 1811), pp. 568-69. It was convincingly demonstrated by G. G. A. Caussin de Perceval, "Mémoire sur l'Optique de Ptolémée," pp. 20-23. Cf. Enrico Narducci, "Intorno ad una traduzione italiana, fatta nel secolo decimoquarto, del trattato d'Ottica d'Alhazen," pp. 31-40.

11. Ibn Abi Usaibi'a's sketch is available in a German translation by Eilhard Wiedemann, "Ibn al Haitam, ein arabischer Gelehrter," pp. 147-78. On Alhazen's life, see also A. I. Sabra, "Ibn al-Haytham, Abu 'Ali al-Hasan ibn al-Hasan," *Dictionary of Scientific Biography*, 6: 189-210; Wiedemann, "Ueber das Leben von Ibn al Haitam und al Kindi," *Jahrbuch für Photographie und Reproduktionstechnik* 25 (1911): 7-11; and Lufti M. Sa'di, "Ibn-al-Haitham (Alhazen), Medieval Scientist," *University of Michigan Medical Bulletin* 22 (1956): 249-57.

12. The most useful bibliographies of Alhazen's works are Giorgio Nebbia, "Ibn al-Haytham nel millesimo della nascita," pp. 174-201; and Wiedemann, "Ibn al Haitam," pp. 161-75. The extant works, along with manuscripts, editions, and translations, are conveniently listed by Sabra, "Ibn al-Haytham," pp. 204-9. I will avoid duplicating information supplied by Sabra, and the reader should consult his article for additional bibliography concerning optical works in the list shortly to be given.

13. Both the Arabic and the Latin texts of this work are extant, and there is also a fourteenth-century Italian translation by Guerruccio di Cione Federighi (Vatican City, Bibl. Apostolica, MS Vat. Lat. 4595). An English translation of the Arabic text is in preparation by A. I. Sabra. On the medieval Latin translation, see this volume, pp. 209-10. The Latin text was published by Friedrich Risner along with the *Perspectiva* of Witelo as *Opticae thesaurus Alhazeni Arabis libri septem. . . . Item Vitellonis Thuringopoloni libri X* (Basel, 1572); for an analysis of this edition, as well as a list of the medieval Latin manuscripts, see the facsimile reprint with an introduction by David C. Lindberg (New York, 1972), pp. xxvi-xxxiv; see also Lindberg, *Catalogue of Optical Manuscripts*, pp. 17-19. On the Italian translation, see Narducci, "Intorno ad una traduzione italiana," pp. 1-4; Graziella Federici Vescovini, "Contributo per la storia della fortuna di Alhazen in Italia," pp. 17-49.

14. In addition to the translations listed by Sabra, a portion of this work has been translated into English from the medieval Latin text in Edward Grant, *Source Book in Medieval Science*, pp. 413-17.

15. There is an English translation (from Baarmann's German translation of the Arabic text) not listed by Sabra in Thomas H. Shastid, "History of Ophthalmology," pp. 8701-17. There is also an inferior English translation of this and other optical works of Alhazen in *Ibn al-Haitham: Proceedings of the Celebration of* [*the*] *1000th Anniversary* [*of His Birth*], ed. Hakim Mohammed Said.

16. See A. I. Sabra, "The Authorship of the *Liber de crepusculis*, an Eleventh-Century Work on Atmospheric Refraction," pp. 77-85.

17. See the bibliographies of Nebbia and Wiedemann.

18. Nebbia, "Ibn al-Haytham," pp. 179-80, treats these three chapters as separate works; Sabra, "Ibn al-Haytham," p. 204, points out that they are but chapters of the work on the art of medicine.

19. On the dating of Alhazen's optical works, see Matthias Schramm, *Ibn al-Haythams Weg zur Physik*, pp. 274-85; Sabra, "Ibn al-Haytham," pp. 204-9. On the chronology of Avicenna's works, see A. J. Arberry, "Avicenna: His Life and Times," in *Avicenna: Scientist and Philosopher*, ed. G. M. Wickens (London, 1952), pp. 21-25.

20. See above, chap. 3.

21. See this volume, pp. 109-12, 114-16.

22. The Latin translation omits the first three chapters of the Arabic text; see Eilhard Wiedemann, "Zu Ibn al Haitams Optik," p. 4. The most useful studies of Alhazen's optical thought are Leopold Schnaase, *Die Optik Alhazens*; Hans Bauer, *Die Psychologie Alhazens*; Vasco Ronchi, *Storia della luce*, pp. 35-47 (translated into French by J. Taton as *Histoire de la lumière*, pp. 33-45, and into English by V. Barocas as *The Nature of Light*, pp. 45-57); H. J. J. Winter, "The Optical Researches of Ibn al-Haitham," pp. 190-210; Matthias Schramm, "Zur Entwicklung der physiologischen Optik in der arabischen Literatur," pp. 291-99; Schramm, *Weg zur Physik*; Vescovini, *Studi sulla prospettiva*, chap. 7; Crombie, "Mechanistic Hypothesis," pp. 17-22; David C. Lindberg, "Alhazen's Theory of Vision and Its Reception in the West," pp. 322-35; Lindberg, "Introduction" to fascimile reprint of *Opticae thesaurus*, pp. xiv-xix; Roshdi Rashed, "Optique géométrique et doctrine optique chez Ibn Al Haytham," pp. 271-98; Sabra, "Ibn al-Haytham," pp. 190-97; Sabra, "The Physical and the Mathematical in Ibn al-Haytham's Theory of Light and Vision"; M. Nazif, "Ibn al-Haitham's Work on Optics," pp. 285-93.

23. *De aspectibus*, ed. Risner in *Opticae thesaurus* (henceforth cited simply as *De aspectibus*), bk. 1, chap. 1, sec. 1, p. 1. I have compared Risner's text with early Latin manuscripts, and unless otherwise stated have found no differences in substance. The subdivisions of the text that I have termed "sections" have frequently been regarded by historians as propositions or theorems. In fact, they do not exist in either the original Arabic text or the medieval Latin translation, where the only division of the text is into books and chapters; moreover, they are not propositional in form, but simply represent the additional subdivision of the text added by the sixteenth-century editor, Friedrich Risner.

24. *De aspectibus*, bk. 1, chap. 1, sec. 1, p. 1. The afterimage had been discussed earlier by Ptolemy and Alexander of Aphrodisias; see *L'Optique de Claude Ptolémée*, ed. Albert Lejeune, p. 66; Saint-Pierre, "La physique de la vision," pp. 56-59, 240, 271-72.

25. *De aspectibus*, bk. 1, chap. 1, sec. 1, p. 1.

26. Ibid. Alhazen makes it clear that color produces a similar effect in the eye.

27. Cf. Ronchi, *Nature of Light*, p. 46.

28. See above, p. 40.

29. More exactly, he has demonstrated that light and color affect the visual power. However this, in Alhazen's view, was tantamount to an effect on its seat, the eye.

30. Bk. 1, chap. 5, sec. 14, p. 7.

31. Although the first three chapters of the Arabic text were not translated into Latin, their

content is ascertainable from Kamal al-Din al-Farisi's commentary; see Wiedemann, "Zu Ibn al Haitams Optik," pp. 25-26. In addition, Prof. A. I. Sabra has made available to me the following excerpt from his English translation of the Arabic text of chap. 3 (Istanbul, MS Fatih 3212, fols. 25v-26r): "All this being so, the light shining from the self-luminous body into the transparent air therefore radiates from every part of the luminous body facing that air; and the light in the illuminated air is continuous and coherent; and it issues from every point on the luminous body in every straight line that can be imagined to extend in the air from that point." A similar claim is made in the Latin text, bk. 1, chap. 5, sec. 19, p. 10. On the original enunciation of this point by al-Kindi, see above, p. 30.

32. *De aspectibus*, bk. 1, chap. 5, sec. 14, p. 7.

33. Ibid. I have emended Risner's text slightly by reference to the manuscripts.

34. On the Galenic theory, see above, pp. 10-11, 38-41.

35. *De aspectibus*, bk. 1, chap. 5, sec. 23, p. 14. I have translated the greater part of sec. 23 in Grant, *Source Book*, pp. 403-5.

36. *De aspectibus*, bk. 1, chap. 5, sec. 23, p. 14.

37. See just above.

38. *De aspectibus*, bk. 1, chap. 5, sec. 23, p. 14.

39. Ibid.

40. See above, p. 49. This is one of several similarities between Alhazen's refutation of the extramission theory and that of Avicenna; see just below.

41. *De aspectibus*, bk. 1, chap. 5, sec. 23, p. 14.

42. It is implicit to Alhazen's argument that the substance issuing from the eye is continuous. Cf. Avicenna's similar argument, above, p. 45.

43. Cf. Avicenna, above, p. 44.

44. See chap. 6, pp. 114-16, 118.

45. *De aspectibus*, bk. 1, chap. 5, sec. 23, p. 14. The term "something (*aliquid*)" was omitted by Risner but is found in all the manuscripts I have examined. After the final word of the quotation, British Museum, MS Royal 12.G.VII, fol. 7v, adds "vel a visu." Alhazen's position in this argument is similar to al-Kindi's; see above, p. 26.

46. *De aspectibus*, bk. 1, chap. 5, sec. 23, p. 14.

47. *De aspectibus*, bk. 1, chap. 5, sec. 23, pp. 14-15.

48. *Storia della luce*, p. 42. Despite strong and repeated criticism, Ronchi repeats the same claim in the augmented English translation of this work, *Nature of Light*, p. 52.

49. For a fuller discussion of Risner's edition, see my "Introduction" to the facsimile reprint, pp. xxviii-xxxii.

50. *De aspectibus*, p. 15: "Visio videtur fieri per συναύγειαν, id est receptos simul et emissos radios."

51. For example, Aetius uses the term to describe Plato's theory of sight, 4. 13. 11 (Hermann Diels, *Doxographi Graeci*, 3d ed., p. 404). Cf. Charles Mugler, *Dictionnaire historique de la terminologie optique des grecs*, p. 379. Apparently Plato did not himself employ this term; see Arthur Erich Haas, "Antike Lichttheorien," p. 393, n. 100.

52. *De aspectibus*, bk. 1, chap. 5, sec. 23, p. 15.

53. Ibid. The specific arrangement that Alhazen appears to have in mind is that of forms on the surface of the glacial humor, by means of the visual pyramid; see this volume, pp. 73-74.

54. Ibid., sec. 23, p. 15.

55. On the status of visual rays or radial lines, cf. al-Kindi, above, pp. 24-26.

56. *De aspectibus*, bk. 1, chap. 4, sec. 13, p. 7: "All that we have said concerning the tunics of the eye and their composition has already been stated by the anatomists in the books of anatomy."

57. Alhazen's anatomical scheme is presented most fully in bk. 1, chap. 4, sec. 4, pp. 3-4; English translations of this section from both the Arabic and Latin texts are found in Stephen Polyak, *The Retina*, pp. 109-11. See also *De aspectibus*, bk. 1, chap. 4, secs. 5-13, pp. 4-7; bk.

1, chap. 6, secs. 33-35, pp. 20-22. Figures 7-9, from Arabic and Latin manuscripts of *De aspectibus*, allegedly represent Alhazen's conception of the anatomy of the visual system; actually, none of them is an accurate portrayal. A better (but still imperfect) idea of what Alhazen had in mind can be gained from the figure reconstructed by M. Nazif; see Nazif, "Ibn al-Haitham's Work on Optics," p. 287; same figure reproduced by Sabra, "Ibn al-Haytham," p. 192.

58. This is not true, of course, for the optic nerve actually enters the eye slightly to the nasal side of the central axis. However, it was the universally held belief of anatomists until the seventeenth century.

59. The term "glacial humor" used without further qualification ordinarily refers to the anterior glacial humor or crystalline lens, and I will so employ the expression. I will generally avoid using the term "lens" because it may connote the focusing properties of lenses, which have no place in Alhazen's theory of vision.

60. Alhazen departs from Hunain in supposing that the arachnoid membrane surrounds both the crystalline and vitreous humors; see above, pp. 36, 55.

61. See Schramm, "Entwicklung der physiologischen Optik," p. 292.

62. See *De aspectibus*, bk. 1, chap. 4, secs. 6-12, pp. 4-6. It is true, of course, that Gaien and others had already described the eye in quasi-geometrical terms, but Alhazen went considerably further than any predecessor; cf. Galen, *Usefulness of Parts*, trans. May, pp. 467-71.

63. Since the uvea is perforated by the pupil, its eccentricity is of no matter; indeed, its perforation is the result of its eccentricity. On the eccentricity of the posterior surface of the glacial humor, see pp. 81, 85.

64. This point has been made by Schramm, "Entwicklung der physiologischen Optik," pp. 295-96. Cf. Crombie, "Mechanistic Hypothesis," p. 22.

65. *De aspectibus*, bk. 1, chap. 5, sec. 16, p. 8. The English translation is reprinted from Grant, *Source Book*, p. 399. On the fate of this argument in the West, see David C. Lindberg, "Lines of Influence in Thirteenth-Century Optics, p. 80.

66. *Ten Treatises*, trans. Meyerhof, p. 4; cf. Galen, *Usefulness of Parts*, trans. May, pp. 463-64; *De placitis* 7. 6, *Opera omnia*, ed. Kühn, 5: 635.

67. *De aspectibus*, bk. 1, chap. 4, sec. 4, p. 4. The emphasis is, of course, mine. Alhazen expresses this same view more confidently in bk. 1, chap. 6, sec. 33, p. 21, where he writes: "The optic nerve . . . is hollow, and visual spirit from the brain runs through it and reaches the glacial humor and gives it the power of sensibility."

68. See above, chap. 1.

69. *De aspectibus*, bk. 1, chap. 5, sec. 25, p. 15.

70. Ibid., bk. 2, chap. 2, sec. 18, p. 35. The same conclusion can be gathered from the claim made during a discussion of the completion of vision in the common nerve, that the form of light illuminates and the form of color colors the visual spirits in the optic nerve and that by this means the *ultimum sentiens* perceives the light and color of the visible object; ibid., sec. 16, p. 35.

71. Ibid., bk. 1, chap. 5, sec. 25, p. 15. English translation quoted from Grant, *Source Book*, p. 401. The same description holds true, of course, for the form of color.

72. In the passage quoted just above, the term "suffering" is ambiguous and might be taken as a reference simply to "reception" or "passivity." However, Alhazen makes himself quite clear later in the passage, where he writes: "And the action of all light on the eye is of the same kind and is diversified only according to more and less. Since all are of one kind, and the operation of strong lights is a kind of pain (*dolor*), all actions of light are a kind of pain" (*De aspectibus*, bk. 1, chap. 5, sec. 26, pp. 15-16; full passage translated in Grant, *Source Book*, p. 401).

73. *De aspectibus*, bk. 1, chap. 5, sec. 19, p. 10. Notice that in this passage Alhazen applies the principle of punctiform analysis to nonluminous bodies. In other passages, Alhazen often speaks of a single form for an entire body (or so the Latin translation has it), but this in no way compromises his position; if it is sometimes easier to speak of form in the singular, it remains in reality that all points of a body radiate independently.

74. Ibid., sec. 14, pp. 7-8.

75. See this volume, pp. 58-59.

76. Ptolemy's analysis of refraction has been examined in detail by Albert Lejeune, *Recherches sur la catoptrique grecque*, pp. 152-75.

77. I.e., the perpendicular to the interface between the two media at the point of refraction.

78. Alhazen briefly summarizes the qualitative features of refraction in *De aspectibus*, bk. 1, chap. 5, sec. 17, p. 9. He treats refraction more fully in bk. 7.

79. *De aspectibus*, bk. 1, chap. 5, sec. 18, p. 9.

80. Ibid., sec. 19, pp. 10-12. Because the cornea and the anterior surface of the glacial humor are concentric, a ray that is perpendicular to one of them is perpendicular to the other, and there is no refraction of the rays making up the pyramid until they encounter the posterior surface of the glacial humor.

81. This might seem to be an instance of the modern conception of a real optical image and thus the first application of that conception to image-formation in the eye. However, Alhazen does not say that a picture is formed, and such a conception cannot justifiably be read into his work. Sabra writes that "the 'distinct form' he succeeded in realizing inside the eye is apparent only to the sensitive faculty; it is not a visibly articulate image such as that produced by a pinhole camera" ("Ibn al-Haytham," p. 193). Cf. Kepler, chap. 9 of this volume.

82. *De aspectibus*, bk. 1, chap. 5, sec. 18, p. 10.

83. Ibid.

84. Ibid., bk. 7, chap. 2, sec. 8, p. 241. See my discussion of Alhazen's theory of refraction, "The Cause of Refraction in Medieval Optics," pp. 25-29, from which the passage here quoted (with two minor changes) is reprinted.

85. Although in the Latin version of *De aspectibus* the efficacy of perpendicular rays is explained principally in terms of their greater strength, the Arabic version reveals another explanation (which is also obscurely hinted at in the Latin translation), namely, that (in Sabra's phrase) "there exist certain privileged directions in the lens *as a sensitive body*" ("The Physical and the Mathematical"; cf. Sabra, "Ibn al-Haytham," p. 193). The point is made most clearly in bk. 2, chap. 1, sec. 4, p. 26 (of Risner's Latin edition), where it is stated that sentient bodies, such as the glacial humor, do not receive forms in the same manner as do nonsentient transparent bodies and consequently that the path of a form in the glacial humor should be different from that in an ordinary transparent body—and where (after "exiguunt" in line 6) the Arabic text adds that forms proceed in the glacial humor according to the extension of the parts of the sensitive body. For a full elucidation of this point it is clear that one must have the Arabic text at his disposal and therefore that we must await the appearance of Sabra's edition and translation; however, I wish to express gratitude to Professor Sabra for discussing the matter with me and for devoting part of an afternoon to comparing the Arabic and Latin texts at this point.

86. Bk. 1, chap. 2, sec. 2, p. 2.

87. *De aspectibus*, bk. 7, chap. 6, sec. 37, pp. 268-69. Alhazen does not explain why he permitted the inconsistency to remain. Perhaps he would have been prepared to maintain that the theory of book 1 simply represents a simplified version and that book 7 introduces the full complexity; but the fact remains that book 1 gives no hint of revisions to come. As an experimental demonstration of his new position on vision by refracted rays, Alhazen points out that a small object placed before one eye (with the other eye closed) does not obstruct vision of things behind it; see *De aspectibus*, bk. 7, chap. 6, sec. 37, p. 269; cf. Sabra, "Ibn al-Haytham," pp. 193-94.

88. *De aspectibus*, bk. 7, chap. 6, sec. 37, p. 269.

89. Ibid., chap. 5, secs. 18-19, pp. 253-56.

90. On its history, see Colin M. Turbayne, "Grosseteste and an Ancient Optical Principle," pp. 467-72.

91. Alhazen is quite explicit on the matter. After the passage quoted above (n. 88), he continues: "Formae ergo omnium visibilium quae opponuntur parti superficiei visus que

opponitur foramini uveae et existunt in hac parte superficiei visus refringuntur in diaphanitate tunicarum visus, et perveniunt ad membrum sensibile, quod est humor glacialis, et comprehenduntur a virtute sensibili per lineas rectas que continuant centrum visus cum ipsis visibilibus, scilicet quod forma cuiuslibet puncti cuiuslibet visi oppositi superficiei visus quae opponitur foramini uveae existit in universo superficiei huius partis, et refringitur a tota superficie, et pervenit ad humorem glacialem. Et tunc ille humor sentit formam ad se venientem, et virtus sensibilis comprehendit omnia quae perveniunt ad glacialem ex forma visus puncti super unam lineam continuantem centrum visus cum illo puncto."

92. Whether or not the backward extension of ray *AE* eventually intersects *OB* (i.e., in front of the eye, in the direction of *O*) will depend on the densities of the various media, the radius of curvature of surface *AD*, and the magnitude of distance *OB*.

93. See chap. 9, pp. 189-90.

94. See above pp. 2-3, 6-9, 42-56.

95. On the visual pyramid, see above, pp. 12-13, 15-17, 24-30.

96. Alhazen's theory also has the advantage of escaping the problems associated with spaces between the rays. Since every visible point emits its own ray, there is no possibility of a point becoming invisible because it happens to fall between two adjacent rays of the visual pyramid.

97. Roshdi Rashed, "Le 'Discours de la lumière' d'Ibn al-Haytham (Alhazen)," pp. 207, 233; J. Baarmann, "Ablandlung über das Licht von Ibn al-Haitam," pp. 198-99, 235. Cf. S. Pines, "What Was Original in Arabic Science?" in *Scientific Change*, ed. A. C. Crombie (New York, 1963), p. 200.

98. Rashed, "Optique géométrique," p. 273. Cf. Sabra, "Ibn al-Haytham," p. 191.

99. See this volume, p. 71. On the terminology employed in the Arabic text, see Schramm, *Weg zur Physik*, pp. 203-5, 216-17; Sabra, "Ibn al-Haytham," p. 191.

100. Alhazen insists, however, that transmission of forms through a transparent medium in no way alters the medium, for otherwise rays could not cross without mixing: "Therefore let us say that transparent substances are not changed (*immutantur*) by colors, nor are they [transparent substances] altered (*alterantur*) by them with a fixed alteration; but the property of color and light (*lux*) is that their forms extend [through transparent substances] along straight lines" (*De aspectibus*, bk. 1, chap. 5, sec. 28, p. 17). Some of my thoughts on this subject I owe to Stephen McCluskey.

101. *De aspectibus*, bk. 4, chap. 3, sec. 18, p. 112. Translation quoted, with one minor change, from Grant, *Source Book*, p. 418. On Alhazen's physical theory of reflection, see also Rashed, "Optique géométrique," pp. 281 ff. Before Alhazen, the reflection of light had been compared to that of a projectile by Hero of Alexandria, Ptolemy, and the author of the Pseudo-Aristotelian *Problemata*; see Saint-Pierre, "La physique de la vision," pp. 350-52; Ptolemy, *L'Optique*, ed. Lejeune, p. 98, n. 32.

102. See A. I. Sabra, "Explanation of Optical Reflection and Refraction," p. 552. In the Latin translation, the obstacle is referred to simply as a thin board (*tabula*); cf. Lindberg, "Cause of Refraction," p. 26.

103. *De aspectibus*, bk. 4, chap. 4, sec. 18, p. 113. Translation quoted from my "Introduction" to the reprint of the Risner edition, p. xviii.

104. See above, pp. 40-41.

105. *De aspectibus*, bk. 2, chap. 1, sec. 2, p. 25.

106. Ibid., sec. 3, pp. 25-26; sec. 5, p. 26. Alhazen argues that the interface between the glacial and vitreous humors must be either plane or spherical and (if spherical) of moderately large radius; cf. this volume, p. 126. Alhazen does not, so far as I can discover, specify which humor (the glacial or vitreous) is the denser, although his theory appears to require that the vitreous have the greater density.

107. *De aspectibus*, bk. 1, chap. 6, sec. 33, p. 21; bk. 2, chap. 1, secs. 5-6, pp. 26-27. The visual spirit has the same density or transparency as the vitreous humor so that forms pass from the one to the other without refraction; see ibid., sec. 6, p. 27.

108. Ibid., secs. 2-6, pp. 25-26.

109. Ibid., chap. 1, sec. 6, pp. 26-27; chap. 2; sec. 16, p. 34.

110. Alhazen discusses the operation of the *ultimum sentiens* and the completion of vision in *De aspectibus*, bk. 1, chap. 5, secs. 25-26, pp. 15-16; bk. 2, chap. 1, secs. 1-8, pp. 24-29; bk. 2, chap. 2, sec. 16, pp. 34-35. I will not delve further into Alhazen's psychology of visual perception, for to trace the history of that problem would require a full monograph in itself. The best sources on Alhazen's psychology are Vescovini, *Studi sulla prospettiva*, chap. 7, pp. 113-35; and Hans Bauer, *Psychologie Alhazens*. For a good medieval summary of Alhazen's psychological doctrines, see John Pecham's discussion, in Lindberg, *Pecham*, pp. 135-49.

There appears to be some uncertainty about the location of the *ultimum sentiens* in Alhazen's theory. Alhazen maintains quite explicitly, as I have indicated, that forms or their impressions are perceived by the *ultimum sentiens* when they reach the common nerve. However, there are also passages where he speaks of the propagation of forms to the anterior part of the brain and explicitly identifies this region of the brain as the seat of the *ultimum sentiens* (see, for example, *De aspectibus*, bk. 1, chap. 5, sec. 26, p. 16; bk. 2, chap. 2, sec. 16, p. 34). The difficulty is resolved by recognizing that Alhazen does not in these passages distinguish the anterior region of the brain from the nerves leading to the optic chiasma and the chiasma itself; all are of the same substance, and Alhazen treats them as a unit.

111. See chap. 4, n. 107.

112. *De aspectibus*, bk. 2, chap. 1, secs. 2-6, pp. 25-27; chap. 2, sec. 16, pp. 34-35.

113. See Vescovini, *Studi sulla prospettiva*, pp. 118-19.

114. *De aspectibus*, bk. 1, chap. 5, sec. 30, p. 17; cf. bk. 2, chap. 1, sec. 4, p. 26. The same point had been made by Alexander of Aphrodisias; see Saint-Pierre, "La physique de la vision," p. 240.

115. *De aspectibus*, bk. 2, chap. 1, sec. 5, p. 26; see also sec. 4, p. 26.

116. *De aspectibus*, bk. 2, chap. 1, sec. 5, p. 26.

117. Ibid., sec. 3, p. 26; cf. ibid., sec. 4, p. 26. Nevertheless, Alhazen appears to hold that in fact rays continue to move along refracted, but otherwise rectilinear, paths at least until they reach the optic nerve; he argues, in bk. 2, chap. 1, sec. 8, p. 29, that "forma illius rei visae perveniet ad superficiem glacialis secundum rectitudinem linearum radialium; deinde extenduntur formae ab ista superficie secundum rectitudinem linearum radialium etiam quousque perveniant ad superficiem vitrei; deinde punctum axis extendetur ab ista superficie secundum rectitudinem axis quousque perveniat ad locum gyrationis concavi nervi, et omnia puncta residua refringuntur super lineas secantes lineas radiales et consimilis ordinationis quousque perveniant ad locum gyrationis concavi nervi." Deviation from rectilinearity must occur, then, primarily (if not entirely) in the optic nerve.

118. The refraction occurring at the interface between the glacial and vitreous humors is symmetrical about the axis of the visual pyramid and thus does not deviate the beam as a whole; and in any case, it does not affect the central ray of the visual pyramid, which is incident on the interface perpendicularly. Although Alhazen is vague on the matter, there seems to be no way of escaping the conclusion that the nonrectilinearity of light in the optic nerve is a function of the medium. However, in his treatise *On Light* (composed after *De aspectibus*), Alhazen argues that rectilinear propagation is a property of light and that the only function of the medium is to oppose or admit light; see Rashed, "Discours de la lumière," p. 213; Baarmann, "Abhandlung über das Licht," p. 212. The same difficulty appears in Alhazen's theory of refraction; see Lindberg, "Cause of Refraction," p. 26.

119. *De aspectibus*, bk. 2, chap. 1, secs. 2-6, pp. 25-27; ibid., chap. 2, sec. 16, pp. 34-35. On the nature of the radiation propagated from the visible object to the glacial humor, see above, pp. 78-80.

120. See above, pp. 54-56.

121. *De aspectibus*, bk. 1, chap. 5, sec. 25, p. 15.

122. Ibid., sec. 27, p. 16.

123. Ibid., bk. 2, chap. 1, sec. 5, p. 26.

124. On the sensitivity of the visual spirits, see ibid., chap. 2, sec. 16, p. 34: "Spiritus visibilis est sentiens per totum, quoniam virtus sensitiva est per totum istius corporis."

125. On Alhazen's doctrine of certification, see *De aspectibus*, bk. 2, chap. 1, secs. 7-9, pp. 27-30; chap. 3, secs. 64-66, pp. 67-69; Vescovini, *Studi sulla prospettiva*, pp. 122-24.

126. *De aspectibus*, bk. 1, chap. 5, sec. 18, pp. 9-10; above, pp. 74-78.

127. *De aspectibus*, bk. 2, chap. 3, sec. 64, p. 67; cf. Vescovini, *Studi sulla prospettiva*, p. 122.

128. *De aspectibus*, bk. 2, chap. 1, sec. 8, p. 29.

129. Ibid., secs. 8-9, pp. 29-30. Alhazen also points out that the closer a ray is to the axis, the less its refraction and the clearer the perception it produces.

130. Ibid., chap. 3, sec. 42, p. 57; sec. 65, p. 68.

131. See especially p. 78, above; Sabra, "Ibn al-Haytham," pp. 191-93. In his work *On Light*, Alhazen maintained explicitly that optics is both a mathematical and a physical science; see Rashed, "Discours de la lumière," p. 205; Baarmann, "Abhandlung über das Licht," p. 197; see also Rashed, "Optique géométrique," pp. 272 ff. The more mathematical aspects of Alhazen's optical thought appear in books 4-7 of *De aspectibus*.

132. Part of Kamal al-Din al-Farisi's commentary has been rendered into German by Wiedemann, "Zu Ibn al Haitams Optik." On Kamal al-Din, see also Rashed, "Le modèle de la sphère transparente." On Alhazen's influence (and lack thereof) in Islam, see H. J. J. Winter and W. 'Arafat (trans.), "A statement on Optical Reflection and 'Refraction' Attributed to Nasir ud-Din at-Tusi," pp. 141-42; Sabra, "Ibn al-Haytham," p. 196; Rashed "Le modèle de la sphère transparente," p. 110, n. 1.

133. On Alhazen's influence in the West, see Lindberg, "Alhazen's Theory of Vision," pp. 330-41; Lindberg, *Pecham*, pp. 24-25, 34 ff.; and "Introduction" to the reprint of the Risner edition, pp. xxi-xxv. I must forcefully reject the contention of Vasco Ronchi that Alhazen's views were ignored, rejected, or misunderstood in the West; see Ronchi, *Nature of Light*, pp. 57, 62-69, 83, 87.

134. See chap. 9.

135. It seems clear that the elements out of which a theory is constructed represent as great a creative effort as the theory constructed out of them.

Chapter 5 The Origins of Optics in the West

1. The Roman handbook tradition has been ably discussed by William H. Stahl, *Roman Science*, especially chap. 5.

2. On Seneca, see ibid., pp. 98-100; John Clarke, *Physical Science in the Time of Nero: Being a Translation of the Quaestiones naturales of Seneca*, pp. xxi-xxxi; E. Phillips Barker, "Seneca," *Oxford Classical Dictionary*, pp. 827-28.

3. Stahl, *Roman Science*, p. 99, argues for the influence of Posidonius; Saint-Pierre, "La physique de la vision," pp. 112-13, defends the influence of Chrysippus in addition to that of Posidonius.

4. *Quaestiones naturales* 1. 5, trans. Clarke, p. 23.

5. Ibid. 1. 3, trans. Clarke, p. 19. Saint-Pierre, "La physique de la vision," pp. 110-11, has pointed out that remarks in the *Quaestiones naturales* 2. 8-9, on the tension of the air as a medium for the propagation of sight reveal Stoic influence.

6. On Pliny, see Stahl, *Roman Science*, chap. 7.

7. *Natural History* 11. 54. 143; 11. 55. 151.

8. Ibid. 11. 54. 147-48.

9. Ibid. 11. 55. 149, trans. H. Rackham, 3: 527.

10. Ibid. 33. 45. 128, trans. Rackham, 9: 97.

11. On Solinus, see Stahl, *Roman Science*, pp. 136-41.

12. *Collectanea rerum memorabilium*, ed. Theodore Mommsen, p. 26. Pliny makes similar remarks in his *Natural History* 7. 2. 16-18. Both Pliny and Solinus are quoted on this subject by Roger Bacon, *Opus maius*, pt. 5.2, dist. 1, chap. 3, ed. Bridges, 2: 91, as providing a possible

explanation of double vision. The evil eye (*fascinum*) is a theme that runs through European folklore (often emerging in the intellectual tradition as well) from antiquity to the present; see, for example, Frederick T. Elworthy, *The Evil Eye: An Account of This Ancient and Widespread Superstition* (London, 1895); Robert C. Maclagan, *Evil Eye in the Western Highlands* (London, 1902); Siegfried Seligmann, *Der böse Blick und verwandtes*, 2 vols. (Berlin, 1910); Seligmann, *Die Zauberkraft des Auges und das Berufen* (Hamburg, 1922); A. M. Hocart, "The Mechanism of the Evil Eye," *Folklore* 49 (1938): 156-57. On the double pupil, see Walton Brooks McDaniel, "The *Pupula Duplex* and Other Tokens of an 'Evil Eye' in the Light of Ophthalmology," *Classical Philology* 13 (1918): 335-46.

13. Virtually nothing else is known about Chalcidius; see Stahl, *Roman Science*, pp. 142 ff.; Stahl, "Chalcidius," *Dictionary of Scientific Biography*, 3: 14-15.

14. For a list of manuscripts, see *Timaeus a Calcidio translatus*, ed. J. H. Waszink and P. J. Jensen, pp. cvii-cxxviii.

15. Ibid., pp. 249-54.

16. See Robert Grosseteste, *De iride*, in Grant, *Source Book*, p. 389; Bartholomaeus Anglicus, *De rerum proprietatibus*, bk. 3, chap. 17, pp. 62-63; Lindberg, *Pecham*, p. 25. Cf. *Damianos Schrift über Optik*, ed. Richard Schöne, pp. 2-3.

17. Chalcidius, *Commentary*, in *Timaeus a Calcidio translatus*, ed. Waszink and Jensen, p. 256.

18. Ibid., pp. 256-57.

19. *De trinitate* 11. 4. 4, trans. Marcus Dods and Arthur W. Haddan, p. 146.

20. The issuance of radiation from the eyes (the power of sight remaining behind) is not the same as the power of sight itself issuing forth; the latter is the Galenic conception.

21. *De genesi ad litteram*, ed. Joseph Zycha, p. 23. This passage would later be quoted by John Pecham, *Tractatus de perspectiva*, ed. David C. Lindberg, pp. 37-38; and cited by Albert the Great, *De homine*, qu. 19, in *Opera omnia*, ed. A. Borgnet, 35: 169. Cf. also Pseudo-Grosseteste, *Summa philosophie*, in *Die philosophischen Werke des Robert Grosseteste*, ed. Ludwig Baur, p. 498. On Augustine's theory of vision, see also his *De musica* 6. 5. 10, in J.-P. Migne, *Patrologiae cursus completus, series latina*, vol. 32, col. 1169.

22. M.-D. Chenu, *Nature, Man, and Society in the Twelfth Century*, trans. Jerome Taylor and Lester K. Little, p. 14. Chenu is one of the best sources on the intellectual life of the twelfth century; see also Charles H. Haskins, *The Renaissance of the Twelfth Century*; E. J. Dijksterhuis, *The Mechanization of the World Picture*, trans. C. Dikshoorn (Oxford, 1961), pp. 116-25.

23. Chenu, *Nature, Man, and Society*, p. 10; see also Haskins, *Renaissance*, chap. 10.

24. A portion of this generalization has been defended by Alexandre Koyré, who has argued that advances in theory do not ordinarily follow from advances in methodology; see "The Origins of Modern Science: A New Interpretation," *Diogenes* 16 (Winter 1956): 1-22.

25. The reevaluation of William's relationship to the school of Chartres and Bernard's teaching is due to R. W. Southern, *Medieval Humanism and Other Studies* (Oxford, 1970), pp. 61-85. But see also the reply of Nikolaus Häring, "Chartres and Paris Revisited," in *Essays in Honor of A. C. Pegis*, ed. J. R. O'Donnell (Toronto, 1974), pp. 268-329. On William, see also Reginald L. Poole, *Illustrations of the History of Medieval Thought and Learning*, 2d ed. (London, 1920), pp. 298-310; Tullio Gregory, *Anima mundi: La filosofia di Guglielmo di Conches e la scuola di Chartres*; Giorgio Rialdi, *Il De philosophia mundi (XII sec.): L'Autore, la storia, il contenuto medico* (Genoa, 1965); J.-M. Parent, *La doctrine de la création dans l'école de Chartres*, pp. 15-16, 115-21; Guillaume de Conches *Glosae* [*sic*] *super Platonem*, ed. Edouard Jeauneau, pp. 9-16; Chenu, *Nature, Man, and Society*, passim; Dijksterhuis, *Mechanization*, pp. 119-23; Ynez Violé O'Neill, "William of Conches' Descriptions of the Brain"; and A. Vernet, "Un remaniement de la *Philosophia* de Guillaume de Conches."

26. *Glossae*, ed. Jeauneau, pp. 236-37. In the following account, I follow chiefly the *Glossae*, for it is here that William presents his fullest discussion of vision.

27. Ibid., p. 237; cf. *Dragmaticon*, published as *Dialogus de substantiis physicis* (Strasburg, 1567), pp. 281-82.

28. *Glossae*, p. 237. Cf. *Dragmaticon*, pp. 282, 289; *Philosophia mundi*, in J.-P. Migne, *Patrologiae cursus completus, series latina*, vol. 172, col. 95 (there attributed to Honorius of Autun).

29. *Glossae*, pp. 237-38; *Dragmaticon*, p. 285; *Philosophia mundi*, col. 95.

30. *Glossae*, p. 238; *Philosophia mundi*, col. 95. On the evil eye, see this chapter, n. 12.

31. *Dragmaticon*, p. 283.

32. Ibid., p. 285.

33. *Glossae*, pp. 238-39; *Dragmaticon*, pp. 280-81; *Philosophia mundi*, col. 95. William refers his reader to the *Isagoge* of Johannitius and the *Pantegni* of Constantinus for a fuller description of the eye. The *Pantegni* (actually by Haly Abbas) was first translated by Constantinus Africanus late in the eleventh century, and its availability to William is thus unproblematic. Translation of the *Isagoge*, on the other hand, has been traditionally attributed to Marc of Toledo (on the basis of his own statement); as Marc flourished at the beginning of the thirteenth century, this appears to present serious problems. However, d'Alverny and Vajda have argued that Marc's statement cannot be taken at face value, for the *Isagoge* was in circulation in a Latin version by the end of the eleventh century; see M.-Th. d'Alverny and Georges Vajda, "Marc de Tolède, traducteur d'Ibn Tumart," *Al-Andalus* 16 (1951): 109-112. William's reliance on the *Pantegni* for some of his terminology has been demonstrated by Ynez Violé O'Neill, "William of Conches and the Cerebral Membranes," pp. 13-21.

34. *Glossae*, pp. 243-45.

35. The dichotomy is certainly not complete, since William had had contact with the newly translated learning, and Adelard had Platonic tendencies. Nevertheless, William relied chiefly on traditional sources for his philosophy, whereas Adelard was far more eclectic and made his mark principally as a translator of Arabic works into Latin.

36. On Adelard's life and works, see Franz Bliemetzrieder, *Adelhard von Bath*; Charles H. Haskins, *Studies in the History of Mediaeval Science*, chap. 2; Marshall Clagett, "Adelard of Bath," *Dictionary of Scientific Biography*, 1: 61-64.

37. *Die Quaestiones naturales des Adelardus von Bath*, ed. Martin Müller, qu. 23, pp. 27, 30.

38. Ibid., p. 30.

39. Ibid., pp. 30-31.

40. See above. If there was indeed influence, it was probably from Adelard to William, since Adelard appears to have written his *Questiones naturales* in the 1130s or sooner (and no later than 1142), whereas William's *Dragmaticon* dates from the late 1140s. On the dating of the two works (a somewhat intricate matter), see Haskins, *Studies*, pp. 26-27; Lynn Thorndike, *History of Magic*, 2: 44-49; Bliemetzrieder, *Adelhard von Bath*, pp. 72-82; Gregory, *Anima mundi*, pp. 3, 7; *Glossae*, ed. Jeauneau, p. 10; Reginald L. Poole, "The Masters of the Schools of Paris and Chartres in John of Salisbury's Time," *English Historical Review* 35 (1920): 334.

41. *Questiones naturales*, ed. Müller, qu. 24, p. 31.

42. Ibid., qu. 25, pp. 31-32.

43. Ibid., p. 32.

44. Ibid., qu. 26, pp. 32-33.

45. Ibid., qu. 27, pp. 33-34.

46. Ibid., qu. 28, p. 34. I have here borrowed some phrases from the English translation of the *Questiones naturales* by Hermann Gollancz, in Barachya Hanakdan, *Dodi Ve-Nechdi . . . to which is added the first English Translation from the Latin of Adelard of Bath's Questiones Naturales*, p. 122. In general, however, Gollancz's translation is too defective to be usable.

47. On the source of Grosseteste's optical thought, see A. C. Crombie, *Robert Grosseteste and the Origins of Experimental Science*, pp. 116-17; Ludwig Baur, *Die Philosophie des Robert Grosseteste*, pp. 110-11.

48. Crombie has argued that Grosseteste made use of Ptolemy's *Optica*; however, Ptolemy is not cited in Grosseteste's works, and I do not believe that Grosseteste could have presented his primitive "law of refraction" if he had seen Ptolemy's discussion of the same subject. See

Crombie, *Grosseteste*, pp. 116-17; Crombie, "Mechanistic Hypothesis," pp. 24-26. R. W. Hunt has argued that a reference to Alhazen's *Perspectiva* appears in a concordance drawn up by Grosseteste and Adam Marsh from works available to them (R. W. Hunt, "The Library of Robert Grosseteste," in *Robert Grosseteste, Scholar and Bishop*, ed. D. A. Callus, pp. 123-25). However, Hunt admits that Alhazen's name does not appear in the concordance (but only the expression *Perspectiva*), and he does not explain how he knows that this particular *Perspectiva* was Alhazen's; it should be noted, moreover, that Alhazen's work was usually entitled *De aspectibus* in the thirteenth century, whereas there were other works circulating at the time (e.g., Euclid's *De visu*) which might be designated *Perspectiva*. In any case, S. Harrison Thomson ("Grosseteste's Topical Concordance of the Bible and the Fathers," *Speculum* 9 [1934]: 139-44) has argued that the concordance dates from the years 1235-43 (after Grosseteste had completed his works on optics), and it is abundantly clear from internal considerations that Grosseteste's optical works were written without knowledge of Alhazen's *De aspectibus*.

49. On Grosseteste's life, see Francis S. Stevenson, *Robert Grosseteste, Bishop of Lincoln* (London, 1899); D. A. Callus, "Robert Grosseteste as a Scholar," in Callus, *Grosseteste*, pp. 1-11; Crombie, *Grosseteste*, pp. 44-45.

50. Callus claims that "the central figure in England in the intellectual movement of the first half of the thirteenth century was undoubtedly Robert Grosseteste" ("Grosseteste as a Scholar," p. 1). Grosseteste's influence on the scientific thought of the thirteenth century is one of the main themes of Crombie's *Grosseteste*, esp. pp. 10-15, 135-39, 147-52, 162-72, 186-201, 213-15.

51. D. E. Sharp, *Franciscan Philosophy at Oxford in the Thirteenth Century*, p. 10, notes that "in him [Grosseteste] the Augustinian thought first encounters the philosophy of Aristotle."

52. See ibid., p. 9; Robert Grosseteste, *On Light*, trans. Clare C. Riedl, p. 1.

53. The best sources on the philosophy (or "metaphysics") of light are Clemens Baeumker, *Witelo, ein Philosoph und Naturforscher des XIII. Jahrhunderts*, pp. 357-467; Baur, *Philosophie des Grosseteste*, pp. 76-109; Charles C. McKeon, *A Study of the Summa philosophiae of the Pseudo-Grosseteste*, pp. 85-97; Otto von Simson, *The Gothic Cathedral*, pp. 50-55; Crombie, *Grosseteste*, pp. 104-16, 128-31; Franz-Norbert Klein, *Die Lichtterminologie bei Philon von Alexandrien und in den hermetischen Schriften*; Vescovini, *Studi sulla prospettiva*, pp. 7-32; and Franz Rosenthal, *Knowledge Triumphant: The Concept of Knowledge in Medieval Islam* (Leiden, 1970), pp. 155-93.

54. McKeon, *Study of the Summa philosophiae*, p. 157; cf. Plato *Republic* 4. 508d-509a.

55. In the *City of God* 10. 2, Augustine quotes John 1: 6-9 as his authority for this point. Cf. *De trinitate* 13. 1.

56. Pecham, *Tractatus de perspectiva*, ed. Lindberg, p. 28; cf. Etienne Gilson, *The Christian Philosophy of Saint Augustine*, trans. L. E. M. Lynch (New York, 1960), p. 78; Eugène Portalié, S.J., *A Guide to the Thought of St. Augustine*, trans. Ralph J. Bastian (Chicago, 1960), p. 110; McKeon, *Study of the Summa philosophiae*, p. 158.

57. *De trinitate* 12. 15. 24, trans. Dods and Haddan, p. 164.

58. On Grosseteste's epistemology of light, see Lawrence E. Lynch, "The Doctrine of Divine Ideas and Illumination in Robert Grosseteste, Bishop of Lincoln"; Crombie, *Grosseteste*, pp. 128-31.

59. Crombie, *Grosseteste*, p. 130.

60. *Die philosophischen Werke des Robert Grosseteste*, ed. Ludwig Baur, p. 138.

61. Emile Bréhier, *The Philosophy of Plotinus*, trans. Joseph Thomas (Chicago, 1958), pp. 43-52.

62. Grosseteste, *On Light*, trans. Riedl, p. 10. On Grosseteste's cosmogony of light, see also Crombie, *Grosseteste*, pp. 106-9; Baur, *Philosophie des Grosseteste*, pp. 85-93; McKeon, *Study of the Summa philosophiae*, pp. 160-63; Lynch, "Doctrine of Divine Ideas," pp. 164-66.

63. Plotinus, *The Enneads* 5. 1. 6, trans. Stephen MacKenna, p. 374. Cf. Crombie, *Grosseteste*, pp. 104-5; McKeon, *Study of the Summa philosophiae*, pp. 85-95.

64. Avicebron, *Fons vitae*, 3, 52, ed. Clemens Baeumker, p. 196.

65. See above, p. 19.

66. The terminology was not new, for the term "species," with the connotation of visual form or image, had been current in the West for centuries; if there was any novelty, it was in the detail added by Grosseteste (and later Bacon) regarding the process of "multiplication." On medieval usages of the term "species," see Pierre Michaud-Quantin, *Etudes sur le vocabulaire philosophique du Moyen Age*, pp. 113-50.

67. Quoted from Grant, *Source Book*, pp. 385-86.

68. Ibid., p. 385.

69. Grosseteste, *De natura locorum*, quoted from Crombie, *Grosseteste*, p. 110.

70. John 1: 4-9.

71. *Dionysius the Areopagite on the Divine Names and the Mystical Theology*, trans. C. E. Rolt, p. 94. On the Dionysian tradition in the twelfth century (and its utilization by Suger of Saint-Denis in the creation of Gothic architecture), see von Simson, *Gothic Cathedral*, pp. 102-7.

72. *De veritate*, in *Philosophischen Werke*, ed. Baur, p. 142.

73. Crombie, *Grosseteste*, p. 131.

74. *Opus maius*, pt. 4, ed. Bridges, 1: 216-17. For a later example of the theology of light, see "Tractatus de luce Fr. Bartholomaei de Bononia," ed. Irenaeus Squadrani.

75. Crombie, *Grosseteste*, pp. 104, 109, 116, 128, 135-36.

76. *Opus tertium*, in *Opera quaedam hactenus inedita*, ed. J. S. Brewer, p. 37. In another passage Bacon strains credibility, arguing that optical marvels are useful for conversion of the infidel; for just as optical marvels cannot be grasped immediately by all men but must first be believed, so in matters of faith man must be taught first to believe in order that he may later understand; see *Part of the Opus Tertium of Roger Bacon*, ed. A. G. Little, pp. 41-42.

77. My translations of *De lineis* and *De iride* are available in Grant, *Source Book*, pp. 385-91; another translation of *De iride* appears in "Robert Grosseteste on Refraction Phenomena," trans. Bruce S. Eastwood. The important secondary sources on Grosseteste's optical work are Baur, *Philosophie des Grosseteste*, pp. 109-30; Crombie, *Grosseteste*, chap. 5; Colin M. Turbayne, "Grosseteste and an Ancient Optical Principle"; Carl B. Boyer, *The Rainbow*, pp. 88-94; Bruce S. Eastwood, "Grosseteste's 'Quantitative' Law of Refraction"; Eastwood, "Mediaeval Empiricism: The Case of Grosseteste's Optics"; Eastwood, "Robert Grosseteste's Theory of the Rainbow"; David C. Lindberg, "Roger Bacon's Theory of the Rainbow: Progress or Regress?" pp. 238-41.

78. Richard C. Dales, "Robert Grosseteste's Scientific Works," pp. 394-401.

79. *De iride*, quoted from Grant, *Source Book*, p. 389. The two quotations are from Aristotle's *De generatione animalium* 5. 1. 781a1-2, 781b2-13. However, at this point Aristotle had not in actuality defended an extramission theory, but only seemed to do so as a result of the mistranslation of Michael Scot, whose version Grosseteste is here following; see Eastwood, "Grosseteste on Refraction Phenomena," p. 197; S. D. Wingate, *The Mediaeval Latin Versions of the Aristotelian Scientific Corpus, with Special Reference to the Biological Works* (London, 1931), p. 78.

80. Quoted from Crombie, *Grosseteste*, p. 114.

81. On this law, see Eastwood, "Grosseteste's 'Quantitative' Law."

82. For an elaboration of this point, see Lindberg, "Bacon's Theory of the Rainbow," pp. 238-41; for a different point of view, see Eastwood, "Grosseteste's Theory of the Rainbow." No two scholars have yet understood the details of Grosseteste's scheme to the satisfaction of each other.

83. For a general treatment of the translations, see Haskins, *Renaissance*, chap. 9; J. M. Millas-Vallicrosa, "Translations of Oriental Scientific Works (to the End of the Thirteenth Century)," in *The Evolution of Science*, ed. Guy S. Métraux and François Crouzet (New York, 1963), pp. 128-67 (published originally in Spanish as "La corriente de las traducciones cientificas de origen oriental hasta fines del siglo XIII," *Journal of World History* 2 [1954]: 395-428); F. Gabrieli, "The Transmission of Learning and Literary Influences to Western Europe," in *Cambridge History of Islam*, ed. P. M. Holt, Ann K. S. Lambton, and Bernard Lewis, vol. 2

(Cambridge, 1970), pp. 851-89.

84. A preliminary approach to some of these questions has been made by Sir Hamilton Gibb, "The Influence of Islamic Culture on Medieval Europe," *Bulletin of the John Rylands Library*, vol. 38 (1955-56), reprinted in *Change in Medieval Society*, ed. Sylvia Thrupp (New York, 1964), pp. 155-67. See also George F. Hourani, "The Medieval Translations from Arabic to Latin Made in Spain," *The Muslim World* 62 (1972): 97-114.

85. The "steady flow" at the end of the thirteenth century consisted almost entirely of translations from the Greek, and ultimately it led to a new burst of Greco-Latin translating activity during the Renaissance.

Chapter 6 The Optical Synthesis of the Thirteenth Century

1. "Perspectivist" may seem a barbaric expression, but there has long been a need in English for a term to denote the practitioner of *perspectiva*, and this word seems preferable to the sixteenth-century terms "perspectiver" and "perspectivian." Henceforth I will use the expression without apology and without quotation marks. I will mean by it precisely what late medieval writers meant by the term *perspectivus*, namely, a member of the mathematical tradition in optics, which included Euclid, Ptolemy, al-Kindi, Alhazen, Roger Bacon, Witelo, John Pecham, and others.

2. On Albert's life and works, see T. M. Schwertner, O.P., *St. Albert the Great* (New York, 1932); James A. Weisheipl, O.P., "Albert the Great (Albertus Magnus), St.," *New Catholic Encyclopedia*, 1: 254-58; William A. Wallace, O.P., "Albertus Magnus, Saint," *Dictionary of Scientific Biography*, 1: 99-103; G. Meersseman, O.P., *Introductio in opera omnia B. Alberti Magni O.P.* (Bruges, 1931); Franz Pelster, S.J., *Kritische Studien zum Leben und zu den Schriften Alberts des Grossen* (Freiburg, i.B., 1920); Heribert C. Scheeben, *Albert der Grosse: Zur Chronologie seines Lebens* (Vechta, 1931).

3. Weisheipl, "Albert the Great," p. 258, dates *De homine* to the period 1244-48, though without explaining the grounds for this choice. Lottin argues at length that Albert had drafted the *Summa de creaturis* (containing *De homine*) before 1243 (O. Lottin, O.S.B., "Problèmes concernant la 'Summa de creaturis' et le Commentaire des Sentences de Saint Albert le Grand," *Recherches de théologie ancienne et médiévale* 17 [1950]: 328). For the dates of Albert's remaining works, see Weisheipl's summary.

4. In addition, an extract from qu. 21 of *De homine* circulated widely under the title *Questio de forma resultante in speculo*, or *Katoptrik*. For further detail, including a list of manuscripts, see Lindberg, *Catalogue of Optical Manuscripts*, pp. 16-17. An early fifteenth-century list attributes to Albert a *Summa perspectivae* and a commentary on Alhazen's *De aspectibus*; however, since many works were spuriously attributed to Albert (indeed, Bacon's *Perspectiva* is attributed to Albert in at least one manuscript—Munich, Bayerische Staatsbibl. MS CLM 453) there is no reason to take this fifteenth-century list seriously; see Heribert C. Scheeben, "Les écrits d'Albert le Grand d'après les catalogues," *Revue thomiste* 36 (1931): 282, 290. Albert's work on optics has been briefly analyzed by Methodius Hudeczek, O.P., "De lumine et coloribus [according to Albert the Great]"; Giuseppe Ovio, "Cenni d'ottica fisiologica [according to Albert the Great]"; Crombie, *Grosseteste*, pp. 191-200; Carl B. Boyer, *The Rainbow*, pp. 94-97; William A. Wallace, O.P., *The Scientific Methodology of Theodoric of Freiberg*, pp. 141-43.

5. *De homine*, qu. 22 (part 2 of the *Summa de creaturis*), in *Opera omnia*, ed. A. Borgnet, 35: 210-28.

6. *De anima*, bk. 2, tract. 3, chap. 9, ed. Clemens Stroick, in *Opera omnia*, ed. Bernhard Geyer, vol. 7, pt. 1, pp. 111-12. *De sensu*, tract. 1, chaps. 5-14, in *Opera omnia*, ed. Borgnet, 9: 9-35.

7. *De homine*, qu. 21-22, ed. Borgnet, pp. 181-83, 203-10, 213-15; *De sensu*, tract. 1, chaps. 5, 10, ed. Borgnet, pp. 10, 24-25; *De anima*, bk. 2, tract. 3, chap. 7, ed. Stroick, pp. 108-10. Albert employs a variety of expressions to denote this alteration of the medium,

including *forma, species, lumen, color*, and *radius*.

8. *De homine*, qu. 20, ed. Borgnet, pp. 171–72; *De sensu*, tract. 1, chaps. 5, 10, 14, ed. Borgnet, pp. 10, 25, 35.

9. *De animalibus*, bk. 1, tract. 2, chap. 7, in *Opera omnia*, ed. Borgnet, 11: 50–52.

10. Ed. Borgnet, pp. 215–28; cf. *De sensu*, tract. 1, chaps. 6–8, ed. Borgnet, pp. 11–21.

11. *De homine*, qu. 20, ed. Borgnet, p. 172; *De anima*, bk. 2, tract. 3, chap. 12, ed. Stroick, p. 117; *De sensu*, tract. 1, chaps. 4, 10, ed. Borgnet, pp. 7–8, 25–26.

12. *De homine*, appendix to qu. 22, ed. Borgnet, p. 223; *De sensu*, tract. 1, chap. 5, ed. Borgnet, p. 10. In the latter passage, Albert writes: "Quaedam autem novella et fatua invenit, non opinio, sed insania quorumdam dicentium nos videre, et intus suscipientes et extra mittentes." It appears that Albert could not be referring here to Roger Bacon, since *De sensu* was completed before 1260, and Bacon probably did not engage in serious optical work until after 1260; on the dating of Bacon's works, see p. 108. On the intromission-extramission theory of Bartholomeus Anglicus, see chap. 6, n. 26. Albert knew at least Grosseteste's *Commentary on Aristotle's Posterior Analytics* (see William A. Wallace, *Causality and Scientific Explanation*, 1: 66–67), and there he would have found a brief statement of the combined intromission-extramission theory; see above, chap. 5.

13. I will make no attempt to provide exact citations to the passages in which Albert uses each of the various works, since these are plentiful and are easily discovered by a perusal of his writings.

14. However, Albert attributed Averroes's *De sensu* to al-Farabi; see chap. 3, n. 67.

15. These works are cited in *De homine*, ed. Borgnet, pp. 199, 212, 214, 216–17, 224. On the exact identity of the Euclidean works, see my *Catalogue of Optical Manuscripts*, nos. 79E–79K, 79C, 79D, and 80, respectively.

16. See *De sensu*, tract. 1, chaps, 9–10, ed. Borgnet, pp. 24–27, where Borgnet's "Haceuben Huchaym" is apparently a misreading of "Hacen ben Huchaym"; as manuscripts of Alhazen's *De aspectibus* commonly give the author's name as "Hacen [or Alhacen] filii Huchaym [or Hucayn or Hucaym]," there can be no doubt about the identity of the treatise. See also the *Quaestiones de animalibus*, ed. Ephrem Filthaut, O.P., in *Opera omnia*, ed. Geyer, 12: 98, where the expression *auctor Perspectivae* can refer to nobody except Alhazen.

17. *De animalibus*, bk. 1, tract. 2, chap. 3, ed. Borgnet, pp. 37–38, 41–42; Borgnet's text reads "Constabulini." On the identification of "Constabulini" as Costa ben Luca, see Nicholas H. Steneck, "Albert the Great on the Classification and Localization of the Internal Senses," p. 206.

18. Polemon and Loxus are cited in *De animalibus*, bk. 1, tract. 2, chap. 3, ed. Borgnet, pp. 37–38, 41–42. Since their views on sight had been discussed in the *Physiognomonia* of Pseudo-Apuleius, a work specifically mentioned by Albert (ibid., p. 39), the latter was in all probability Albert's source. See *Physiognomonia*, ed. Valentin Rose, in *Anecdota Graeca et Graecolatina*, 1: 116–32.

19. *De homine*, qu. 21, ed. Borgnet, p. 176. The specific work cited is Isaac's *De diffinitionibus*.

20. Favoring such an interpretation, over that of English Franciscan and continental Dominican science, is the fact that there is no evidence to associate the optical work of either Bacon or Pecham with Oxford or England (both appear to have pursued optical studies primarily while on the continent) apart from familiarity with Grosseteste's work (which was also known by Albert the Great, who never set foot on English soil); moreover, an examination of Little's list of thirty-three lecturers to the Franciscan school at Oxford between 1229 and 1300 (four seculars and twenty-nine friars) reveals that none except Grosseteste and Pecham ever distinguished himself in optics or mathematical science; see A.G. Little, "The Franciscan School at Oxford in the Thirteenth Century," *Archivum Franciscanum Historicum* 19 (1926): 803–74. On the Dominican side there is the troublesome fact that few affinities are to be found among the Dominicans who wrote on optics.

21. On Bacon's life, see Stewart Easton, *Roger Bacon and His Search for a Universal Science*;

Theodore Crowley, O.F.M., *Roger Bacon: The Problem of the Soul in His Philosophical Commentaries* (Louvain/Dublin, 1950); A. C. Crombie and J. D. North, "Bacon, Roger," *Dictionary of Scientific Biography*, 1: 377-85.

22. The Latin text of the *Opus maius*, accompanied by *De multiplicatione specierum*, was published in the eighteenth century by Samuel Jebb, *Fratris Rogeri Bacon ordinis minorum Opus majus* (London, 1733). A new edition by John H. Bridges, *The Opus Majus of Roger Bacon*, appeared in 2 vols. (Oxford, 1897); it was reissued in 3 vols. with additions and corrections (London, 1900). I have used this latter, 3-volume edition. A careless and inadequate English translation of the *Opus maius* is available in *The Opus Majus of Roger Bacon*, trans. Robert B. Burke. *De speculis comburentibus* appears in *Rogerii Bacconis Perspectiva*, ed. I. Combach; for a partial analysis of its contents, see David C. Lindberg, "A Reconsideration of Roger Bacon's Theory of Pinhole Images." On the identity and composition of *De multiplicatione specierum*, see Lindberg, "Lines of Influence in Thirteenth-Century Optics," pp. 69-71. A good general bibliography of Bacon's works is found in A. G. Little, "Roger Bacon's Works."

23. See above, chap. 5.

24. On the relationship between Grosseteste and Bacon, see Easton, *Roger Bacon*, pp. 89-91, 206-9; cf. Ludwig Baur, "Der Einfluss des Robert Grosseteste auf die wissenschaftliche Richtung des Roger Bacon"; Crombie, *Grosseteste*, pp. 139-40. Easton points out that there is no reason to believe that Grosseteste's lectures to the Oxford Franciscans were on anything except theology. On the disposition of Grosseteste's books, see R. W. Hunt, "The Library of Robert Grosseteste," in *Robert Grosseteste, Scholar and Bishop*, ed. D. A. Callus, pp. 130-32; one cannot be absolutely certain about the bequeathal of Grosseteste's books to the Franciscans or about the inclusion of his own writings among them, but both appear to be strong probabilities.

25. On *De radiis stellarum*, see above, p. 19. Two other works containing claims of the same sort are Avicebron's *Fons vitae* (see above, p. 97) and the Pseudo-Aristotelian *Liber de causis* (upon which Bacon lectured during his Parisian career).

26. On Bartholomeus, see Thomas Plassmann, O.F.M., "Bartholomaeus Anglicus," *Archivum Franciscanum Historicum*, 12 (1919): 68-109. Plassmann argues that *De proprietatibus rerum* was completed about 1250 but was based on earlier work. Although Bartholomeus completed this treatise in Germany, it may have circulated rapidly among friends, Franciscans, or Paris alumni. As for content, *De proprietatibus rerum* is based to a significant degree on Alhazen's *De aspectibus*, as well as Grosseteste's work and many other sources, and attempts a reconciliation of the intromission and extramission theories of vision that anticipates, in a number of respects, Bacon's later synthesis; see *De proprietatibus rerum*, bk. 3, chap. 17 (Frankfurt, 1601), pp. 62-66.

27. On Bacon's attitude toward Albert, see Easton, *Roger Bacon*, pp. 210-31; Bacon notes specifically that the "unnamed master" (whom Easton convincingly demonstrates to be Albert) is ignorant of *perspectiva*, which implies that Bacon was reading Albert's optical works and found them lacking in the more mathematical aspects (generally connoted by the term "*perspectiva*"). On Albert's possible attitude toward Bacon, see A. G. Little, "On Roger Bacon's Life and Works," pp. 8-9.

28. I know of no Western optical writer before Bacon to cite Ptolemy's *Optica*, and only Bartholomeus Anglicus and Albert the Great preceded him in the use of Alhazen's *De aspectibus*.

29. *Opus maius*, p. 5. 1, dist. 6, chap. 2, ed. Bridges, 2: 37-38. Cf. Alhazen's account, above, chap. 4.

30. Bacon distinguishes two parts of the glacial humor, the anterior and the interior; the former corresponds to the crystalline lens, the latter to the vitreous humor. However, unless otherwise noted, I will continue to employ the expression "glacial humor" to denote the anterior glacial humor or crystalline lens. See *Opus maius*, pt. 5. 1, dist. 2, chap. 3, ed. Bridges, 2: 17.

31. Ibid., dist. 6, chap. 1, ed. Bridges, 2: 35.

32. Ibid., dist. 5, chap. 2; dist. 7, chap. 1, ed. Bridges, 2: 32-33, 47-49. It is by no means correct to suggest, as Crombie does (*Grosseteste*, p. 154), that Bacon conceives the rays to be "focused" on the end of the optic nerve. In fact, there is no focusing at all, if by "focusing" is

meant the convergence at a single point in the eye of rays that originate from a single point in the visual field; moreover, the convergence of rays from different points in the visual field (i.e., the convergence of the visual pyramid to an apex, which is what Crombie really has in mind) occurs not on the end of the optic nerve, but in the common nerve (i.e., the junction of the two optic nerves in the optic chiasma). Thus Bacon is perfectly faithful to Alhazen except for two minor revisions: first, he is more explicit than Alhazen on the freedom of species to disregard the law of rectilinear propagation once inside the optic nerve (so as to follow its tortuous path); second, he suggests that the species issuing from different points in the visual field converge to a point in the common nerve, whereas Alhazen insisted only that the forms maintain their proper spatial arrangement through the optic nerve to the common nerve. Cf. Alhazen, this volume, pp. 80-83.

33. I have copied fig. 13 from London, British Museum, MS Royal 7.F.VIII, fol. 61v (13th c.); and British Museum, MS Harley 80, fol. 8r (15th c.). Fig. 14 is copied from the same two MSS, fols. 54v and 4v, respectively, with minor corrections and with omission of some of the lettering. See Crombie, "Mechanistic Hypothesis," p. 20, for a photographic reproduction of the Royal MS version of fig. 14; and *Opus maius*, ed. Bridges, 2: 24, 48, for other versions (albeit not perfectly accurate) of both figures.

34. The circular arc through which visible object *AL* passes in fig. 14 does not represent an actual part of the eye, but merely the forward edge of the circle used to define the posterior surface of the glacial humor. This entire drawing is a highly idealized representation of Bacon's view of ocular structure.

35. *Opus maius*, pt. 5.3, dist. 2, chap. 1, ed. Bridges, 2: 147-48. Cf. Alhazen, above, pp. 76-78.

36. *Opus maius*, pt. 5.1, dist. 8-9, ed. Bridges, 2: 54-74. Cf. Alhazen, *De aspectibus*, bk. 1, chap. 5, sec. 16, p . 8; bk. 1, chap. 7, pp. 22-23; bk. 2, chap. 2, sec. 20, pp. 36-37.

37. *Opus maius*, pt. 5.1, dist. 1, chap. 3, ed. Bridges, 2: 6. Cf. Alhazen, *De aspectibus*, bk. 2, chap. 2, sec. 15, p. 34.

38. *Opus maius*, pt. 5.1, dist. 2-3, ed. Bridges, 2: 15-25; actually there is one exception, for the uvea is eccentric to the rest; but this is of no matter since it is perforated by an aperture. For the views of Constantine (i.e., Hunain), Avicenna, and Alhazen on ocular anatomy, see this volume, pp. 34-36, 51, 67-69.

39. *Opus maius*, pt. 5.1, dist. 4, chap. 2, ed. Bridges, 2: 27; cf. Alhazen's argument, this volume, p. 69.

40. In the *Opus tertium*, he wrote, with regard to optics: "I have determined not to imitate one author; rather, I have selected the most excellent opinions from each" (*Un fragment inédit de l'Opus tertium de Roger Bacon*, ed. Pierre Duhem, p. 75).

41. Bacon was convinced that God had originally revealed the entirety of human knowledge to ancient sages, from whom it passed through such philosophers as Pythagoras, Plato, Aristotle, and Avicenna to his own era; see *Opus maius*, pt. 1, chap. 14; pt. 2, chaps. 9-14, ed. Bridges, 1: 44-59.

42. N. W. Fisher and Sabetai Unguru, "Experimental Science and Mathematics in Roger Bacon's Thought," *Traditio* 27 (1971): 378.

43. *Opus maius*, pt. 4, dist. 2, chap. 1, ed. Bridges, 1: 111. The translation is reprinted from Grant, *Source Book*, p. 393, Cf. Grosseteste, above, p. 98.

44. *De mult. spec.*, pt. 1, chap. 1, ed. Bridges, 2: 409.

45. *Opus maius*, pt. 5.1, dist. 9, chap. 4, ed. Bridges, 2: 71-72; translation reprinted from Grant, *Source Book*, p. 394. Cf. Bacon's *De multiplicatione specierum*, where the same doctrines are spelled out in more detail, especially pt. I, chap. 3, ed. Bridges, 2: 433-38. On Bacon's doctrine of the multiplication of species, see also Crombie, *Grosseteste*, pp. 145-47; Vescovini, *Studi sulla prospettiva*, pp. 57-62; and Sebastian Vogl, "Roger Bacons Lehre von der sinnlichen Spezies und vom Sehvorgange," pp. 205-22.

46. See this volume, pp. 7-8.

47. *De mult. spec.*, pt. 2, chap. 9, ed. Bridges, 2: 494-96.

48. Ibid., pt. 1, chap. 1, ed. Bridges, 2: 410.

49. Ibid., pp. 409-10. Cf. Pierre Michaud-Quantin, *Etudes sur le vocabulaire philosophique du Moyen Age*, pp. 126-27.

50. *Opus maius*, pt. 5.1, dist. 5, chap. 1, ed. Bridges, 2: 31.

51. Ibid., pp. 31-32.

52. Ibid., pt. 5.1, dist. 7, chap. 2, ed. Bridges, 2: 49.

53. See ibid., chap. 4, ed. Bridges, 2: 52, for the claim that the eye sees by means of the species of its own visual power.

54. Ibid., chap. 2, ed. Bridges, 2: 49. Reference here is to Aristotle's *De generatione animalium* 5. 1. 780a5-15, 780b34-781a14, where Aristotle says no such thing; however, the passage had been mistranslated by Michael Scot, with most pernicious results for the attempt to understand Aristotle's true position: cf. chap. 5, n. 79.

55. *Opus maius*, pt. 5.1, dist. 7, chap. 4, ed. Bridges, 2: 52; see also the corrections in ibid., 3: 138.

56. Ibid., pt. 5.1, dist. 7, chap. 4, ed. Bridges, 2: 52.

57. Ibid., chap. 3, ed. Bridges, 2: 50-51.

58. For an elaboration of these data and the whole question of Bacon's influence upon Pecham and Witelo, see my "Lines of Influence," pp. 72-77; cf. Crombie, *Grosseteste*, pp. 165, 213-16.

59. Two passages in Bacon's works confirm that his friends and brothers knew at least something of the content of his writings. In the *Opus minus* he wrote: "For my superiors and brothers, disciplining me with hunger, kept me under close guard and would not permit anyone to come to me, fearing that my writings would be divulged to others than to the chief pontiff and themselves" (from a portion of the *Opus minus* not presently extant, but quoted in Anthony Wood's life of Bacon and translated by Easton, *Roger Bacon*, p. 134). In his apology to the pope for his delay in sending his writings, Bacon noted that when he received the pope's request he had as yet "composed nothing, except that I had sometimes compiled in cursory fashion certain chapters now on one science and now on another at the request of friends. . . . And those things that I wrote I no longer have [in my possession], for because of their imperfection I did not take care of them" (F. A. Gasquet, "An Unpublished Fragment of a Work by Roger Bacon," *English Historical Review* 12 [1897]: 500).

60. It is possible that Bacon and Witelo briefly overlapped in Paris during the 1250s, but there is no reason to believe that either of them had yet become interested in optics; see my "Lines of Influence," pp. 72-73.

61. Witelo might have obtained Bacon's works through his friendship with William of Moerbeke, who was papal confessor and the one who encouraged Witelo to undertake optical studies. We know relatively little about Witelo's life. He was born in Silesia, probably in the vicinity of Breslau, and studied at the universities of Paris and Padua before appearing at the papal court late in 1268 or early in 1269. For a fuller account, see Aleksander Birkenmajer, *Etudes d'histoire des sciences en Pologne*, pp. 361-404; and an abbreviated version of the same work in "Witelo e lo studio di Padova"; see also my introduction to the reprint of the 1572 edition of Alhazen's and Witelo's optical works, pp. vii-ix, and my "Witelo," *Dictionary of Scientific Biography*, forthcoming. On Witelo see also Clemens Baeumker, *Witelo, ein Philosoph und Naturforscher des XIII. Jahrhunderts*; and Birkenmajer's "Etudes sur Witelo," first issued in the form of brief summaries in *Bulletin international de l'Académie polonaise des sciences et des lettres*, and recently published in full in his *Etudes d'histoire des sciences en Pologne*.

62. John Pecham was born in the early or mid-1230s and studied the arts at Oxford and Paris. After entering the Franciscan order, he received the theological doctorate at Paris in 1269. Thereafter he lectured at Paris, Oxford, and the Papal University in Viterbo and Rome, and became archbishop of Canterbury in 1279; he died in 1292. Pecham wrote two books on optics, the *Tractatus de perspectiva* and the *Perspectiva communis*. On Pecham's life and works, see Decima L. Douie, *Archbishop Pecham* (Oxford, 1952); Lindberg, *Pecham*, pp. 3-11; Lindberg, "Pecham, John," *Dictionary of Scientific Biography*, 10: 473-76.

63. Possibly also Witelo; see my "Lines of Influence," pp. 77-83; and Pecham, *Tractatus de perspectiva*, ed. David C. Lindberg, p. 14, n. 26.

64. *Perspectiva communis*, bk. 1, prop. 46, in Lindberg, *Pecham*, pp. 130-31. Pecham never referred to Alhazen by name, but only as *auctor Perspective*.

65. Ibid., p. 20.

66. I have elaborated on these themes in my "Alhazen's Theory of Vision and Its Reception in the West," pp. 336-40; and *Pecham*, pp. 34-40.

67. See Lindberg, "The Cause of Refraction in Medieval Optics," pp. 32-33.

68. *Perspectiva communis*, bk. 1, prop. 40, in Lindberg, *Pecham*, p. 125. Cf. Bacon, above, pp. 110-11.

69. *Perspectiva communis*, bk. 1, prop. 46, in Lindberg, *Pecham*, pp. 128-29.

70. Ibid., pp. 34-35.

71. Giovanni Battista Della Porta, *De refractione optices parte libri novem*, p. 76; Joseph Priestly, *The History and Present State of Discoveries Relating to Vision, Light, and Colours*, p. 22; Boyer, *Rainbow*, p. 103.

72. On Witelo's optical work, see Crombie, *Grosseteste*, pp. 214-32; Boyer, *Rainbow*, pp. 103-7; Vescovini, *Studi sulla prospettiva*, pp. 132-35; Lindberg, "Witelo"; Lindberg, "The Theory of Pinhole Images from Antiquity to the Thirteenth Century," pp. 164-76; Lindberg, "Cause of Refraction," pp. 30-34; Lindberg, "Introduction," p. xix-xxi. On Witelo's mathematical work, see Sabetai Unguru, "Witelo as a Mathematician: A Study in XIIIth Century Mathematics Including a Critical Edition and English Translation of the Mathematical Book of Witelo's *Perspectiva*"; also Unguru, "Witelo and Thirteenth-Century Mathematics: An Assessment of His Contributions."

73. I owe this translation largely to my research assistant, Kent Kraft, and my colleague, Fannie J. LeMoine; it is rendered from Baeumker's critical edition of the preface in *Witelo*, p. 128, lines 6-17. For another portion of Witelo's preface, see Crombie, *Grosseteste*, p. 215.

74. Crombie, *Grosseteste*, p. 215. Cf. Birkenmajer, *Etudes d'histoire des sciences*, pp. 276-78, 342.

75. Birkenmajer, "Robert Grosseteste and Richard Fournival," *Medievalia et Humanistica* 5 (1948): 36; cf. Crombie, *Grosseteste*, pp. 214-15.

76. Birkenmajer, *Etudes d'histoire des sciences*, p. 284.

77. Ibid, p. 292; cf, pp. 288-93.

78. See above, p. 19, and the works by Vescovini and Thorndike there cited. M.-Th. d'Alverny's forthcoming edition of *De radiis* may shed new light on the question.

79. There is a slight possibility that Witelo also had access to Pecham's *Perspectiva communis*, though I have argued that influence between Witelo and Pecham was more likely in the other direction; see my "Lines of Influence," pp. 77-83.

80. The following data are gathered from my *Catalogue of Optical Manuscripts*.

81. Leopold Delisle, *Le cabinet des manuscrits de la Bibliothèque Nationale*, vol. 3 (Paris, 1881), p. 90.

82. Joseph Aschbach, *Geschichte der Wiener Universität im ersten Jahrhunderte ihres Bestehens* (Vienna, 1865), pp. 121, 155, 201, 339, 365, 453 (also see the index for references to other lectures on perspective). Hastings Rashdall, *The Universities of Europe in the Middle Ages*, ed. F. M. Powicke and A. B. Emden (London, 1936), 1: 449, n. 3. Cracow, Bibl. Jagiellońska, MS 1840, fols. 179r-219r (containing Pecham's *Perspectiva communis*) is based on lectures given at the University of Vienna in 1444.

83. Rudolf Helssig, *Die wissenschaftlichen Vorbedingungen für Baccalaureat in artibus und Magisterium im ersten Jahrhundert der Universität*, in *Beiträge zur Geschichte der Universität Leipzig im fünfzehnten Jahrhundert* (Leipzig, 1909), pp. 17, 21, 37, 46, 56. *Registrum epistolarum fratris Johannis Peckham archiepiscopi Cantuariensis*, ed. Charles T. Martin, vol. 3 (London, 1885), p. lx.

84. On von Czechel, see the explicit to the *Perspectiva communis* in Cracow, Bibl. Jagiellońska, MS 1929, fol. 85r; cf. Baeumker, *Witelo*, p. 185. On Brudzewski's lectures, see Albertus de Brudzewo, *Commentariolum super theoricas novas planetarum Georgii Purbachii*, ed. L. A. Birkenmajer (Cracow, 1900), pp. xxx, 131. On the use of the *Perspectiva communis* at Cracow, see also Stefan Sweiżawski, "La philosophie à l'Université de Cracovie des origines au XVIe siècle," *Archives d'histoire doctrinale et littéraire du Moyen Age* 30 (1963): 97.

85. Lindberg, *Pecham*, p. 14. Radanus is probably Bartholomaeus de Radom, who received the M.A. degree from the University of Cracow in 1427; see Charles H. Lohr, S.J., "Medieval Latin Aristotle Commentaries: Supplementary Authors," *Traditio* 30 (1974): 125. The catalog of MSS in the Jagellonian Library, *Catalogus codicum manuscriptorum Bibliothecae Universitatis Jagellonicae*, ed. W. Wisłocki, 2 vols. (Cracow, 1877-81), also lists treatises by Henricus, Jacobus, Joannes, and Nicolaus de Radom, but none of an optical sort.

86. Guy Beaujouan, "Motives and Opportunities for Science in the Medieval Universities," in *Scientific Change*, ed. A. C. Crombie (New York, 1963), p. 221.

87. Baeumker, *Witelo*, pp. 184-86. Charles M. Coffin, *John Donne and the New Philosophy* (New York, 1958), p. 30.

88. William A. Wallace, O.P., "The 'Calculatores' in Early Sixteenth-Century Physics," *British Journal for the History of Science* 4 (1968-69): 225-26.

89. Lindberg, *Pecham*, p. 31.

90. For substantial lists of such citations, see ibid., pp. 31-32; Lindberg, "Introduction," pp. xxi-xxv.

Chapter 7 Visual Theory in the Later Middle Ages

1. For a list of extant manuscripts of these treatises, see my *Catalogue of Optical Manuscripts*, pp. 47-54, 57-62.

2. The *Quaestiones perspective* of Dominicus de Clavasio is extant in a single manuscript, Florence, Bibl. Nazionale, MS Conv. soppr. J.X.19, fols. 44r-55v (ca. 1400). The first and sixth questions have been edited by Graziella Federici Vescovini, "Les questions de 'perspective' de Dominicus de Clivaxo"; for an analysis of the treatise, see Vescovini, *Studi sulla prospettiva*, pp. 204-11. Henry of Langenstein and Blasius of Parma will be dealt with later in this chapter.

3. For example, bk. 1, qu. 13 of Blasius of Parma's *Questiones super perspectivam* is given over to a discussion of contingent angles.

4. For a fuller description of the Scholastic method, see M.-D. Chenu, O.P., *Toward Understanding Saint Thomas*, trans. A.-M. Landry and D. Hughes (Chicago, 1964), pp. 93-96.

5. On these and other medieval optical treatises, see my *Catalogue of Optical Manuscripts*.

6. Blasius's treatise is extant in sixteen MSS, Henry's in only five; however, Henry's *Questiones* had the advantage of being printed in 1503 (in Valencia). See my *Catalogue of Optical Manuscripts*, pp. 43-44, 62-63.

7. On Henry's life, see Otto Hartwig, *Henricus de Langenstein dictus de Hassia. Zwei Untersuchungen über das Leben und Schriften Henrichs von Langenstein* (Marburg, 1857); Joseph Aschbach, *Geschichte der Wiener Universität*, vol. 1 (Vienna, 1965), pp. 366-402; F. W. E. Roth, "Zur Bibliographie des Henricus Hembuche de Hassia," *Centralblatt für Bibliothekswesen*, Beihefte 2, vol. 1 (1888), pp. 97-118; Lynn Thorndike, *A History of Magic and Experimental Science*, 3: 472-510; and Nicholas Steneck, *Science and Creation: Henry of Langenstein on Genesis* (forthcoming), chap. 1.

8. Henricus Denifle and Aemilius Chatelain, eds., *Auctarium chartularii Universitatis Parisiensis*, vol. 1 (Paris, 1894), cols. 284, 289.

9. As Steneck will show in *Science and Creation*, Henry also included a great deal of natural philosophy in his *Lectures on Genesis*.

10. The following analysis is based on the forthcoming critical edition of the text by H. L. L. Busard, to be published with my introduction by the Österreichische Akademie der Wissen-

schaften in Vienna. I have consulted the manuscripts frequently, especially Erfurt, Wissenschaftliche Bibl., MS Ampl. F.380, fols. 29r-40v (14th c.); and Florence, Bibl. Nazionale, MS Conv. soppr. J.X.19, fols. 56r-85v (ca. 1400); as well as the printed edition, in *Preclarissimum mathematicarum opus in quo continentur perspicacissimi mathematici Thome Brauardini arismetica et eiusdem geometria necnon et sapientissimi Pisani carturiensis perspective ... cum acutissimis Ioannis d'Assia* [i.e., Henry of Langenstein] *super eadem perspectiva questionibus annexis* (Valencia, 1503), fols. 47r-65r. For convenience I will cite this printed text even when it diverges slightly from the critical edition or manuscript readings on which my analysis depends. On Henry's optical work, see Vescovini, *Studi sulla prospettiva*, pp. 165-93; David C. Lindberg, "The Theory of Pinhole Images in the Fourteenth Century," pp. 308-16.

11. *Preclarissimum opus*, fols. 47r-48v.

12. The other explanation is reflection or refraction in the atmosphere surrounding the luminous object; see ibid., fol. 51v.

13. This figure is found in the Erfurt MS, fol. 31v, and the Florence MS, fol. 63r. The smaller circle represents the crystalline humor, though its precise diameter and location cannot be determined from the manuscript drawings. The manuscripts do not distinguish *K* from *K' or E* from *E'*.

14. Although the text reads "toward the center of the eye" (*Preclarissimum opus*, fol. 51v), the manuscript copies of the figure have the rays actually reaching the center. This discrepancy is probably the result of error in copying the figure, but as Henry does not give sufficient information to permit rectification of the drawing, I have no alternative but to leave it as in the manuscripts.

15. Ibid., fol. 58r. Question 9, as Henry initially enunciates it, is ambiguous ("Utrum omne quod videtur recte videatur"), and I have taken this restatement ("An in medio uniformi visibilia irrefracte videantur") from a dozen lines farther down.

16. See this volume, p. 85; Lindberg, *Pecham*, pp. 119-21.

17. This figure is found in Erfurt MS F.380, fol. 35v, and in Florence MS J.X.19, fol. 72v.

18. Members of the medical tradition, such as Galen, Hunain, and Avicenna, had universally regarded the crustalline humor as lenticular or round; see this volume, pp. 34, 51; Rudolph E. Siegel, *Galen on Sense Perception*, pp. 44-45. Members of the perspectivist tradition, however, had been more ambiguous. Alhazen and Witelo had both argued at one point that the crystalline is spherical or lenticular and at another that the interface between the crystalline and vitreous humors is either plane or spherical, without choosing one or the other and also without specifying whether the sphericity would be concave or convex (as seen by the incoming ray)—though it would seem to follow clearly enough from the laws of refraction that only a concave surface would prevent intersection of the rays. It is also true, of course, that if the crystalline humor is spherical or lenticular the interface between it and the vitreous humor is concave. Bacon had clearly maintained the concavity of this interface, and Pecham had said nothing about its shape. See this volume, pp. 67 and 244 (n. 106); Alhazen, *De aspectibus*, bk. I, chap. 3, sec. 4, and bk. 2, chap. 1, sec. 3, pp. 3, 25-26; Witelo, *Perspectiva*, bk. 3, theor. 4, 23, pp. 85, 95-96; Bacon, *Opus maius*, pt. 5.1, dist. 3, chap. 3, and dist. 7, chap. 1, ed. Bridges, 2: 24, 47-48; Lindberg, *Pecham*, pp. 115, 123-25. In Henry's defense, it must be noted that the eye-diagrams in a number of manuscripts of Pecham's *Perspectiva communis* show this interface to be convex; see Lindberg, *Pecham*, plates 2*b* and 3*b*.

19. *Preclarissimum opus*, fol. 58v.

20. Ibid.

21. Ibid. Fig. 17 is copied from Erfurt MS F.380, fol. 35v, and Florence MS J.X.19, fol. 72v. The drawing is quite primitive in both manuscripts, and I can claim only to have captured the rough sense of it. Only two refractions are shown, either because of the difficulty of arranging the drawing to include all four refractions or because Henry wished to represent an instance of a ray that fails to reach the visual power.

22. Question 10 does not appear in the printed edition of Henry's *Questiones*; it is found in Erfurt MS F.380, fols. 35v-36r; and in Florence MS J.X.19, fols. 72v-74r.

23. For an illustration of this scheme, see Lindberg, *Pecham*, p. 251, n. 105.

24. This figure is mine rather than Henry's. Henry does not indicate whether we are to consider refraction of the rays in the aqueous and crystalline humors, and for the purposes of illustration I have excluded such refraction.

25. Erfurt MS F.380, fol. 35v.

26. However, it is possible that some confusion might have arisen from Henry's custom (following Alhazen, Pecham, and others) of regarding the crystalline and vitreous humors as the anterior and interior parts, respectively, of the glacial humor. On Alhazen, see above, p. 84. Cf. Oresme, below, chap. 7, n. 67.

27. An intention is that which acts upon and can be grasped by the interior senses or the intellect (as opposed to the five exterior senses). See Vescovini, *Studi sulla prospettiva*, pp. 64–69, 80–85; also consult the entry for this word in *The Oxford English Dictionary*. The twenty-two visible intentions commonly distinguished by perspectivists are listed by Pecham in bk. 1, prop. 55 of the *Perspectiva communis*; see Lindberg, *Pecham*, p. 135. The printed edition of Henry's *Questiones* also omits question 11; for the text, see Erfurt MS F.380, fols. 36r–37r; and Florence MS J.X.19, fols. 74r–76v.

28. Here Henry expressly denies Ockhamist teaching; see this volume pp. 141–42.

29. Erfurt MS F.380, fol. 36v; Florence MS J.X.19, fol. 75v: "In concavitatibus nervorum deferentium spiritus sensitivos descendentium a cerebro oportet esse corpus dyaphanum ad multiplicationem specierum aptum illuminatum, quod corpus terminatur ad organum exteriorum sensuum; vel oportet quod in illis concavitatibus sint subtilissima corpora et clarissima, scilicet spiritus, continue a cerebro fluentes usque vel versus exteriora et postea quodam motu refluentes, in quibus spiritibus refluentibus simulacra sensibilium recepta ad communem sensum vel ad ymaginationem."

30. On Blasius's life and works, see Edward Grant, "Blasius of Parma," *Dictionary of Scientific Biography*, 2: 192–95; Thorndike, *History of Magic*, 4: 65–79; Graziella Federici Vescovini, "Problemi di fisica aristotelica in un maestro del XIV [secolo]: Biagio Pelacani da Parma," *Rivista di filosofia* 51 (1960): 180–85.

31. Grant, "Blasius of Parma," p. 193.

32. Ferrara, Bibl. Comunale Ariostea, MS Classe II, n. 380, fol. 48r. The first ten questions of book 1 of Blasius's *Questiones super perspectivam* have been edited from Vienna, Österreichische Nationalbibl. MS 5447, by Franco Alessio, "Questioni inedite di ottica di Biagio Pelacani da Parma"; questions 14 and 16 of book 1 and question 3 of book 2 have been edited from Florence, Bibl. Medicea Laurenziana, MS Plut. 29 cod. 18, by Graziella Federici Vescovini, "Le questioni di 'Perspectiva' di Biagio Pelacani da Parma," pp. 207–43. Unfortunately, neither MS is dependable, and we badly need a critical edition; I have constantly checked the texts of Alessio and Vescovini against the Ferrara MS. For analysis of Blasius's optical work, see Vescovini, *Studi sulla prospettiva*, pp. 239–67; Vescovini, "Le questioni di 'Perspectiva,'" pp. 163–206; Lindberg, "Pinhole Images in the Fourteenth Century," pp. 316–23.

33. Vescovini, *Studi sulla prospettiva*, chap. 12, has attempted to discover a coherent visual theory by pulling together the contents of Blasius's *Questiones super perspectivam*, *Questiones in libros de anima*, and *Questiones in libros metheororum*. This may be a useful endeavor, but from the fact that the resulting doctrine (drawn mainly from the *Questiones in libros de anima*) is largely psychological and epistemological in orientation, we are not entitled to conclude (as Vescovini does, p. 246) that the perspectivist tradition has taken a major turn toward psychological and epistemological concerns. It would not be in the least out of character for a fourteenth-century scholar such as Blasius to behave as an Aristotelian while commenting on Aristotelian texts and as a perspectivist while commenting on works from the perspectivist tradition, and if Vescovini wishes to sustain her position she must demonstrate it from Blasius's *Questiones super perspectivam* alone. See below, chap. 7, n. 43.

34. "Questioni inedite," ed. Alessio, pp. 81–88. The argument against the existence of species is based on the principle of economy: "When an object causes vision, it is superfluous to posit the concurrence of species" (p. 85).

35. Ibid., p. 83.

36. Ibid., p. 84; see Vescovini, *Studi sulla prospettiva*, pp. 245-46.

37. "Questioni inedite," ed. Alessio, p. 83.

38. Ibid., p. 91.

39. Ferrara, Bibl. Comunale Ariostea, MS Classe II, n. 380, fol. 19r-v.

40. Ibid., fol. 21v.

41. Ibid., fol. 21r; cf. Oxford, Bodleian Lib., MS Canon. Misc. 363, fols. 89v-90r. The text here translated is not found in any individual manuscript, as it is based on a collation of seven manuscripts; the two cited above will suffice to locate the passage.

42. Qu. 2, 3, and 6, respectively, of Henry's *Questiones*; bk. I, qu. 3, 4, and 2, respectively, of Blasius's *Questiones*.

43. It must be added that in the *Questiones super perspectivam* of Henry and Blasius we can also perceive a certain shift away from traditional perspectivist concerns toward physical and psychological issues. This point has been made by Vescovini in chaps. 9 and 12 of her *Studi sulla prospettiva*, but somewhat too strongly. In the first place, her case rests on the implicit assumption that Henry and Blasius each had a unified, monolithic philosophy of nature, articulated more or less consistently in each of their works—which enables her to supply what is not stated in the *Questiones super perspectivam* by drawing on other works of Henry and Blasius (see above, chap. 7, n. 33). Moreover, she tends to overlook or underrate the presence of many of these same physical and psychological issues in the works of the thirteenth-century perspectivists. Thus while I agree that there was a shift in emphasis within the perspectivist tradition (an exact assessment of which must await a more detailed analysis of the relevant texts), I do not believe it amounted to a major change of direction.

44. For a listing of such commentaries, see Charles H. Lohr, S.J., "Medieval Latin Aristotle Commentaries," *Traditio* 23 (1967): 313-413; 24 (1968): 149-245; 26 (1970): 135-216; 27 (1971): 251-351; 28 (1972): 281-396; 29 (1973): 93-197; 30 (1974): 119-44. Optical discussions might also appear in commentaries on other Aristotelian works; see, for example, Nicole Oresme's *Questiones super de celo*, bk. I, qu. 20, ed. and trans. Claudia Kren, "The *Questiones super de celo* of Nicole Oresme," doctoral dissertation (University of Wisconsin, 1965), 1: 314-40.

45. On Buridan, see Ernest A. Moody, "Buridan, Jean," *Dictionary of Scientific Biography*, 2: 603-8; Edmond Faral, "Jean Buridan, maître ès arts de l'Université de Paris," *Histoire littéraire de la France*, vol. 38 (Paris, 1949), pp. 462-605.

46. For a discussion of the psychological aspects of Buridan's optical thought, see Vescovini, *Studi sulla prospettiva*, pp. 145-63; see pp. 155-57 on the nature of *lux*, *lumen*, and color.

47. Published in *Questiones et decisiones physicales insignium virorum*, ed. George Lockert, fol. 14va. On the authenticity of the *Questiones breves super librum de anima*, see Faral, "Jean Buridan," pp. 561-69.

48. *Questiones et decisiones*, fol. 14vb.

49. Ibid., fol. 14va.

50. Ibid., fol. 15ra.

51. Ibid., fol. 19rb.

52. Ibid.

53. Ibid., fol. 29va.

54. The eyelids also give off their own *lumen*, since every color except black participates to some degree in *lux*; see ibid., fol. 29vb.

55. The word that I have translated "reflected" (or "reflects" or "reflection") here and following is *refrangere* (or *refractio*). It is apparent that at this point Buridan does not wish to distinguish between reflection and refraction, but employs the term generically to denote breaking of the ray by either means. Forced to choose between the two in my translation, I have in each case chosen reflection, though in some instances, such as this first one, my decision might be disputed.

56. *Questiones et decisiones*, fol. 29vb. The same conclusion is reached in Buridan's *Questiones de anima*, bk. 2, qu. 24, ibid., fol. 19va-vb.

57. For a biographical sketch, see Nicole Oresme, *De proportionibus proportionum* and *Ad pauca respicientes*, ed. Edward Grant (Madison, Wis., 1966), pp. 3-10; Marshall Clagett, "Oresme, Nicole," *Dictionary of Scientific Biography*, 10: 223-30. See the latter for a list of Oresme's works.

58. See above, p. 100.

59. "Nicole Oresme and the Marvels of Nature," ed. and trans. Bert Hansen, p. 123. I have made several minor alterations in Hansen's translation.

60. "Nicole Oresme on Light, Color, and the Rainbow," ed. and trans. Stephen C. McCluskey, pp. 130-31. Here and in quotations below, I have slightly modified McCluskey's translation. For an analysis of Oresme's optical work, see ibid.; also Vescovini, *Studi sulla prospettiva*, pp. 195-204. A short optical treatise, *Questio de apparentia rei*, attributed to Oresme, has been edited and analyzed by Linda B. Watson; I am grateful to her for making available to me her unpublished work.

61. See this volume, chap. 5, n. 79.

62. "Oresme on Light, Color, and the Rainbow," ed. McCluskey, p. 132. What Oresme states here is that the visual power is bounded not by a lower limit (the maximum of those things that fall below the limit of vision) but by an upper limit (the minimum of those things that are above the limit of vision), and that vision must therefore be a passive power. Now if this maximum and minimum are taken to refer to the size of visible objects viewed from a given distance (as was customary in such discussions), the passage makes no sense, for sight as a passive power is limited by the largest of those objects that are too small to be seen at a given distance. Moreover, Oresme's conclusion here is in opposition to his later accounts in the *Questiones super de celo*, bk. 1, qu. 20 (ed. Kren, 1: 296-352); and *Le livre du ciel et du monde*, ed. and trans. A. D. Menut and A. J. Denomy (Madison, Wis., 1968), bk. 1, chap. 29, lines 74-91, p. 195. It seems then that Oresme was guilty of confusion at this point, or else that the text, as it has come down to us, is defective. On the question of maxima and minima, see Curtis Wilson, *William Heytesbury: Medieval Logic and the Rise of Mathematical Physics* (Madison, Wis., 1960), pp. 57-69; on Oresme's discussion of the matter, see McCluskey's commentary, "Oresme on Light, Color, and the Rainbow," pp. 405-8, n. 11.

63. "Oresme on Light, Color, and the Rainbow," ed. McCluskey, pp. 134-35.

64. Ibid., pp. 136-37.

65. Ibid., pp. 158-59.

66. Ibid., pp. 182-83.

67. Ibid., pp. 164-65. I am not certain what Oresme means by "the inner eye"—possibly the vitreous humor. If so, this remark may reflect the influence of Aristotle, who stressed the transparent watery substance of the eye as that which sees (see above, p. 9); it is possible also that the influence of Alhazen is involved (see above, p. 129).

68. "Oresme on Light, Color, and the Rainbow," ed. McCluskey, pp. 176-77.

69. For biographical notes on Versoris and an analysis of his works, see Pierre Duhem, *Le système du monde*, vol. 10 (Paris, 1959), pp. 111-31; Lohr, "Medieval Latin Aristotle Commentaries," *Traditio* 27 (1971): 290.

70. Joannes Versoris, *Questiones super parva naturalia cum textu Aristotelis*, qu. 3, fols. 2v-3r.

71. Ibid., qu. 4, fols. 3v-4r.

72. On the doctrine of the evil eye, see chap. 5, n. 12.

73. Versoris, *Questiones super parva naturalia*, fol. 3va.

74. Ibid., fol. 3vb.

75. Ibid.

76. Ibid.

77. Ibid.

78. For a listing of extant sentence commentaries, see Friedrich Stegmüller, *Repertorium Commentariorum in Sententias Petri Lombardi*, 2 vols. (Würzburg, 1947); Victorin Doucet, O.F.M., "Commentaires sur les Sentences: Supplément au Répertoire de M. Frédéric Stegmueller," *Archivum Franciscanum Historicum* 47 (1954): 88–170, 400–427.

79. *Commentarium in Primum*[*-quartum*] *librum Sententiarum*, 2: 180–85. Peter also discusses vision in the prologue of this same work, in the context of intuitive and abstractive cognition; cf. Philotheus Boehner, "*Notitia intuitiva* of Non Existents according to Peter Aureoli, O.F.M."

80. *In Petri Lombardi Sententias theologicas commentariorum libri IIII*, vol. 1, fols. 139r–140r. This discussiun has been analyzed by Anneliese Maier, "Das Problem der 'Species sensibiles in medio' und die neue Naturphilosophie des 14. Jahrhunderts," pp. 429–32. In other questions of the same work, Durandus inquires "whether light [*lumen*] is body" and "whether light [*lumen*] has real or intentional existence in the medium" (fols. 154v–155v).

81. *Gregorius Ariminensis super primo et secundo Sententiarum*, fols. 89v–99v, 36r–44r, respectively.

82. Ockam's birth was closer to 1280 than to 1290, according to Boehner; see William of Ockham, *Philosophical Writings: A Selection*, ed. Philotheus Boehner, O.F.M. (Edinburgh, 1957), pp. xi–xii. I have relied principally on Boehner for the following biographical sketch.

83. I have employed the edition of Lyons, 1495, bks. 2–4 of which contain some spurious sections. However, the portions from these books that I have used (bk. 2, qu. 17–18) have been shown to be genuine by Philotheus Boehner, "The Notitia intuitiva of Non-Existents according to William Ockham," p. 242.

84. Sebastian J. Day, *Intuitive Cognition: A Key to the Significance of the Later Scholastics*, pp. 143–48.

85. Ibid., pp. 149–58.

86. For further discussion, see Day's detailed analysis. See also Boehner, "*Notitia intuitiva* of Non Existents according to Peter Aureoli"; the same author's "The Notitia intuitiva of Non-Existents according to Ockham"; and C. K. Brampton, "Scotus, Ockham and the Theory of Intuitive Cognition."

87. A *habitus* is "an acquired ability to do certain things not possessed before or the ability to do them with greater ease or perfection than before" (Oswald Fuchs, *The Psychology of Habit according to William Ockham*, p. 1).

88. See Boehner, "The Notitia intuitiva of Non-Existents according to Ockham," pp. 223, 231–36.

89. For the Latin text, see Day, *Intuitive Cognition*, p. 189.

90. Ibid., p. 190. The claim is more than false; it is self-contradictory, since by definition intuitive cognition truthfully informs one regarding the existence of the object.

91. For the Latin text, see Fuchs, *Psychology of Habit*, p. 43. The middle clause ("which is capable of affecting sense and can be perceived by sense") is a free translation of "quae est passio vel passibilis qualitas quae potest sentiri a sensu."

92. Ibid.

93. This question has been analyzed at length by Maier, "Das Problem der 'Species sensibiles in medio,'" pp. 433–44.

94. William of Ockham, *In quattuor libros Sententiarum*, bk. 2, qu. 18, sec. D, sig. F7r.

95. Ibid. Ockham's position is that if light is present at any point in the medium, illumination will emanate in all directions from that point.

96. Maier, "Das Problem der 'Species sensibiles in medio,'" pp. 441–44.

97. For an effective summary of Ockham's position, see ibid., p. 444.

98. It should not be thought that my survey of the various categories of optical literature exhaustively covers the places where one might find optics treated or utilized. On the contrary, optics as a mode of thought pervaded medieval philosophy and theology, and treatises on virtually any subject might provide occasion for a discussion of visual theory or the use of an

optical analogy or metaphor. For some illustrations of fourteenth-century use of optics in seemingly nonoptical contexts, see Edith Sylla, "Medieval Quantifications of Qualities: The 'Merton School,'" *Archive for History of Exact Sciences* 8 (1971): 9-39. It should also be noted that I have omitted medical works from discussion in this chapter on the grounds that medical authors of the fourteenth and fifteenth centuries presented only brief and relatively trivial discussions of the physiology of sight and, moreover, that they will be dealt with in the following chapter.

99. But see the reservations expressed above, chap. 7, n. 43.

100. I recognize that an analysis of other issues, such as the nature of light and color, might yield different systems of classification.

101. Martin Grabmann, *Die Geschichte der scholastischen Methode*, vol. 2 (Freiburg i.B., 1911), pp. 13-27; Palémon Glorieux, "Sentences (Commentaires sur les)," *Dictionnaire de théologie catholique*, vol. 14, pt. 2 (Paris, 1941), cols. 1871-77; Chenu, *Understanding Saint Thomas*, pp. 79-96; Chenu, *Nature, Man, and Society in the Twelfth Century*, trans. Jerome Taylor and Lester K. Little, pp. 291-300.

102. Chenu, *Understanding Saint Thomas*, pp. 86-87. An important aspect of the quest "for a deeper understanding," of which Chenu writes, was the attempt to apply new conceptual languages to old problems—the language of intension and remission of forms, the calculus of proportions, the mathematics of the infinite, the theory of supposition, and the conceptual apparatus of Ockham's nominalism, to name a few. These new "languages" and their impact on natural philosophy are skillfully summarized by John E. Murdoch, "Philosophy and the Enterprise of Science in the Later Middle Ages," in *The Interaction between Science and Philosophy*, ed. Yehuda Elkana (Atlantic Highlands, N.J., 1974), pp. 51-74.

103. Other forms are briefly described by Lohr, "Medieval Latin Aristotle Commentaries," *Traditio* 23 (1967): 313.

Chapter 8 Artists and Anatomists of the Renaissance

1. See, for example, John White, *The Birth and Rebirth of Pictorial Space*, 2d ed., pp. 57-69; Erwin Panofsky, *Renaissance and Renascences in Western Art* (New York, 1969), pp. 136-39.

2. White, *Birth and Rebirth*, pp. 58-66; Miriam S. Bunim, *Space in Medieval Painting and the Forerunners of Perspective* (New York, 1940), pp. 139-43. For a much fuller discussion of the Arena Chapel frescoes, see James H. Stubblebine, *Giotto: The Arena Chapel Frescoes* (New York, 1969).

3. The words are Lorenzo Ghiberti's, quoted by Richard Krautheimer and Trude Krautheimer-Hess, *Lorenzo Ghiberti*, p. 14. Similar statements were commonplace among Renaissance artists.

4. Antonio di Tuccio Manetti, *The Life of Brunelleschi*, ed. and trans. Howard Saalman and Catherine Enggass (University Park, Pa., 1970), p. 42. On Brunelleschi's perspective drawings see also Samuel Y. Edgerton, Jr., "Brunelleschi's First Perspective Picture"; Edgerton, *The Renaissance Rediscovery of Linear Perspective*, chap. 10; White, *Birth and Rebirth*, pp. 113-21; Alessandro Parronchi, *Studi su la dolce prospettiva*, pp. 226-95 (which should be read along with the review by Samuel Y. Edgerton, Jr., in *Art Bulletin* 49 [1967]: 77-80); Joan Gadol, *Leon Battista Alberti*, pp. 25-32; Giulio Carlo Argan, "The Architecture of Brunelleschi and the Origins of Perspective Theory in the Fifteenth Century"; and R. Wittkower, "Brunelleschi and 'Proportion in Perspective.'"

5. After Parronchi, *Studi*, plate 91.

6. Manetti, *Life of Brunelleschi*, p. 44.

7. Ibid.

8. It has been claimed both by the translator of Alberti's treatise, John Spencer, and by one of Alberti's most recent biographers, Joan Gadol, that Alberti's decision to employ a pyramid (with square or rectangular base) rather than the ancient and medieval cone represented an

important methodological innovation. In fact, "pyramid (*pyramis*)" was the standard medieval term for designating such a figure without specifying the shape of its base, and there is no reason to suppose that Alberti intended any more than to continue the medieval tradition. (The medieval Latin term *conus* meant point or apex.) See Leon Battista Alberti, *On Painting*, trans. John R. Spencer, 2d ed., p. 103; Gadol, *Alberti*, pp. 29–31.

9. Alberti, *On Painting*, p. 48.

10. Ibid., p. 46. This passage is found only in the Latin version of *Della Pittura* (*On Painting*), probably written several years after the Italian version. For a cogent defense of this reversal of the order traditionally assigned the two versions, see Maria Picchio Simonelli, "On Alberti's Treatises of Art and Their Chronological Relationship," *Yearbook of Italian Studies, 1971* (Florence, 1972), pp. 82–92.

11. Alberti, *On Painting*, p. 47, once again only in the Latin version.

12. Ibid., pp. 51–52.

13. Ibid., pp. 68–69.

14. Ibid., p. 69.

15. The best of the innumerable analyses of Alberti's perspective construction is Edgerton, *Renaissance Rediscovery of Linear Perspective*, chap. 2. See also Edgerton, "Alberti's Perspective: A New Discovery and a New Evaluation"; Cecil Grayson, "L. B. Alberti's 'Costruzione Legittima'"; Gadol, *Alberti*, pp. 37–53; Erwin Panofsky, "Das perspektivische Verfahren Leone Battista Albertis"; Panofsky, "Die Perspektive als 'symbolische Form'"; White, *Birth and Rebirth*, pp. 121–26; Alessandro Parronchi, "La 'costruzione legittima' è uguale alla 'costruzione con punti di distanza'"; Parronchi, *Studi*, pp. 296–312; William M. Ivins, Jr., *On the Rationalization of Sight*; and Spencer's notes to his translation of Alberti's *On Painting*, pp. 110–17.

16. Figures 20*a* and 20*b* are taken, with minor changes, from Edgerton, *Renaissance Rediscovery of Linear Perspective*, p. 45. Alberti gives no figures and only an elliptical description.

17. Edgerton, "Brunelleschi's First Perspective Picture," pp. 176–78; Edgerton, *Renaissance Rediscovery of Linear Perspective*, pp. 134–35.

18. Edgerton, "Brunelleschi's First Perspective Picture," p. 177.

19. Alberti, *On Painting*, p. 83. Leonardo da Vinci may have had something similar in mind when he wrote that "the mirror is the master of painters" (*The Notebooks of Leonardo da Vinci*, trans. Edward MacCurdy, p. 240).

20. Il Filarete, *Filarete's Treatise on Architecture*, trans. John R. Spencer, 1: 305.

21. Parronchi, *Studi*, pp. 240 ff.; cf. Edgerton's review of Paronchi's *Studi*, p. 78. Edgerton points out in *Renaissance Rediscovery of Linear Perspective*, p. 63, that Michele Savonarola, grandfather of Girolamo, noted the tendency of painters in his native Padua to study *perspectiva*.

22. In *Della pittura* Alberti writes: "Therefore, the distance and the position of the central ray are of greatest importance to the certainty of sight" (*On Painting*, p. 48). Ptolemy, Damianos, al-Kindi, and Avicenna all argued that vision is clearer near the central ray of the visual pyramid, but certainty entered the matter only with Alhazen and his followers. See this volume, pp. 26–28, 85, 235 (n. 79); Albert Lejeune, *Euclide et Ptolémée*, pp. 35–36; Bacon, *Opus maius*, pt. 5.1, dist. 7, chap. 4, ed. Bridges, 2: 53; Lindberg, *Pecham*, pp. 121–23. On Alberti's debt to medieval optics, see also Edgerton, *Renaissance Rediscovery of Linear Perspective*, chap. 5.

23. To be precise, Edgerton associates Alberti's construction with prop. 10 of Theon of Alexandria's recension of Euclid's *Optica*. However, Theon's work seems not to have been in circulation until later, and prop. 11 of *De visu* (which had an enormous circulation during the Middle Ages and Renaissance) has essentially the same content. Edgerton's argument is found in his "Alberti's Perspective," p. 373; cf. Decio Gioseffi, *Perspectiva artificialis*, p. 19. On a possible fifteenth-century version of Theon's recension, see Lindberg, *Catalogue of Optical Manuscripts*, pp. 46–47.

24. Wilfred R. Theisen, ed., "The Medieval Tradition of Euclid's *Optics*," p. 197. The same

proposition and the same accompanying figure would also have been available to Alberti in Witelo's *Perspectiva*, bk. 4, theor. 37, p. 135.

25. Krautheimer and Krautheimer-Hess, *Ghiberti*, pp. 307–8. The standard edition of Ghiberti's *Commentarii* is *Lorenzo Ghibertis Denkwürdigkeiten* (*I Commentarii*), ed. Julius von Schlosser, 2 vols.

26. Others cited with regard to optics are Aristotle, Ptolemy, al-Kindi, Avicenna, Averroes, Constantinus Africanus, and the authors of the *Libro dello specchio* (Euclid) and *Libro dei crepuscoli* (Abhomadi Malfegeyr). However, Vescovini has argued that these citations are based largely on Bacon; see Graziella Federici Vescovini, "Contributo per la storia della fortuna di Alhazen in Italia," pp. 21–22. Vescovini has also demonstrated (pp. 21–49) how extensively Ghiberti relied on Bacon and Alhazen (whose *De aspectibus* Ghiberti used in an Italian translation, Biblioteca Apostolica Vaticana, MS Vat. Lat. 4595). On the optical content of the *Third Commentary*, see also Arturo Castiglioni, "Il trattato dell'ottica de Lorenzo Ghiberti"; White, *Birth and Rebirth*, pp. 126–30; Parronchi, *Studi*, pp. 313–48; Gezienus Ten Doesschate, *De derde Commentaar van Lorenzo Ghiberti in Verband met de middeleeuwsche Optiek*, pp. 4–9; and Kathryn Bloom "Lorenzo Ghiberti's Space in Relief: Method and Theory."

27. *Birth and Rebirth*, p. 127. Cf. Ghiberti, *I Commentarii*, ed. von Schlosser, 1: 126–43.

28. White, *Birth and Rebirth*, pp. 129–30.

29. There has been debate on the extent to which Albertian perspective is embodied in Ghiberti's own artistic compositions. Krautheimer, *Ghiberti*, pp. 321–34, argues that there was an "intricate give and take" between Alberti and Ghiberti and that Ghiberti's later artistic work closely followed the theories set forth by Alberti. As for perspective, Krautheimer writes: "One recalls that in the *Isaac* and *Joseph* panels Ghiberti employed a system of linear perspective which followed almost trait for trait the prescription given in the first book of *Della Pittura*" (p. 323). Parronchi, on the other hand, has argued that Ghiberti's artistic creations reflect a system of perspective that is anti-Albertian; see Parronchi, *Studi*, pp. 313–48.

30. Since the *Third Commentary* was never completed, it is possible that in later sections Ghiberti intended to provide an explicit discussion of the relationship between visual theory and the problem of perspective.

31. I have in mind Piero della Francesca, Jean Pélerin Viator, Antonio di Piero Averlino (Il Filarete) (d. ca. 1465), Leonardo da Vinci, and Giacomo Barozzi da Vignola (d. 1573). For their perspective constructions, see L. Brion-Guerry, *Jean Pélerin Viator*, pp. 18–115; Il Filarete, *Treatise on Architecture*, trans Spencer, 1: 301–3; M. H. Pirenne, "The Scientific Basis of Leonado da Vinci's Theory of Perspective"; White, *Birth and Rebirth*, pp. 207–15; Carlo Pedretti, "Leonardo on Curvilinear Perspective"; Pierre Francastel, "La perspective de Léonard de Vinci et l'expérience scientifique au XVIe siècle"; Corrado Maltese, "Per Leonardo prospettico"; and Timothy K. Kitao, "Prejudice in Perspective: A study of Vignola's Perspective Treatise."

32. Above, chap. 7, n. 10.

33. See the quotation from Alberti, above, p. 149. Similar remarks were made by Il Filarete, *Treatise on Architecture*, trans. Spencer, 1: 301; and Viator, *De artificiali perspectiva* (Toul, 1509), chap. 1, reprinted in Ivins, *On the Rationalization of Sight*, sig. A2r.

34. No less a scholar than James Ackerman has written: "Renaissance science, even when concerned with phenomena too distant to be seen, conformed to common sense experience, and it was just for this reason that an artist like Leonardo, trained from childhood in observation, could be a more original scientist than physicists and anatomists of this time, who looked rather at ancient books than at the world around them" ("Science and Visual Art," in *Seventeenth Century Science and the Arts*, ed. Hedley Howell Rhys [Princeton, 1961], p. 650).

35. There are scores of studies of Leonardo's life and works. See, for example, Kenneth Clark, *Leonardo da Vinci: An Account of His Development as an Artist* (Cambridge, 1939). For a recent study, oriented toward Leonardo's scientific work, see V. P. Zubov, *Leonardo da Vinci*, trans. David H. Kraus (Cambridge, Mass., 1968).

36. *Notebooks*, trans. MacCurdy, p. 58.

37. Ibid., p. 57.

38. "Léonard de Vinci, ingénieur et savant," in *Léonard de Vinci et l'expérience scientifique au XVIe siècle*, p. 20.

39. "Léonard et ceux qu'il n'a pas lus," in ibid., pp. 46-47. Cf. Alexandre Koyré's more cautious statement in the "Rapport final" of the same conference, ibid., p. 239.

40. "Leonardo da Vinci: Mechanics," *Dictionary of Scientific Biography*, 8: 215-16.

41. Leonardo's Italian version of Pecham's proemium, accompanied by an English translation, is provided in *The Literary Works of Leonardo da Vinci*, ed. and trans. Jean Paul Richter, 3rd ed., 1: 117, par. 13. Leonardo also quotes the first proposition of Pecham's *Perspectiva communis*; see ibid., 1: 124, par. 39. Leonardo informs us that a copy of Witelo's work was to be found in the library of "Sancto Marco" in Florence, where he apparently was able to use it. For Leonardo's references to Witelo, see the indexes to the volumes edited by MacCurdy and Richter. I do not believe that the often-stated view that Leonardo relied heavily on Roger Bacon can be sustained. Leonardo does indeed refer to Bacon by name, but not in connection with optics, and the case for Leonardo's dependence on Bacon has rested wholly on inconclusive parallels among their optical writings—parallels that could equally well be taken as evidence of the influence of Pecham and Witelo. For a long but uncritical study of Leonardo's sources, see Edmondo Solmi, "Le fonti dei manoscritti di Leonardo da Vinci," *Giornale storico della letteratura italiana*, suppl. 10-11 (1908), pp. 1-344. For a more balanced approach, see Eugenio Garin, "Il problema delle fonti del pensiero di Leonardo," in Garin, *La cultura filosofica del rinascimento italiano* (Florence, 1961), pp. 388-401.

42. Leonardo's manuscripts and their facsimile editions are listed in Kate T. Steinitz, *Manuscripts of Leonardo da Vinci: Their History, with a Description of the Manuscript Editions in Facsimile* (Los Angeles, 1948). The greatest quantity of optical material is found in MSS A, C, D, and F of the Bibliothèque de l'Institut de France (Paris), the Codex Atlanticus of the Biblioteca Ambrosiana in Milan, and MS Ital. 2038 of the Bibliothèque Nationale in Paris.

43. Portions of both manuscripts are translated in the volumes edited by MacCurdy and Richter. MS D has been translated (almost entirely) by Nino Ferrero, "Leonardo da Vinci: Of the Eye"; and (entirely) by Donald S. Strong, "Leonardo da Vinci on the Eye."

44. The best work is Strong's. Also useful are Edmondo Solmi, *Nuovi studi sulla filosofia naturale di Leonardo da Vinci*, pp. 138-218; J. Playfair McMurrich, *Leonardo da Vinci the Anatomist*, pp. 217-27; Anna Maria Brizio, *Razzi incidenti e razzi refressi*; Giovanni Perrod, "La diottrica oculare di Leonardo da Vinci"; W. Elsässer, "Die Funktion des Auges bei Leonardo da Vinci"; Vasco Ronchi, "L'optique de Léonard de Vinci"; Ronchi, "Leonardo e l'ottica"; Ronchi, "A Scarcely Known Aspect of the Activity of Leonardo da Vinci in the Field of Optics"; Kenneth D. Keele, "Leonardo da Vinci on Vision"; Keele, "Leonardo da Vinci's Physiology of the Senses"; and Willem van Hoorn, *As Images Unwind*, pp. 115-47. Some of these studies must be used with caution.

45. Leonardo recognizes that the medium must first be illuminated: "The instant the atmosphere is illuminated it will be filled with an infinite number of species which are produced by the various bodies and colors assembled in it" (*Literary Works*, trans. Richter, 1: 135). I have substituted the term "species" for Richter's "images," since the Italian original reads *spetie*.

46. The figure is from Paris, Bibl. Nationale, MS Ital. 2038, fol. 6v; cf. *Literary Works*, ed. Richter, I, 137.

47. *Notebooks*, trans. MacCurdy, p. 965.

48. *Literary Works*, trans. Richter, 1: 138. Again I have made minor alterations in Richter's translation.

49. Ibid., 1: 139, translation slightly modified.

50. Ibid. I have corrected Richter's figure by reference to a facsimile of the original in *The Drawings of Leonardo da Vinci in the Collection of Her Majesty the Queen at Windsor Castle*, ed. Kenneth Clark, 2d ed. revised with the assistance of Carlo Pedretti, vol. 3 (London, 1969),

fol. 19150v. Leonardo has frequently been credited with discovering the principles of the pinhole camera or *camera obscura*. In fact, Leonardo had a fair knowledge of the phenomena of the *camera* (which he could easily have acquired by reading medieval treatises on the subject) but no more than an elementary understanding of the theory behind them. Leonardo knew that light radiates in straight lines through small apertures (as revealed in fig. 23), that at a distance behind the aperture equal to the distance of the object before the aperture the dilation of the radiant pyramid is as wide as the object, and that a pinhole camera produces inverted images. He also knew that in radiation through large apertures, the radiating beam has the shape of the aperture just beyond the aperture, the shape of the luminous source far beyond the aperture. But Leonardo presented no theory to explain the phenomena of large apertures. See *Notebooks*, trans. MacCurdy, pp. 233, 952, 960-61. On the ancient and medieval theory of radiation through apertures, see my three articles on pinhole images.

51. Strong, "Leonardo on the Eye," p. 50.

52. *Literary Works*, trans. Richter, 1: 136.

53. *Notebooks*, trans. MacCurdy, p. 962.

54. *Literary Works*, trans. Richter, 1: 131. I have substituted the term "similitudes" for Richter's "images." Cf. *Notebooks*, trans. MacCurdy, p. 993.

55. *Literary Works*, trans. Richter, 1: 134. Cf. *Notebooks*, trans. MacCurdy, p. 994.

56. *Notebooks*, trans. MacCurdy, pp. 217, 258, 982-83.

57. Ibid., p. 217.

58. Ibid., p. 237.

59. One evidence of this process of intromission to which Leonardo frequently refers is the after-image; see ibid., pp. 241-42, 951.

60. Ibid., p. 234. The leaf, fol. 270vc, was provisionally dated to 1492 by Carlo Pedretti, *Studi Vinciani: Documenti, analisi e inediti leonardeschi* (Geneva, 1957), p. 278. However, in private correspondence Professor Pedretti has informed me that he would now prefer a date of 1487-90 and would go no later than 1491.

61. *Notebooks*, trans. MacCurdy, p. 236.

62. Ibid., p. 235.

63. Ibid.

64. *Literary Works*, trans. Richter, 1: 171: "So long as the air is free from grossness or moisture they [the lines from the eye and the solar and other luminous rays] will preserve their direct course, always carrying the image of the object that intercepts them back to their point of origin. And if this is the eye, the intercepting object will be seen by its colour, as well as by form and size." This passage is from the Windsor anatomical drawings, fol. 19148v, which dates from ca. 1483-85; on the dating of this leaf, see *Drawings of Leonardo at Windsor Castle*, ed. Clark and Pedretti, 3: 56.

65. *Notebooks*, trans. MacCurdy, p. 982; from the Windsor drawings, fol. 19148v.

66. *Literary Works*, trans. Richter, 1: 139-40. The first bracketed expression is mine; in its place Richter has "[of the eye]." For medieval uses of this same argument, see above, pp. 45, 64.

67. *Notebooks*, trans. MacCurdy, pp. 195, 197, 231, 251; *Literary Works*, trans. Richter, 2: 99; Strong, "Leonardo on the Eye," pp. 62-63.

68. A passage of the treatise *On the Eye* (written about 1508) that might seem to refer to visual rays (Strong, "Leonardo on the Eye," p. 84) is in fact referring merely to the fringes or rays that appear to surround a luminous body viewed through squinted eyes.

69. Kenneth Keele, who has done much recent writing on the subject, states that "Leonardo saw vision as the result of the percussion of light, first on the eye, then on the optic nerve, up which its waves of percussion flow" ("Physiology of the Senses," p. 47). Cf. van Hoorn, *As Images Unwind*, pp. 120-22; Domenico Argentieri, "Leonardo's Optics," in *Leonardo da Vinci* (Istituto Geografico de Agostini-Novara) (New York, 1956), pp. 405-8.

70. *Literary Works*, trans. Richter, 1: 140, quoting from MS A, fol. 9v. A similar claim from

Codex Atlanticus, fol. 373r-v, is quoted by Keele, "Physiology of the Senses," pp. 38-39: "Every body situated within the luminous air fills the infinite parts of this air circle-wise with its images, and goes lessening its images throughout equidistant space. . . . The stone flung into the water becomes the centre of various circles, and these have their centre at the point percussed. The air in the same way is filled with circles the centres of which are the sounds and voices formed within them."

71. See above, pp. 79-80. It is true that Leonardo frequently refers to the "percussion" (*percussione*) of radiation, but I am aware of no passage in which we cannot accurately translate this by the term "incidence." And there is no passage, to my knowledge, in which Leonardo affirms that light is a percussion wave. If it should be argued that the term *percussione* has unavoidable mechanical connotations, it could be replied that so do the Latin terms *fractio* and *reflexio* employed by upholders of traditional, nonmechanistic conceptions of light to denote refraction and reflection. In short, the use of mechanical terminology to describe the phenomena of radiation does not demonstrate belief in the mechanistic nature of the radiated entity. Dr. Kenneth Keele and Kim Veltman, whose knowledge of the Leonardo manuscripts far surpasses my own, are still investigating these matters, and their labors should shed important light on the problem. I am grateful to them for an enjoyable afternoon discussing Leonardo and a fruitful correspondence.

72. See Strong, "Leonardo on the Eye," pp. 97, 131; and *Literary Works*, trans. Richter, passim.

73. *Literary Works*, trans. Richter, 2: 99. I have altered Richter's translation at several points, and because of the crucial nature of the passage I quote in full the Italian text: "Dico il vedere essere operato da tutti li animali mediante la luce; . . . impero chè chiaro si comprende, i sensi ricievendo le similitudini delle cose non mandano fori di loro alcuna virtù; anzi mediante l'aria, che si trova infra l'obietto e 'l senso, incorpora in sé le spetie delle cose, e per lo contatto, che à col senso, le porgie a quello."

74. Van Hoorn, *As Images Unwind*, pp. 146-47, argues that Leonardo does away with species, but nothing could be further from the truth.

75. *Literary Works*, trans. Richter, 1: 141; *Notebooks*, trans MacCurdy, p. 233. My translation is roughly a composite of the translations of Richter and MacCurdy. However, where my translation reads "emanation," the Italian text reads "virtù spirituali"; but this cannot be what Leonardo meant, for the argument makes sense only if this emanation or virtue is interpreted in corporeal terms—so that through such emanation the body becomes diminished in size. The analogy of the lodestone to sight had earlier been expressed by Filarete, *Treatise on Architecture*, trans. Spencer, 1: 301.

76. *Notebooks*, trans, MacCurdy, p. 232; cf. *Literary Works*, trans. Richter, 1: 120, par. 21.

77. Strong, "Leonardo on the Eye," pp. 54-57. Elmer Belt has claimed that Leonardo's placement of the crystalline sphere in the center of the eye was influenced by his dissection of the eye of an ox (*boeuf*), which (according to Belt) has the crystalline lens central to the eye; however, a friend of mine expert on bovine anatomy assures me that this is not the case—that the placement of the crystalline lens in cattle is basically identical to that in humans. See Elmer Belt, "Les dissections anatomiques de Léonard de Vinci," in *Léonard de Vinci et l'expérience scientifique*, p. 206.

78. Strong, "Leonardo on the Eye," pp. 89-110. Strong notes that many of Leonardo's drawings show the optic nerve penetrating the crystalline sphere; for two of these, see Keele, "Physiology of the Senses," p. 46. This is the most bizarre aspect of Leonardo's anatomy of the eye, but I suggest that its origin can be easily understood. The medieval perspectivists usually conceived the crystalline and vitreous humors (sometimes referred to as the anterior glacial humor and the interior glacial humor, respectively) as together forming a sphere. They never failed to distinguish between the two parts of this glacial sphere, recognizing that the two had different consistencies and different functions; they sometimes even specified that the crystalline or anterior glacial humor was lenticular in shape. Now Leonardo took this sphere, regarded it as a single homogeneous humor, and designated it by the names that medieval perspectivists had

reserved for its respective parts ("the vitreous or crystalline sphere," he calls it). This is an understandable, if lamentable, mistake. The probable explanation of Leonardo's supposition that the albugineous (our aqueous) humor is concentric with this glacial sphere and surrounds it on all sides is that whereas medieval texts in the perspectivist tradition clearly specify that the albugineous is situated in front of the crystalline humor, the drawings of the eye in many of those same texts show it behind the vitreous humor; see Lindberg, *Pecham*, p. 115 and plates 1-4. As for the penetration of the optic nerve through the albugineous humor to the posterior surface of the glacial sphere, in some of these same drawings the optic nerve is shown extending through the albugineous humor to the back of the vitreous humor. The same misconception could have arisen from (or been reinforced by) a misunderstanding of a passage in Alhazen's *De aspectibus* (bk. I, sec. 5, p. 4) or Witelo's *Perspectiva* (bk. 3, theor. 25, p. 97), where it is argued that the optic nerve extends to the circumference of the glacial humor (*ad circumferentiam glacialis*). What Alhazen and Witelo meant, of course, was the back of the vitreous (or interior glacial) humor rather than the crystalline (or anterior glacial) humor; but Leonardo, having merged the two glacial humors and placed them in the center of the eye, had no choice (if he were to follow Alhazen and Witelo) but to extend the optic nerve through the albugineous humor until it encountered the rear surface of this glacial sphere.

79. See above, pp. 36, 42; Lindberg, *Pecham*, pp. 123, 251.

80. *Notebooks*, trans. MacCurdy, p. 232. Cf. ibid., pp. 243-44, 248-49; Strong, "Leonardo on the Eye," pp. 64-65, 70-71.

81. *Notebooks*, trans. MacCurdy, p. 251. Cf. ibid., pp. 252, 257; Strong, "Leonardo on the Eye," pp. 63-64. From the latter passage it appears that Leonardo's idea may have stemmed from the dependency of the fringes that surround a point of light on the size of the pupillary aperture—though on that hypothesis one would expect the relationship between aperture size and the perceived size of the object to be inverse.

82. Strong, "Leonardo on the Eye," p. 57.

83. Ibid., pp. 45, 77. This new convergence also explains for Leonardo why pictures drawn according to the rules of artificial perspective appear too small—i.e., because in actual vision, refraction of the rays produces magnification; see ibid., p. 46.

84. In fact, as medieval perspectivists well knew, the size of the image is a function solely of the radius of curvature and the locations of the object and observer. It could therefore be argued, on the basis of the visual theory of the perspectivists, that the size of the pupillary opening would determine how large a portion of the image is perceived by the visual power, but surely not that the pupillary opening would affect the size of the perceived portions. It follows from Leonardo's theory that all visible objects viewed through a pupil of given diameter should appear identical in size.

85. It appears to follow from Leonardo's theory that vision is produced by the peripheral rays of the visual pyramid (see fig. 24), irrespective of their angle of incidence on the eye and without regard for other rays from the same point incident on other parts of the eye or rays from other points incident on the same part of the eye. This conclusion is reinforced by another proposition from the treatise *On the Eye*, in which Leonardo insists that if object *G* (fig. 38) is seen by pupil *R*, it will be seen through pyramid *GR*; however, if the pupil is opened to its full width *MF*, the object will be seen through the cylinder, which cuts line *AB* (representing the cornea?) in a larger area, with the result that the object will appear larger (Strong, "Leonardo on the Eye," p. 71). Thus Leonardo neither eliminates the multiplicity of rays (emanating from a single point) through refraction, as the medieval perspectivists had done, nor returns the rays to a single point of focus in the eye, as Kepler would do; though he makes the general claim that every point of the eye receives an image of the whole object (ibid., p. 50), Leonardo occupies himself in each particular case with the peripheral rays and ignores the others.

86. Ibid., p. 53.

87. Ibid. I have slightly altered Strong's translation.

88. Ibid., pp. 78, 89-90; *Notebooks*, trans. MacCurdy, p. 237.

89. Strong, "Leonardo on the Eye," pp. 59-60, 113.

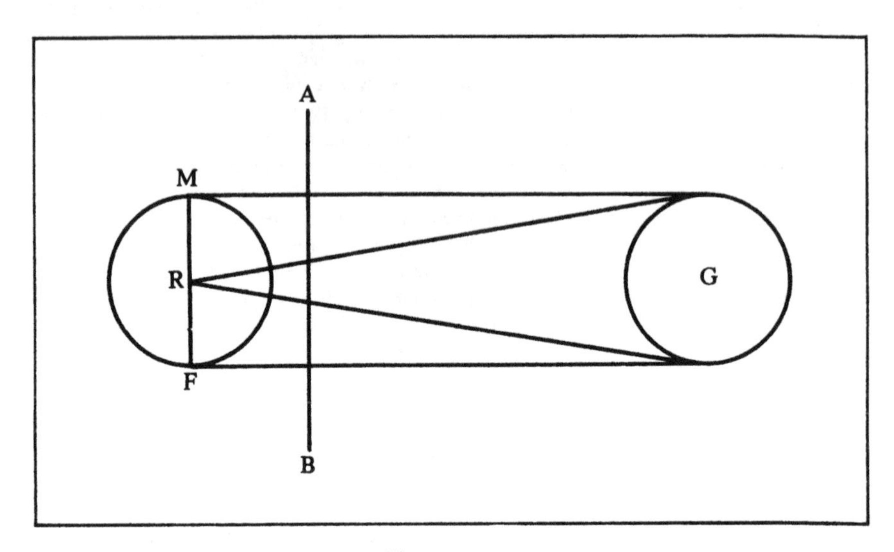

Fig. 38 (see note 85)

90. Ibid., p. 68.

91. Ibid., p. 69; cf. *Notebooks*, trans. MacCurdy, p. 996. This important conclusion was first expressed (so far as extant documents reveal) by Archimedes in *The Sandreckoner*; see *The Works of Archimedes*, trans. Thomas L. Heath, 2d ed. (Cambridge, 1912), p. 224; Albert Lejeune, "La dioptre d'Archimède." It was subsequently repeated by Alhazen, *De aspectibus*, bk. 7, chap. 6, sec. 37, p. 269. Although Leonardo here agrees with Archimedes and Alhazen, in an earlier manuscript he maintained the opposite position; see *Literary Works*, trans. Richter, 1: 131, par. 52; *Notebooks*, trans. MacCurdy, p. 993.

92. For the unique instance, see Strong, "Leonardo on the Eye," p. 73. For the others, see ibid., pp. 53-54, 56-57, 73-74. It is ironic that Leonardo has chosen as the sensitive part of the eye the blind spot, i.e., the part of the posterior surface of the eye that we know to be insensitive.

93. Leonardo wrote: "The vitreous [i.e., crystalline] sphere is placed in the middle of the eye in order to reinvert the species which intersect inside the small hole of the pupil" (ibid., p. 56).

94. Leonardo was interested by the fact that rays from various parts of the visual field were not mingled and confused as a result of intersection. He wrote: "The intersections of the images at the entrance of the pupil do not mingle one in another in that space where this intersection unites them; and this is evident [by contrast with the event when] rays of the sun pass through two panes of glass in contact one with another, the one of these being blue and the other yellow, [for] the ray that penetrates them does not assume the hue of blue or yellow but of a most beautiful green. And the same process would occur with the eye if the images yellow and green in colour should come to mingle one with the other at the intersection which they make within themselves at the entrance of the pupil, but as this does not happen such a mingling does not exist" (*Notebooks*, trans. MacCurdy, p. 260). Of course, Leonardo is here stating a conclusion that was commonplace among the perspectivists; see Alhazen, *De aspectibus*, bk. 1, chap. 5, sec. 29, p. 17.

95. Strong, "Leonardo on the Eye," p. 56.

96. Keele, "Physiology of the Senses," pp. 42-43; *Literary Works*, trans. Richter, 2: 100. On the interior faculty of judgment, see Alhazen's *ultimum sentiens*, this volume, pp. 81-84; and

Pecham's *iudicium interior*, Lindberg, *Pecham*, pp. 38, 118-19. For an extended discussion of medieval doctrines of the internal senses, which I have generally omitted from discussion in this book, see Nicholas H. Steneck, "The Problem of the Internal Senses in the Fourteenth Century."

97. *Notebooks*, trans MacCurdy, p. 110; *Literary Works*, trans. Richter, 2: 101.

98. See Steinitz, *Manuscripts of Leonardo*, pp. 41-43; *Notebooks*, trans. MacCurdy, pp. 41-47.

99. Human dissection was rare before the fourteenth century and uncommon then. Whereas Mondino, at the beginning of the fourteenth century, could claim to have performed several dissections, Berengario da Carpi, at the beginning of the sixteenth, could claim hundreds; see the introduction to *The Fasciculo di medicina*, ed. and trans. Charles Singer, 1: 45-50; Jacopo Berengario da Carpi, *A Short Introduction to Anatomy*, trans. L. R. Lind, p. 35. On anatomical dissection in the medieval medical curriculum, see also Charles Singer, *A Short History of Anatomy and Physiology from the Greeks to Harvey* (New York, 1957), pp. 71-74; Vern L. Bullough, "Medieval Bologna and the Development of Medical Education," *Bulletin of the History of Medicine* 32 (1958): 204-7; and Nancy G. Siraisi, *Arts and Sciences at Padua: The Studium of Padua before 1350* (Toronto, 1973), pp. 168-70.

100. Singer, *Short History*, p. 90.

101. On Mondino, see ibid., pp. 74-86; George Sarton, *Introduction to the History of Science*, 3: 842-45; and Singer's introduction to the *Fasciculo di medicina*, vol. 1.

102. *Fasciculo di medicina*, 1: 53-54.

103. *Anothomia* (Pavia, 1478), reprinted in facsimile in *Anatomies de Mondino dei Luzzi et de Guido Vigevano*, ed. Ernest Wickersheimer, pp. [44]-[47]. I have also consulted Charles Singer's translation and facsimile text of the *Fasciculo di medicina* (which contains an Italian version of Mondino's *Anatomia*); and the fourteenth-century Latin manuscript reproduced in facsimile in Mondino dei Luzzi, *Anatomia*, ed. Lino Sighinolfi.

104. The anterior capsule of the crystalline lens; see above, p. 55.

105. *Anothomia*, ed. Wickersheimer, p. [47]: "Et hic humor est magis versus partem anteriorem quam humor vitreus in quo locatur." Cf. *Fasciculo di medicina*, ed. and trans. Singer, 1: 96. Certain other versions, as the Italian version edited by Sighinolfi, p. 183, omit this claim altogether.

106. *Anothomia*, ed. Wickersheimer, p. [46].

107. Ibid.

108. Ibid., p. [44].

109. But Galen had also done this; see Galen, *On the Usefulness of the Parts of the Body*, trans. Margaret T. May, pp. 490-98.

110. The works of Mondino, Lanfranc, Henry of Mondeville, and Guy de Chauliac all had an enormous circulation. Mondino's *Anatomia* went through at least forty-six printed editions before 1600, including translations into French, Italian, and Flemish; Lanfranc's *Chirurgia major* was translated into English twice and also into French, Spanish, German, and Hebrew; Henry of Mondeville's *Chirurgia* was translated into French in his own lifetime, into Provençal and Dutch in the fourteenth century, and into English in the fifteenth century; and Guy de Chauliac's *Chirurgia magna* was rendered into French, Provençal, Catalan, Italian, Dutch, English, Irish, and Hebrew. See Wickersheimer's notes to his edition of Mondino's *Anatomia*, pp. 86-91; Sarton, *Introduction to the History of Science*, 2: 1080-81; 3: 870, 1961. For descriptions of ocular anatomy by the three surgeons mentioned, see *Lanfrank's "Science of Cirurgie,"* ed. Robert von Fleischhacker, p. 241; *Die Anatomie des Heinrich von Mondeville*, ed. J. L. Pagel, pp. 37-42; *The Cyrurgie of Guy de Chauliac*, ed. Margaret S. Ogden, 1: 42-43.

111. Gabriele Zerbi, *Liber anathomie corporis humani*, pt. 1, fol 123r. Regarding arguments for the existence of only four tunics, see above, pp. 36, 67.

112. Zerbi, *Liber anathomie*, pt. 1, fols. 123r-128r.

113. Mondino regarded the crystalline humor as flattened only in front; Zerbi, by contrast, apparently conceived it to be symmetrical, i.e., flattened both front and back.

114. Zerbi, *Liber anathomie*, fols. 128v-129r. However, Zerbi does admit, on the authority of Averroes, that the chief organ of sight might be the *araneu* (arachnoid membrane) rather that the crystalline humor; see this volume, p. 171.

115. Zerbi frequently cites (and even quotes) Pecham's *Perspectiva communis*, and it is clear that he was well acquainted with the perspectivist tradition.

116. On Berengario, see C. D. O'Malley, "Berengario da Carpi, Giacomo" *Dictionary of Scientific Biography*, 1: 617-21; and Lind's introduction to Berengario's *Short Introduction*.

117. The first quotation is from Lind's introduction to Berengario's *Short Introduction*, p. 14, the second from Pierre Huard and M. D. Grmek, "L'oeuvre de Charles Estienne et l'école anatomique Parisienne," a separately printed introduction to the facsimile edition of Charles Estienne's *La dissection des parties du corps humain* . . . (Paris, 1546; reprinted Paris, 1965), p. [1].

118. Lind's introduction to Berengario's *Short Introduction*, p. 8.

119. Berengario admits that some anatomists identify a fourth humor, the "ethereal"; see Berengario, *Short Introduction*, p. 152. This humor was to be mentioned frequently in the sixteenth century.

120. Zerbi, *Liber anathomie*, fol. 129r. On Averroes and the possibility that the *aranea* is the principal sensitive organ, see above, p. 55.

121. Zerbi, *Liber anathomie*, fol. 129r. Zerbi notes, however, that perhaps different visible qualities or properties of an object are perceived by different visual organs—its color, size, and shape by the crystalline humor or *aranea*, its state of motion or rest and its number in the crossing of the optic nerves (ibid., fol. 129v).

122. Ibid., fol. 127v.

123. Above, pp. 54-56.

124. Zerbi, *Liber anathomie*, fol. 129r.

125. The idea that the principal organ of sight must be extremely lucid originated with Galen and was amplified by Hunain and Averroes; see this volume, chap. 3, n. 7; Averroes, *Colliget* (Venice, 1562), bk. 2, chap. 15, fol. 25vb.

126. Zerbi, *Liber anathomie*, fol. 127v. Zerbi seems here to be developing a suggestion made earlier by Galen; see above, p. 40. The physical relationship between the retina and the crystalline humor was frequently discussed during the Middle Ages. Alhazen believed the retina and the anterior capsule of the crystalline humor to constitute a single continuous tunic, which in his *De aspectibus* is designated the *aranea* (see above, p. 67); Alhazen was followed in this by Bacon, Witelo, and Pecham (Bacon, *Opus maius*, bk. 5.1, dist. 4, chap. 3, ed. Bridges, 2: 28-29; Witelo, *Perspectiva*, bk. 3, theor. 4, p. 86; Lindberg, *Pecham*, pp. 114-15). Avicenna distinguished the retina from the *aranea* (here denoting only the capsule of the crystalline humor), but asserted that the latter grew out of the former (*Liber canonis* [Venice, 1507], bk. 3, fen. 3, chap. 1, fol. 203vb). Among anatomists of the Renaissance it was universally acknowledged that the retina is in close contact with the crystalline humor, either because the two are joined at the equator of the crystalline or because the *aranea* is an extension of, or at least is in contact with, the retina. For those who placed the *aranea* and the retina in contact or made them one, it was obviously advantageous to regard the *aranea* as the sensitive ocular organ, since impressions could then pass directly from the *aranea* to the retina and optic nerve; although seldom stated explicitly and certainly not part of Averroes's original scheme, I believe this idea may have motivated many of the discussions of the possible sensitivity of the *aranea*. See (in addition to Zerbi) Alessandro Achillini, *De humani corporis anatomia*, sig. F2v; Charles Estienne, *De dissectione partium corporis humani libri tres*, p. 300; Johannes Jessen, *Anatomia Pragensis*, fol. 122v; Hieronymus Fabricius of Aquapendente, *Tractatus anatomicus triplex, quorum primus de oculo*, pp. 5-6.

127. Achillini, *De humani corporis anatomia*, sig. F2v.

128. *De utilitate particularum* is of course identical with what is now commonly entitled *De usu partium*, translated by Margaret May as *On the Usefulness of the Parts of the Body*. On *De iuvamentis membrorum*, see this volume, chap. 3, n. 6.

129. I have examined it in the *Opera omnia* of Venice, 1541-45, 8 (1545): 365-81. It is also included, under the title *De calore vitali*, with William of Conches's *Dialogus de substantiis physicis* . . . (Strasbourg, 1567).

130. The title page of the Strasbourg, 1567 edition describes it as a book "of uncertain authorship and of the same age" as William of Conches's *Dialogus* (i.e., *Dragmaticon*). The earliest manuscripts that I have located are the following four from the late thirteenth or early fourteenth century: Paris, Bibl. Nationale, MS Lat. 6865, fols. 118(117)v-121(120)r; MS Lat. 11860, fols. 217r-219r; MS Lat. 15456, fols. 147v-150(151)r; and Vatican City, Bibl. Apostolica, MS Palat. Lat. 1097, fols. 114r-116v. I have examined all four manuscripts to confirm their contents and to check the accuracy of the printed text; no clues are there provided regarding the origin of the treatise.

131. Galen, *Opera omnia*, 8: 371: "Dico etiam quod instrumenta omnia sensuum continentur pelliculis, fiuntque sensus in ipsis pelliculis, ut est videre in oculo, cuius pelliculae plenae sunt humore aqueo, et originem habent a cerebri pelliculis. Fitque visus non in crystallino humore oculi, ut Aristoteli visum est, sed in pelliculis oculi, quarum duae procedunt a duabus pelliculis cerebri, et tertia a nervo, qui opticus dicitur." Conceivably the author of *De iuvamentis anhelitus*, when he said that vision occurs "in the tunics of the eye," meant only the *aranea*; however, his use of the plural (tunics), and his description of the tunics (three in number) as the membranes that derive from the two tunics of the brain and the optic nerve, strongly suggest that he had in mind the tunics that surround the posterior portion of the eye, namely, the sclera, *secundina*, and retina.

132. See Huldrych M. Koelbing, *Renaissance der Augenheilkunde*, pp. 69-71; Koelbing gives a good account of Vesalius's contributions to ocular anatomy. For Vesalius's argument regarding the solidity of the nerve, see *De humani corporis fabrica* (Basel, 1543), p. 324. On the doctrine of the hollow nerve, see above, p. 36.

133. Vesalius, *De humani corporis fabrica*, p. 643. For other Renaissance drawings of the eye, see Karl Sudhoff, "Augenanatomiebilder im 15. und 16. Jahrhundert."

134. *De humani corporis fabrica*, pp. 649-50. This passage is translated into German in Koelbing, *Augenheilkunde*, p. 69.

135. *De humani corporis fabrica*, p. 650.

136. Ibid., p. 324: "Quod vero haec tunica . . . praecipuum visus organum a multis habeatur."

137. Ibid., p. 646. Although of no great significance, it is curious that just three years before publication of *De fabrica*, Vesalius, in his first public anatomical demonstration at Bologna (1540), argued that it is the crystalline humor "by which properly the vision occurs"; see *Andreas Vesalius' First Public Anatomy at Bologna, 1540: An Eyewitness Report by Baldasar Heseler, Medicinae Scolaris, Together with His Notes on Matthaeus Curtius' Lectures on Anatomia Mundini*, ed. and trans. Ruben Eriksson (Uppsala, 1959), p. 291.

138. Realdo Colombo, *De re anatomica libri XV*, pp. 219-20.

139. Ibid., p. 219.

140. Georg Bartisch, ΟΦΘΑΛΜΟΔΟΥΛΕΙΑ, *Das ist Augendienst*, fols. 8v-9r; Fabricius of Aquapendente, *Tractatus anatomicus triplex*, pp. 8, 64, tabula 49; Jessen, *Anatomia Pragensis*, p. 117. Bartisch asserts that the two surfaces of the crystalline humor are unequally curved, but he assigns the flattening to the rear rather than the front surface (fol. 7v). Fabricius also argued (p. 6) that the *aranea* completely surrounds the crystalline humor and that it is properly regarded not as a distinct tunic, but merely as the surface of the crystalline humor; on Fabricius, see also Koelbing, *Augenheilkunde*, pp. 76-77.

141. Estienne, *De dissectione*, p. 303: "Humor crystallinus, vitreo innatans, in conum vergens, cuius acies medio optico introrsum respondet." A figure confirms this description. The same description can be found earlier in the *Anothomia* of Hieronymo Manfredi: see "A Study in Early Renaissance Anatomy, with a New Text: The Anothomia of Hieronymo Manfredi (1490)," trans. A. Mildred Westland, p. 120. The same pointed posterior surface of the crystalline humor is shown in Gregorius Reisch, *Margarita philosophica* (Basel, 1517), sig. E1r, though without

accompanying description, and in the 1544 Juntine edition of Avicenna's *Liber canonis* (see Karl Sudhoff, "Augendurchschnittsbilder aus Abendland und Morgenland," p. 20). This strange idea seems to have originated with Avicenna's *Liber canonis*, where the description of the glacial humor (in Gerard of Cremona's Latin translation) concludes with the following remark. "Therefore its posterior surface is briefly constricted so that it will be well covered by the bodies [i.e., humors] that receive it; and these bodies [the receiving humors] are deep, and from the constricted [part] they are spread out, so as to cover it [the glacial humor] well" (*Liber canonis* [Venice, 1507], bk. 3, fen. 3, chap. 1, fol. 203va: "Et propter hoc eius postremum parumper constrictum est ut ipsius coopertura in corporibus fiat bona que ipsum deglutiunt et sunt profunda et dilatata a constricto: ut et ipsa bene cooperiant eum."). The Arabic version, to judge from the French translation of De Koning and the German translation of Hirschberg and Lippert, differs from Gerard's Latin translation only in minor and unimportant respects; see *Trois traités d'anatomie arabes par Muhammad ibn Zakariyya al-Razi, 'Ali ibn al-'Abbas et 'Ali ibn Sina*, trans. P. De Koning, p. 660; *Die Augenheilkunde des Ibn Sina*, trans. J. Hirschberg and J. Lippert, p. 12. However, Andreas Alpago's sixteenth-century Latin translation substitutes *subtiliatum* for Gerard's *constrictum*; see *Avicennae Liber canonis* (Venice, 1555), fol. 220r. This term "constricted" (given by Gerard of Cremona as *constrictum*, by De Koning as *s'amincit*, by Hirschberg and Lippert as *zusammengezogen*, and by Alpago as *subtiliatum*) is exceedingly ambiguous and might easily be interpreted to mean that the rear surface of the glacial humor comes to a point; however, I find it difficult to believe that this was Avicenna's intended meaning, and although I cannot offer a fully satisfactory interpretation of the passage I would suggest that Avicenna may simply have been assigning the posterior surface a thinner covering or capsule than that of the anterior surface (as Alpago's translation suggests and as is indeed the case).

142. Fabricius denied that the retina perceives alterations in the *aranea* or crystalline humor, arguing that the ocular part of visual perception occurs entirely in the crystalline; see *Tractatus anatomicus triplex*, p. 64. Fabricius vigorously defended the principal features of the perspectivist theory of vision, integrated with aspects of the Aristotelian and Galenic theories.

143. André Du Laurens, *Discourse de la conservation de la veue*, trans. Richard Surphlet, p. 32.

144. Ibid., p. 31. Laurens continues, in this work, with a very long analysis of arguments for and against the extramission and intromission theories, concluding finally in favor of the latter; see pp. 37-46.

145. Although *De iuvamentis anhelitus* was not written during the sixteenth century, it received its principal circulation at this time. Although Fabricius defended the perspectivist theory of vision, this did not force a significant departure from the anatomical and physiological aspects of Galen's theory.

146. On Platter's life and work, see J. Karcher, *Felix Platter: Lebensbild des Basler Stadtarztes, 1536-1614* (Basel, 1949); Werner Kolb, *Geschichte des anatomischen Unterrichtes an der Universität zu Basel 1460-1900* (Basel, 1951), pp. 23-28; Koelbing, *Augenheilkunde*, pp. 71-74, 115-24; *Beloved Son Felix: The Journal of Felix Platter, a Medical Student in Montpellier in the Sixteenth Century*, trans. Seán Jennett (London, 1961). It should be noted that despite its influence on Kepler, Platter's visual theory was not widely accepted among physicians (even his own colleagues at Basel) until well into the seventeenth century; see Huldrych M. Koelbing, "Ocular Physiology in the Seventeenth Century and Its Acceptance by the Medical Profession," pp. 222-23.

147. Kolb, *Geschichte des anatomischen Unterrichtes*, p. 26. Koelbing prefers to see Platter's book primarily as a response to Vesalius's *De fabrica*; see Koelbing, *Augenheilkunde*, p. 71; Koelbing, "Felix Platters Stellung in der Medizin seiner Zeit," p. 61. One of Platter's corrections of Vesalius was to argue that the crystalline humor is not situated in the center of the eye.

148. Felix Platter, *De corporis humani structura et usu . . . libri III*, p. 187. For another translation, see Crombie, "Mechanistic Hypothesis," p. 49.

149. Despite what this seems to say, Platter surely recognized that the crystalline humor is between the optic nerve and the pupil—as his diagram of the eye (fig. 28) reveals. What he evidently meant here is simply that the crystalline humor is opposite or facing both the nerve and the pupil.

150. Platter, *De corporis humani structura*, p. 187; see also Crombie's translation, "Mechanistic Hypothesis," p. 49. The Latin text of both this and the preceding quotation, accompanied by a German translation, is given by Koelbing, *Augenheilkunde*, pp. 72-73. Koelbing reproduces the Latin text in facsimile, enabling one to see (in its arrangement) Platter's method of dichotomy.

151. Platter does not even hint at the steps by which he arrived at his theory, but perhaps he was influenced by Vesalius's comparison of the crystalline humor to a looking glass; see this volume, p. 173. A remark of Kepler has been misunderstood as a claim that Platter (in an experimental demonstration of his theory) severed the crystalline humor from the ciliary body, and thus from its connection with the retina, without serious impairment of vision, thereby proving that the crystalline humor is not percipient; see Stephen Polyak, *The Retina*, p. 135. Actually Platter performed no such experiment, nor did Kepler attribute any such to him. What Kepler maintained in the relevant passage is merely that according to Platter (and contrary to traditional belief) the crystalline humor has no connection with the retina and optic nerve, but only with the uvea; no experiment was attributed to Platter, nor any conclusions regarding the insensitivity of the crystalline humor; see this volume, p. 188.

152. Kepler wrote of Platter: "Compare the true means of vision proposed by me with that of Plater [*sic*], and you will see that this most excellent man is no farther from the truth than is compatible with being a man in the medical profession who does not have a working grasp of mathematics" (quoted by Stephen M. Straker, "Kepler's Optics," p. 457).

Chapter 9 Johannes Kepler and the Theory of the Retinal Image

1. See above, pp. 120-21.

2. The Euclidean editions are described in Pietro Riccardi, *Saggio di una bibliografia Euclidea*, 5 pts. (Bologna, 1887-93). On the others, see Lindberg, "Introduction," pp. xxvi, xxviii; Lindberg, *Pecham*, pp. 30, 56-57; Lindberg, *Catalogue of Optical Manuscripts*, pp. 18, 71-72, 79.

3. On Maurolico's biography, see Arnaldo Masotti, "Maurolico, Francesco," *Dictionary of Scientific Biography*, 9: 190-94.

4. There was no edition of 1575, as sometimes alleged; there was, however, a second edition of Lyon (but not Messina), 1613. The 1611 edition (and presumably also the 1613 edition) actually consists of several distinct treatises, of which the *Photismi* is only the first; nevertheless, for convenience I will refer to them all by the short title *Photismi*. On the manuscripts of these works, see Lindberg, *Catalogue of Optical Manuscripts*, pp. 66-67. Maurolico cites Bacon, Witelo, Pecham, the *Optica* and *Catoptrica* of Euclid, and *De speculis comburentibus* of Alhazen (which he attributes, tentatively, to Ptolemy). For a discussion of Maurolico's other scientific works, see Marshall Clagett, "The Works of Francesco Maurolico," *Physis* 16 (1974): 149-98.

5. On the earlier history of the *Camera obscura*, see Joseph Würschmidt, "Zur Theorie der Camera obscura bei Ibn al Haitam"; Eilhard Wiedemann, "Über die Camera obscura bei Ibn al Haitam"; David Lindberg, "The Theory of Pinhole Images from Antiquity to the Thirteenth Century"; Lindberg, "A Reconsideration of Roger Bacon's Theory of Pinhole Images"; Lindberg, "The Theory of Pinhole Images in the Fourteenth Century"; Stephen M. Straker, "Kepler's Optics", pp. 99-222.

6. *Photismi de lumine et umbra* (Naples, 1611), p. 17. See also Henry Crew's English translation, *The Photismi de lumine of Maurolycus*, pp. 28-29. Crew's translation is so generally defective that I have felt compelled to retranslate this and other passages. In every case, I have also checked the text of 1611 against MS 2080 of the Bibl. Governativa in Lucca, fols. 2r-63r. On Maurolico's analysis of the *camera obscura*, see Straker, "Kepler's Optics," pp. 292-99.

7. Contrary to A. C. Crombie (trans.), "Kepler: De modo visionis," p. 141; and Crombie, *Grosseteste*, p. 281. Maurolico discusses sight in *Photismi*, pp. 69-80 (Crew's translation, pp. 105-21). Maurolico's failure to apply the theory of the *camera obscura* to the eye may reflect a belief on his part that the *camera* and the eye are disanalogous—that in the *camera* all rays contribute to the image, whereas in the eye (according to the perspectivist theory, which he holds) only selected rays participate in the visual process. On this disanalogy, see Lindberg, "Pinhole Images in the Fourteenth Century," p. 314, n. 57; and Straker, "Kepler's Optics," pp. 431-32, 445-48, 450.

8. *Photismi*, pp. 72-73 (Crew, p.110).

9. *Photismi*, pp. 69-70 (Crew, pp. 106-7).

10. *Photismi*, p. 74 (Crew, p. 112).

11. *Photismi*, p. 73 (Crew, pp. 110-11).

12. *Photismi*, p. 76 (Crew, p. 115).

13. *Photismi*, p. 76 (Crew, p. 115).

14. *Photismi*, p. 75 (Crew, pp. 113-14).

15. *Photismi*, p. 76 (Crew, p. 115).

16. *Photismi*, p. 77 (Crew, p. 116).

17. *Photismi*, p. 78 (Crew, p. 117).

18. If pressed, Maurolico might have acknowledged that (say) a planoconvex lens could be substituted for a double convex lens. However, his visual theory places great stress on symmetry, and in his examples he always employs symmetrical lenses.

19. See, for example, Vasco Ronchi, *The Nature of Light*, trans. V. Barocas, pp. 99-106. There are, of course, other causes of myopia and hypermetropia than those described by Maurolico.

20. *Photismi*, pp. 69, 80 (Crew, pp. 105, 120).

21. *Photismi*, p. 76 (Crew, p. 115).

22. *Photismi*, p. 75 (Crew, p. 113).

23. That is, he has dismissed all rays that do not enter and emerge from the crystalline humor symmetrically. Surely the principle of symmetry is insufficient justification for this step, even in sixteenth-century terms, for other rays obviously pass through the crystalline humor, and on what grounds can it be maintained that they would not stimulate the visual power?

24. On Della Porta's life, see Louise G. Clubb, *Giambattista Della Porta, Dramatist* (Princeton, 1965), pp. 3-56; and Miller Howard Rienstra, "Giovanni Battista Della Porta and Renaissance Science," pp. 24-148.

25. *Della Porta*, p. 6.

26. Rienstra, "Della Porta," pp. 130-48.

27. Ibid., pp. 83-87.

28. *De refractione optices parte libri novem*, pp. 91-92. Della Porta attempts to prove, in the geometrical argument, that if sight occurs by intromission, an object will appear farther from the eye than it actually is, whereas if sight were to occur by extramission, the object would appear closer to the eye than it actually is; the former, he says, is confirmed by experience. Della Porta acknowledges that animals can see at night, but this, he says, results from an excellence in their visual power not possessed by humans (p. 93).

29. Ibid., pp. 82-83.

30. Ibid., pp. 65-68, 83-86.

31. Ibid., p. 85.

32. Ibid., pp. 83-84, 91-92.

33. *Natural Magick* (an anonymous English translation of the expanded Latin edition of Naples, 1589), pp. 363-64.

34. *De refractione*, p. 91. The same point is made in *Natural Magick*, p. 365.

35. On Risner, see J. J. Verdonk, *Petrus Ramus en de Wiskunde* (Assen, 1966), pp. 66-73; and David C. Lindberg, "Risner, Friedrich," *Dictionary of Scientific Biography*, 11: 468. The

date of Risner's birth is unknown; he died in 1580 or 1581 in his native Herzfeld.

36. For an evaluation of Risner's editorial efforts and a discussion of the influence of his edition, see Lindberg, "Introduction," pp. xxiii-xxv, xxviii-xxxii.

37. *Opticae libri quatuor* (Kassel, 1606). I am aware of only one instance of failure on Risner's part to grasp the intent of either Alhazen or Witelo, this in his belief that Alhazen accepted a process of combined intromission and extramission; see above, pp. 65-66. This misunderstanding did not, however, affect Risner's own theory, for in the *Optica* of 1606 he made no reference whatsoever to the possible extramission of rays and even classified Alhazen among the intromissionists (p. 126).

38. The one notable exception is Willebrord Snell, who extensively annotated his copy of Risner's book; see *Risneri optica cum annotationibus Willebrordii Snellii*, ed. J. A. Vollgraff.

39. Straker, "Kepler's Optics," p. 443; Kepler, *Ad Vitellionem paralipomena, quibus astronomiae pars optica traditur* (Frankfurt, 1604), ed. Franz Hammer in *Gesammelte Werke*, ed. Walther von Dyck and Max Caspar, 2: 181 (hereafter cited simply as *Paralipomena*). For Della Porta's reference to *De refractione* (which he calls "our Opticks"), see *Natural Magick*, p. 365.

40. On Kepler's life, see Max Caspar, *Kepler*, trans. C. Doris Hellman; Owen Gingerich, "Kepler, Johannes," *Dictionary of Scientific Biography*, 7: 289-312.

41. Straker, "Kepler's Optics," pp. 383-84.

42. Ibid., pp. 347-50; 374-83; see also Franz Hammer's "Nachbericht" to his edition of Kepler's *Ad Vitellionem paralipomena*, in *Gesammelte Werke*, 2: 398-400. I must acknowledge my enormous debt to Straker's studies of Kepler's optical work, especially the early work on the *camera obscura*. There are many other analyses of Kepler's optical researches. For visual theory the most useful ones are A. C. Crombie, "Early Concepts of the Senses and the Mind"; Crombie, "Mechanistic Hypothesis," pp. 52-60; Crombie, "Kepler: De modo visionis," pp. 135-44; Hammer, "Nachbericht," pp. 415-18; Huldrych M. Koelbing, "Kepler und die physiologische Optik"; Moritz von Rohr, "Kepler und seine Erklärung des Sehvorganges"; Rohr, "Keplers Behandlung des beidäugigen Sehens"; and Gerard Simon, "A propos de la théorie de la perception visuelle chez Kepler et Descartes."

43. Translated by Straker, "Kepler's Optics," pp. 385-86. For the Latin text, see Kepler, *Paralipomena*, pp. 46-47.

44. On Witelo's and Pecham's theory of radiation through apertures, see Lindberg, "Pinhole Images from Antiquity to the Thirteenth Century," pp. 164-75. Kepler somewhat misrepresents Witelo's position.

45. This aspect of Dürer's and Kepler's work has been admirably discussed by Straker, "Kepler's Optics," pp. 267-71, 390-93.

46. Translated by Straker, ibid., p. 390. For the Latin text see *Paralipomena*, p. 47. Here and elsewhere, Straker's bracketed expressions have been slightly altered.

47. See Straker, "Kepler's Optics," pp. 293-97, 300, 391-92, where it is argued that there is a significant difference between the solutions of Maurolico and Kepler. In fact, the difference between the two theories reduces to the fact that Maurolico superimposed images of the luminous source in a pattern determined by the shape of the aperture, whereas Kepler superimposed images of the aperture in a pattern determined by the shape of the luminous source. Not only is it trivial to show the two approaches to be geometrically equivalent, but Kepler himself clearly recognized this equivalence and, indeed, willingly conceived his own theory according to either scheme; see *Paralipomena*, pp. 51-57.

48. Straker, "Kepler's Optics," pp. 400-1, 411.

49. *Paralipomena*, p. 143. I have borrowed phrases from the translations of Straker, "Kepler's Optics," p. 415; and Crombie, "Kepler: De modo visionis," pp. 144-45.

50. The letter is quoted by Straker, "Kepler's Optics," p. 452.

51. Kepler notes that his friend Johannes Jessen followed Witelo in this (*Paralipomena*, p. 183).

52. Ibid.

53. Ibid., p. 184. Kepler contrasts sight with the perception of warmth, which is indeed an instance of perception by touch.

54. See Witelo, *Perspectiva*, bk. 3, theor. 23, ed. Risner, pp. 95-96. In fact, what Witelo claimed was that the posterior surface was either flat or spherical and, if the latter, of larger radius than the anterior surface of the crystalline humor or the combined crystalline and vitreous humors (which together form a sphere). For a fuller discussion, see this volume, chap. 7, n. 18; cf. Alhazen's view, this volume, chap. 4, n. 106.

55. *Paralipomena*, p. 185. Although Kepler does not say so here, he clearly knew that the gibbosity of the posterior surface of the crystalline humor is greater (i.e., its radius of curvature is smaller) than that of the anterior surface; this is revealed by his own drawing (ibid., p. 151) and by the drawing of Platter, which he reproduces (ibid., p. 159). Thus Kepler could have pointed out that even Witelo's admission that the posterior surface of the crystalline may be curved, but of greater radius than the anterior surface, is contradicted by the anatomical facts.

56. Ibid., p. 185.

57. That is, a man viewing himself in a mirror and raising his right arm will observe a raised left arm in his image. Apparently Kepler is supposing that the visual power in the eye would, according to the perspectivists, view the erect image in the eye from the front, as though an observer were to peer into the eye through the pupil in order to view the image formed within. The medieval perspectivists surely did not face this problem of image-formation within the eye (and hence the question of the direction from which the image is viewed), but it is extremely doubtful that they would have accepted Kepler's interpretation of their theory.

58. *Paralipomena*, p. 185. In fact, this is no remedy at all, as the interchange of right and left (or top and bottom, if you prefer) will still be present if the image is viewed from the front. The source of Kepler's confusion is not clear. Perhaps he believed that in the erect image of the medieval perspectivists there is left-to-right reversal but no inversion, while in the inverted image there is both inversion and left-to-right reversal (and hence no overall transformation). Unfortunately, however, the expression "left-to-right reversal" is ambiguous and in the foregoing sentence has been employed in two different senses; for a fuller discussion of this difficult issue, see Lindberg, *Pecham*, pp. 263-65, n. 123. Finally, it must be noted that Kepler's remarks on this matter are brief and very obscure, and it is not impossible that I have misinterpreted them.

59. *Paralipomena*, p. 183. Proof that the eye can see more than a hemisphere is found, for Kepler, in his experience of seeing "both the sun and my shadow as though they were not opposite but both were situated toward the front," to which he adds that "you fall only a little short of being able to see your own ears" (ibid., p. 157). On the perspectivist theory, see above, pp. 74-78, 111.

60. *Paralipomena*, p. 184.

61. Ibid.

62. Ibid.

63. Ibid., p. 144.

64. Jessen served as intermediary between Kepler and Tycho Brahe in the negotiations that brought Kepler to Prague. Earlier Jessen had been professor of medicine at the University of Prague; he later became its rector. See Caspar, *Kepler*, pp. 109-12, 171-72.

65. *Paralipomena*, pp. 147-48.

66. Ibid., p. 148. Kepler later affirms the shape to be spheroidal (p. 159).

67. Kepler notes the variability of the pupil and recognizes that its purpose is to control the amount of light entering the eye; see ibid., p. 158.

68. Ibid., pp. 149-50.

69. Ibid., p. 157. Pp. 151-62 and 175-79 of the *Paralipomena* can be read in English translation in Crombie's "Kepler: De modo visionis," pp. 147-72. The whole of book 5 on vision can be read in German translation in F. Plehn (trans.), "J. Keplers Behandlung des Sehens." I have profited much from both translations and would quote from Crombie's had I been able to secure permission from the publisher to do so.

70. *Paralipomena*, p. 157.
71. Ibid., pp. 150, 183.
72. Ibid., p. 150. The vitreous, Kepler notes (ibid., p. 157), is surrounded by the hyaloid membrane.
73. Ibid., pp. 151, 158.
74. Ibid., p. 157.
75. Ibid., p. 20.
76. Ibid., pp. 153-54. Cf. Crombie, "Kepler: De modo visionis," p. 152.
77. *Paralipomena*, p. 153.
78. Straker claims that Kepler has taken the "radical step" of inverting the visual pyramid of the perspectivists, which had its base on the visual field and its apex in the observer's eye ("Kepler's Optics," p. 459). The implication is that medieval perspectivists would have objected to, or been surprised by, such an inversion. In fact, the perspectivists would have found no obstacle to accepting Kepler's manner of conceptualizing radiation from the visual field; it was simply that when one eliminated all nonperpendicular rays (as perspectivists did when dealing with visual theory), Kepler's scheme of multiple pyramids having their apexes in the visual field was reduced to the single pyramid of the perspectivists with apex in the eye. Indeed, on occasion (when visual theory was not specifically at issue), medieval perspectivists conceived of the radiation of light in precisely the terms here employed by Kepler; see John Pecham, *Perspectiva communis*, bk. 1, props. 5 (unrevised version) and 7 (revised version), in Lindberg, *Pecham*, pp. 69-70, 75; Roger Bacon, *De multiplicatione specierum*, pt. 2, chaps. 3 and 10, pt. 5, chap. 1, ed. Bridges, 2: 471, 500, 533; Bacon, *De speculis comburentibus* (Lindberg, "Reconsideration of Bacon's Theory of Pinhole Images," pp. 216-17); and Witelo, *Perspectiva*, bk. 2, theor. 19-20, 35, 37-41, ed. Risner, pp. 68-69, 73-76.
79. Pp. 189-90.
80. See Lindberg, *Pecham*, p. 231.
81. The perspectivist theory of refraction can be found in any of the works of the perspectivist tradition; for a brief description, see Grant, *Source Book*, pp. 426-30 (a translation from Bacon); for a comprehensive account, see Alhazen, *De aspectibus*, pp. 233-78.
82. Only Della Porta, among Kepler's predecessors, presented more than the most superficial analysis of nonspherical lenses; and even he dealt principally with transparent spheres. See Della Porta, *De refractione*, pp. 41-64, 175-88. In his later *Dioptrice* (1611), Kepler finally supplied an analysis of thin lenses.
83. *Paralipomena*, pp. 162-63.
84. Ibid., pp. 166-67. I have omitted from the figure certain lines unnecessary for our purposes. Kepler recognizes, from considerations of symmetry, that angle *KBC* equals angle *BCF* and angle *LGH* equals angle *GHI*.
85. Ibid., p. 175. On the hyperbolic shape of this caustic, see ibid., pp. 103, 179.
86. Ibid., pp. 177-78.
87. The only remedy for the indistinctness at *L*, Kepler notes, is to move the paper closer to the sphere at its periphery; but even this, he says, will not completely solve the problem "since the intersections are spread out not only through depth *MN* (which is of little importance) but also in width" (ibid., p. 178).
88. Ibid., p. 154.
89. In *Paralipomena*, p. 148, Kepler appears to affirm the former; in ibid., p. 159, he clearly affirms the latter on the grounds that this is required if the humors of the eye are to have the appropriate focusing properties.
90. Ibid., pp. 151, 154-55.
91. Since the radiation ceases to diverge and begins to converge at the surface of the cornea, it may seem more reasonable to conceive of the first cone terminating and the second cone beginning at this point, rather than at the anterior surface of the crystalline lens. Despite his usual position, Kepler does seem on occasion to conceive the matter in this way, as in *Paralipomena*, p. 154, where he writes: "And so if it were possible for somebody to portray a

series of these rays in a cross-section through the eye, those extending from the top of the cornea to the bottom of the crystalline would form one and the same conical surface." It does not matter, of course, in which of the two ways one conceives the formation of the cones, so long as it is kept in mind that there is little or no refraction as rays pass from the aqueous humor to the crystalline lens.

92. Ibid.

93. *Oeuvres de Descartes*, ed. Charles Adam and Paul Tannery, vol. 6 (Paris, 1902), p. 125.

94. *Paralipomena*, p. 153.

95. Ibid., p. 155. I have borrowed phrases from Crombie's translation, "Kepler: De modo visionis," pp. 154-56.

96. Even the medieval perspectivists, who did not have the benefit of Kepler's argument, found it necessary to reintroduce nonperpendicular rays.

97. Kepler does not indicate the precise role of Platter's teaching in his own discovery of the theory of the retinal image, telling us only that the outcome of his geometrical argument is consistent with Platter's anatomical and physiological doctrines. See *Paralipomena*, p. 187; Straker, "Kepler's Optics," pp. 455-57.

98. Though, to be perfectly accurate, Kepler uses the terms *pictura*, *idolum*, *imago*, and *species* interchangeably; see *Paralipomena*, pp. 151-53, 158.

99. In the theory of the medieval perspectivists, the crystalline humor is stimulated by the perpendicular rays, but no image or picture is formed there; see this volume, chap. 4, n. 81. Cf. Straker, "Kepler's Optics," pp. 460-61.

100. *Paralipomena*, p. 185.

101. *Photismi*, p. 74 (Crew, pp. 111-12).

102. Kepler, *Gesammelte Werke*, 15: 90-91. I have borrowed heavily from Straker's translation in "Kepler's Optics," p. 463.

103. They had failed, in other words, to distinguish between the optical and psychological (or psychophysical) aspects of seeing. It must be recognized that although Kepler's theory of the retinal image gave rise to such a distinction among his successors, and although Kepler himself attempted to distinguish between the optical and the "physical" aspects of sight, Kepler raised no issues of a psychological sort.

104. "Kepler's Optics," p. 464.

105. *Paralipomena*, p. 185; cf. Straker, "Kepler's Optics," pp. 463-64.

106. *Paralipomena*, pp. 151-52. Again I have borrowed phrases from Crombie, "Kepler: De modo visionis," pp. 147-48. Straker calls this passage "satirical" ("Kepler's Optics," p. 464), but there is nothing in the passage to justify such a characterization.

107. *Paralipomena*, p. 152. However, in a marginal note to this passage Kepler backs away from his dogmatic exclusion of the post-retinal transmission from the realm of optics, admitting that "it is this [hidden journey] that brings about vision, from which the name 'optics' is derived; and so it is unjust to exclude it from optics merely because in the limited state of our knowledge it cannot be accommodated within optics" (ibid., p. 152; see also Crombie's translation, "Kepler: De modo visionis," p. 148, from which I have borrowed phrases).

108. *Paralipomena*, p. 152.

109. Ibid. Note that Kepler does not deny the hollowness of the optic nerves; on the history of this doctrine, see this volume, pp. 36, 173.

110. *Paralipomena*, p. 152. Implicit in Kepler's distinction between optics and physics is the belief that in optics, mathematical reasoning must predominate.

111. Ibid., pp. 152-53. The concluding word (*admirabilis*), which I have translated "mysterious," might equally well be translated "worthy of admiration."

112. Ibid., p. 370.

113. Ibid., p. 156. The participation of visual spirit in the process of vision is reaffirmed by Kepler in *Dioptrice*, pp. 23-25; on this, see Koelbing, "Kepler und die physiologische Optik," pp. 240-41.

114. I have discussed the "quasi-optical" aspects of Alhazen's visual theory above, pp. 81-83.

115. In *Dioptrice*, pp. 23-25, Kepler affirms that visual perception ultimately occurs when the common sense receives immaterial species from the eye.

116. This view has been skillfully defended by Straker, "Kepler's Optics," pp. 416, 426, 450-61. It was earlier stated by A. Wolf, *A History of Science, Technology and Philosophy in the 16th and 17th Centuries*, 2d ed. (New York, 1950), 1: 248-49; and by Richard J. Herrnstein and Edwin G. Boring, eds., *A Source Book in the History of Psychology* (Cambridge, Mass., 1965), p. 92.

117. *Paralipomena*, pp. 187-89. Kepler knew only what Della Porta had presented in his *Magia naturalis*.

118. *Paralipomena*, p. 155.

119. Franz Hammer reaches a similar conclusion in his "Nachbericht," p. 416.

120. A brief portion of the passage has been quoted above; see n. 118.

121. See Crombie, "Early Concepts of the Senses," pp. 108-11; Crombie, "Mechanistic Hypothesis," pp. 52-55; Crombie, "Kepler: De modo visionis," p. 142; Ronchi, *Nature of Light*, pp. 91-93. In Ronchi's view, this skillful ray-geometry followed the laying aside of prejudice against inverted images; in Crombie's far more sophisticated view, it followed a conscious strategic decision to separate the answerable questions of ray geometry from the unanswerable "psychophysiological" question of how pictures in the eye are converted into perceptions. Ronchi and Crombie agree, however, that the task facing Kepler was solely one of ray-geometry.

122. When I refer to the perspectivist framework, I have in mind the principle of punctiform analysis, the laws of propagation of light, the requirement that one point in the visual field must stimulate one and only one point within the eye, basic conceptions about the nature of seeing, and the commitment to a methodology that incorporates mathematical, physical, and physiological reasoning.

123. See Crombie, "Mechanistic Hypothesis," p. 54; Crombie, "Early Concepts of the Senses," pp. 108-11; Straker, "Kepler's Optics," pp. 423-24, 450, 482, 498, 521-23, and passim. Straker greatly exaggerates the significance of medieval discussions of the possible nonrectilinear propagation of light and the medieval view of light as a "procreative paradigm of natural activity"; this enables him to view Kepler as the one who finally vindicated the principle of rectilinear propagation and swept away the "procreative paradigm," thereby preparing the ground for "a mathematical physics of light." In fact, the "procreative paradigm," by which Straker means Grosseteste's physics and metaphysics of light, and the tradition that developed from Grosseteste's work, was by no means as ubiquitous as Straker assumes, nor (for that matter) inconsistent with a mathematical physics of light. And medieval discussions of the possible nonrectilinear propagation of light were very rare indeed. John Pecham explored the possibility that light might deviate from rectilinearity in air, but he did so with great reluctance and only when driven to it by the apparent insolubility of the phenomena of the *camera obscura* (see Lindberg, "Pinhole Images from Antiquity to the Thirteenth Century," pp. 167-75). In addition, Alhazen and some of the Western perspectivists were willing to entertain the possibility that forms or species are not bound by the law of rectilinear propagation within the vitreous humor and visual spirit (see this volume, pp. 83, 126, 253 [n. 32]). But except for those brief, reluctant deviations, medieval perspectivists were as firmly committed as Kepler to the principle of rectilinear propagation and, like Kepler, made it the foundation of their visual theory.

Appendix

1. For many of the treatises discussed below, a list of manuscripts and editions, accompanied by a bibliography of the more important secondary literature, can be found in Lindberg, *Catalogue of Optical Manuscripts*.

2. Attributed to Gerard by his own students; see Grant, *Source Book*, p. 37. On Abhomadi's authorship, see A. I. Sabra, "The Authorship of the *Liber de crepusculis*."

3. See, for example, Amable Jourdain, *Recherches critiques sur l'âge et l'origine des traductions latines d'Aristote*, new ed. (Paris, 1843), p. 123; Lucien Leclerc, *Histoire de la médecine arabe*, 1: 521; Moritz Steinschneider, *Die europäischen Übersetzungen aus dem arabischen*, p. 82; Vasco Ronchi, *The Nature of Light*, trans. V. Barocas, pp. 45, 47.

4. For Jordanus's citation, see Marshall Clagett, *Archimedes in the Middle Ages*, vol. 1 (Madison, Wis., 1964), p. 669. Quite apart from questions of dating, there is not a shred of evidence to suggest that either Witelo or Risner was an Arabist or engaged in any translating activity.

5. George Sarton, *Introduction to the History of Science*, 2: 338. Gerard's bibliography is translated in Grant, *Source Book*, pp. 35-38.

6. See my *Catalogue of Optical Manuscripts*, pp. 17-18, for a list of the extant manuscripts. It was George Sarton's conviction that an examination of the manuscripts would almost certainly reveal the name of the translator; see his review of M. Nazif, *Al-Hasan ibn al-Haitham: His Optical Studies and Discoveries* (in Arabic), *Isis* 34 (1942-43): 217. In fact, the manuscripts provide no clues to the translator's identity.

7. Axel Anthon Björnbo, "Über zwei mathematische Handschriften aus dem vierzehnten Jahrhundert," *Bibliotheca Mathematica*, ser. 3, 3 (1902): 63; J. L. Heiberg and Eilhard Wiedemann, "Ibn al Haitams Schrift über parabolische Hohlspiegel," *Bibliotheca Mathematica*, ser. 3, 10 (1909-10): 233. Paris, Bibl. Nationale, MS 9335, which contains *De speculis comburentibus*, is commonly believed to be a collection of Gerard's translations.

8. John E. Murdoch, "Euclides Graeco-Latinus," *Harvard Studies in Classical Philology* 71 (1966): 266-69; Shuntaro Ito, "The Medieval Latin Translation of the Data of Euclid," doctoral dissertation (University of Wisconsin, 1964), p. 38. Ito thinks this person was also the translator of the Greco-Latin version of the *Almagest*; Murdoch argues to the contrary.

9. All of them have been edited by Wilfred R. Theisen, "The Mediaeval Tradition of Euclid's *Optics*."

10. See my *Catalogue of Optical Manuscripts*, nos. 79E-79K, pp. 50-54.

11. Above, n. 8.

12. This work is also found in Paris, Bibl. Nationale, MS 9335; see n. 7, above. Cf. Björnbo, "Über zwei mathematische Handschriften," pp. 63, 71; Theisen, "Euclid's *Optics*," pp. 327-28.

13. Theisen, "Euclid's *Optics*," p. 324, n. 10.

14. Björnbo, "Über zwei mathematische Handschriften," pp. 63, 70; A. A. Björnbo and Sebastian Vogl, "Alkindi, Tideus und Pseudo-Euklid. Drei optische Werke," p. 155. This treatise is also contained in Paris, Bibl. Nationale, MS 9335.

15. Date given in the holograph of Moerbeke's translation, Vatican City, Bibl. Apostolica, MS Ottob. Lat. 1850, fol 61v.

16. Assigned to Gerard by his students; see Grant, *Source Book*, p. 36.

17. Evelyn Jamison, *Admiral Eugenius of Sicily: His Life and Work* (London, 1957), pp. 5, 143; *L'Optique de Claude Ptolémée*, ed. Albert Lejeune, pp. 9*-13*.

18. Björnbo and Vogl, "Drei optische Werke," pp. 151-52.

19. Julius Hirschberg, *Augenheilkunde im Mittelalter*, p. 70; Pierre Pansier, *Collectio ophtalmologica*, fasc. 2, p. 76; Alcoati, *Libre de la figura del uyl, text Català traduït de l'Arab per Mestre Joan Jacme*, ed. Lluis Deztany and Josep M. Simon de Guilleuma (Barcelona, 1933).

20. Leclerc, *Histoire de la médecine arabe*, 2: 498; Sarton, *Introduction to the History of Science*, 1: 677; Charles H. Haskins, *Studies in the History of Mediaeval Science*, pp. 133-34.

21. Bern, Burgerbibl., MS 216, fol. 42v; Prague, Univ. Knihovna, MS 839 (V.B.22), fol. 73r. The latter manuscript was called to my attention by Mlle. M.-Th. d'Alverny. Pansier, *Collectio ophtalmologica*, fasc. 3, pp. 109-91, has argued for Hebrew intermediation, but skepticism has been expressed by Moritz Steinschneider, "Zur Oculistik des 'Isa ben Ali (9. Jahrh.) und des sogenannten Canamusali," pp. 399-400.

22. Steinschneider, *Übersetzungen aus dem arabischen*, p. 8.

23. Ibid., p. 21. For a list of manuscripts of the section on the eye, see my *Catalogue of Optical Manuscripts*, p. 97.

24. See this volume, chap. 3, n. 6.
25. See this volume, chap. 3, n. 58.
26. Steinschneider, *Übersetzungen aus dem arabischen*, p. 25.
27. On this work, see my *Catalogue of Optical Manuscripts*, p. 109.
28. Steinschneider, *Übersetzungen aus dem arabischen*, pp. 5, 26.
29. Alexander of Aphrodisias, *Commentaire sur les météores d'Aristote, traduction de Guillaume de Moerbeke*, ed. A. J. Smet (Louvain, 1968), pp. xi-xiv; L. Minio-Paluello, "Henri Aristippe, Guillaume de Moerbeke, et les traductions latines médiévales des 'Météorologiques' et du 'De Generatione et Corruptione' d'Aristote," *Revue philosophique de Louvain*, ser. 3, 45 (1947): 209.
30. L. Minio-Paluello, "Le texte du 'De anima' d'Aristote: La tradition latine avant 1500," in *Autour d'Aristote: Recueil d'études de philosophie ancienne et médiévale offert à Monseigneur A. Mansion* (Louvain, 1955), pp. 218, 221-26. Moerbeke also translated a portion of John Philoponus's *Commentary on De anima*, but not the section dealing with vision.
31. Minio-Paluello, "Le texte du 'De anima,'" pp. 219, 237-42; Martin Grabmann, *Forschungen über die lateinischen Aristoteles-Übersetzungen des XIII. Jahrhunderts* (*Beiträge zur Geschichte der Philosophie des Mittelalters*, vol. 17, pts. 5-6 [Münster, 1916]), pp. 190-98.
32. J. T. Muckle, C.S.B., "Greek Works Translated Directly into Latin before 1350," *Mediaeval Studies* 5 (1943): 106; George Lacombe, A. Birkenmajer, M. Dulong, and A. Franceschini, *Aristoteles Latinus*, vol. 1 (Rome, 1939), p. 60; S. D. Wingate, *The Mediaeval Latin Versions of the Aristotelian Scientific Corpus, with Special Reference to the Biological Works* (London, 1931), pp. 92-93. Grabmann, *Forschungen*, pp. 198-99, argues that there was also an Arabo-Latin translation by Gerard of Cremona, but Wingate has pointed out (p. 46) that the manuscript in question contains not Aristotle's *De sensu*, but Averroes's epitome of it.
33. Minio-Paluello, "Henri Aristippe," pp. 208, 210, 232-35; Grabmann, *Forschungen*, p. 182; Muckle, "Greek Works," 5 (1943): 106.
34. *Averrois Cordubensis Commentarium Magnum in Aristotelis De anima libros*, ed. F. Stuart Crawford, p. xi; Minio-Paluello, "Le texte du 'De anima,'" p. 237; Lynn Thorndike, *Michael Scot* (London, 1965), pp. 25-26.
35. *Averrois Cordubensis Compendia librorum Aristotelis qui Parva naturalia vocantur*, ed. A. L. Shields and H. Blumberg, pp. xiii-xiv; Wingate, *Mediaeval Latin Versions*, pp. 121-22; Thorndike, *Michael Scot*, p. 25. See also this volume, chap. 3, n. 67.
36. M.-Th. d'Alverny, "Notes sur les traductions médiévales d'Avicenne," *Archives d'histoire doctrinale et littéraire du moyen age*, 19 (1952): 356; S. Van Riet, "La traduction latine du De anima d'Avicenne. Preliminaires à une édition critique," *Revue philosophique de Louvain*, ser. 3, 61 (1963): 600-605; see the latter for speculations regarding a possible role for Gerard of Cremona in the translating process.
37. Muckle, "Greek Works," 5 (1943): 113; Minio-Paluello, "Henri Aristippe," p. 206.
38. Muckle, "Greek Works Translated Directly into Latin before 1350," *Mediaeval Studies* 4 (1942): 40.
39. Muckle, "Greek Works," 5 (1943): 114; Minio-Paluello, "Le texte du 'De anima,'" p. 219.

A SELECTIVE BIBLIOGRAPHY OF SOURCES PERTAINING TO VISUAL THEORY

PRIMARY SOURCES

Achillini, Alessandro. *De humani corporis anatomia*. [Venice, 1521].

Adelard of Bath. *Die Quaestiones naturales des Adelardus von Bath*. Edited by Martin Müller. Beiträge zur Geschichte der Philosophie des Mittelalters, vol. 31, pt. 2. Münster, 1934.

______. *Questiones naturales*. In Berachya Hanakdan, *Dodi Ve-Nechdi (Uncle and Nephew) ... To Which is Added the First English Translation from the Latin of Adelard of Bath's Quaestiones naturales*. Edited and translated by Hermann Gollancz. Oxford, 1920.

Albert the Great (Albertus Magnus). *Opera omnia*. Edited by A. Borgnet. 38 vols. Paris. 1890-99. Vol. 9 (1890), pp. 1-96: *De sensu et sensato*. Vols. 11-12 (1891): *De animalibus*. Vol. 35 (1896): *Summa de creaturis*, part 2 [*De homine*].

______. *Opera omnia*. Edited by Bernhard Geyer. 40 vols. (in progress). Münster, 1951-. Vol. 7, pt. 1 (1968): *De anima*, edited by Clemens Stroick, O.M.I. Vol. 12 (1955), pp. xxxv-xlvii, 77-321: *Questiones de animalibus*, edited by Ephrem Filthaut, O.P.

Alberti, Leon Battista. *On Painting*. Translated by John R. Spencer. 2d ed. New Haven, 1966.

Alcoati. *See* Pansier, Pierre.

Alhazen (Ibn al-Haytham). "Abhandlung über das Licht von Ibn al-Haitam." Translated by J. Baarmann. *Zeitschrift der deutschen morgenländischen Gesellschaft* 36 (1882): 195-237.

______. *De aspectibus*. In *Opticae thesaurus Alhazeni Arabis libri septem* ... Edited by Friedrich Risner. Basel, 1572. Reprint, with an introduction by David C. Lindberg, New York, 1972. MSS Brugge, Stadsbibl., 512, fols. lr-113v (13th c.); Edinburgh, Royal Observatory, 9-11-3 (20), fols. 2r-186r (1269); London, British Museum, 12.G.VII, fols. lr-102v (14th c.).

______. "Le 'Discours de la lumière' d'Ibn al-Haytham (Alhazen): Traduction française critique." Translated by Roshdi Rashed. *Revue d'histoire des sciences et de leurs applications* 21 (1968): 197-224.

______. *Prospettiva* [an Italian version of *De aspectibus*, translated by Guerruccio di Cione Federighi]. MS Vatican City, Bibl. Apostolica, Vat. Lat. 4595, fols. lr-177v (1341).

'Ali ibn al-'Abbas al-Majusi (Haly Abbas). *See* De Koning, P.

'Ali ibn 'Isa (Jesu Haly). *Memorandum Book of a Tenth-Century Oculist for the Use of Modern Ophthalmologists: A Translation of the Tadhkirat of*

Ali ibn Isa of Baghdad. Translated by Casey A. Wood. Chicago, 1936.
______. *See* Hirschberg, Julius; Lippert, J.; and Mittwoch, E.
______. *See* Pansier, Pierre.
Alkindi. *See* Kindi, al-
'Ammar ibn 'Ali al-Mawsili, Abu-l-Qasim. *See* Hirschberg, Julius; Lippert, J.; and Mittwoch, E.
Apuleius [Pseudo-]. *Physiognomonia.* In *Anecdota Graeca et Graecolatina*, ed. Valentin Rose, 1: 61-169. Berlin, 1864.
Aristotle. *Meteorologica.* Translated by H. D. P. Lee. Loeb Classical Library. London, 1952.
______. *On the Soul, Parva naturalia, On Breath.* Translated by W. S. Hett. Loeb Classical Library. Rev. ed. London, 1957.
______. *The Works of Aristotle Translated into English.* Edited by J. A. Smith and W. D. Ross. 12 vols. Oxford, 1908-1952. Vol. 3 (1931): *Meteorologica*, translated by E. W. Webster; *De mundo*, translated by E. S. Forster; *De anima*, translated by J. A. Smith; *Parva naturalia*, translated by J. I. Beare and G. R. T. Ross; *De spiritu*, translated by J. F. Dobson.
Arnald of Villanova. *See* Pansier, Pierre.
Augustine. *De genesi ad litteram.* Edited by Joseph Zycha. In *Corpus scriptorum ecclesiasticorum latinorum*, vol. 28, pt. 1. Vienna, 1894.
______. *De trinitate.* Translated by Marcus Dods and Arthur W. Haddan. In *A Select Library of the Nicene and Post-Nicene Fathers of the Christian Church*, ed. Philip Schaff, vol. 3. Buffalo, 1887.
Aureoli, Peter. *See* Peter Aureoli.
Averroes (Ibn Rushd). *Colliget libri VII.* Venice, 1562. Reprint Frankfurt, 1962.
______. *Commentarium magnum in Aristotelis de anima libros.* Edited by F. Stuart Crawford. Cambridge, Mass., 1953.
______. *Compendia librorum Aristotelis qui Parva naturalia vocantur.* Edited by A. L. Shields and H. Blumberg. Cambridge, Mass., 1949.
______. *Epitome of Parva naturalia.* Translated by Harry Blumberg. Cambridge, Mass., 1961.
Avicebron. *Fons vitae.* Edited by Clemens Baeumker. Beiträge zur Geschichte der Philosophie des Mittelalters, vol. 1. Münster, 1895.
Avicenna (Ibn Sina). *Avicenna Latinus: Liber de anima seu sextus de naturalibus, I-II-III.* Edited by Simone Van Riet. Louvain/Leiden, 1972.
______. *Avicenna's Psychology: An English Translation of the Kitab al-Najat.* Translated by F. Rahman. London, 1952.
______. *A Compendium on the Soul, by Abu-'Aly al-Husayn Ibn 'Abdallah Ibn Sina.* Translated by Edward Abbot van Dyck. Verona, 1906.
______. *Die Augenheilkunde des Ibn Sina.* Translated by Julius Hirschberg and J. Lippert. Leipzig, 1902.
______. "Die Psychologie des Ibn Sina." Translated by S. Laudauer. *Zeitschrift der deutschen morgenländischen Gesellschaft* 29 (1875): 335-418.

______. *Le livre de science*. Translated by Mohammad Achena and Henri Massé. 2 vols. Paris, 1955-58.

______. *Liber canonis*. Translated by Gerard of Cremona. Venice, 1507. Reprint Hildesheim, 1964.

______. *Psychologie d'Ibn Sina (Avicenne) d'après son oeuvre as-Sifa*. Edited and translated by Ján Bakoš. 2 vols. Prague, 1956.

______. *See* De Koning, P.

Bacon, Roger. *De multiplicatione specierum*. In *The Opus Majus of Roger Bacon*, edited by John H. Bridges, 2: 405-552. London, 1900.

______. *The Opus Majus of Roger Bacon*. Edited by John H. Bridges. 3 vols. London, 1900. Reprint Frankfurt, 1964.

______. *The Opus Majus of Roger Bacon*. Translated by Robert B. Burke. 2 vols. Philadelphia, 1928. Reprint New York, 1962.

______. *Opus tertium*. In *Fr. Rogeri Bacon Opera quaedam hactenus inedita*, edited by J. S. Brewer, pp. 3-310. Rerum Britannicarum medii aevi scriptores, no. 15. London, 1859. Reprint New York, 1965.

______. *Part of the Opus tertium of Roger Bacon, Including a Fragment Now Printed for the First Time*. Edited by A. G. Little. Aberdeen, 1912.

______. *Perspectiva*. Edited by I. Combach. Frankfurt, 1614. MSS London, British Museum, Harley 80, fols. lr-33r (15th c.); Royal 7.F.VIII, fols. 47r-98r (13th c.).

______. *Un fragment inédit de l'Opus tertium de Roger Bacon*. Edited by Pierre Duhem. Quaracchi, 1909.

Bartholomaeus Anglicus. *De rerum proprietatibus*. Frankfurt, 1601.

Bartholomaeus de Bononia. "Tractatus de luce Fr. Batholomaei de Bononia." Edited by Irenaeus Squadrani, O.F.M. *Antonianum* 7 (1932): 201-38, 465-94.

Bartisch, George. ΟΦΘΑΛΜΟΔΟΥΛΕΙΑ, *Das ist Augendienst*. Dresden, 1583. Reprint London, 1966.

Berengario da Carpi, Jacopo. *A Short Introduction to Anatomy (Isagogae Breves)*. Translated by L. R. Lind. Chicago, 1959.

Björnbo, Axel Anthon, and Vogl, Sebastian, eds. "Alkindi, Tideus und Pseudo-Euklid. Drei optische Werke." *Abhandlungen zur Geschichte der mathematischen Wissenschaften*, vol. 26, pt. 3 (1912), pp. 1-176.

Blasius of Parma. "Le questioni di 'Perspectiva' di Biagio Pelacani da Parma." Edited by Graziella Federici Vescovini. *Rinascimento*, ser. 2, 1 (1961): 163-243.

______. *Questiones super perspectivam*. MSS Ferrara, Bibl. Comunale Ariostea, classe 2, n. 380, fols. 2r-48r (1390); Florence, Bibl. Medicea Laurenziana, Plut. 29 cod. 18, fols. lr-83r (1428); Oxford, Bodleian Lib., Canon. Misc. 363, fols. lr-193v (ca. 1400); Vatican City, Bibl. Apostolica, Barb. Lat. 357, fols. 61r-108r (1469); Vat. Lat. 2161, fols. lr-40v (15th c.); Venice, Bibl. Nazionale Marciana, Zanetti Lat. 335, fols. lr-97v (1399); Vienna, Österreichische Nationalbibl., 5447, fols. 25r-131r (15th c.).

______. "Questioni inedite di ottica di Biagio Pelacani da Parma." Edited by

Franco Alessio. *Rivista critica di storia della filosofia* 16 (1961): 79-110, 188-221.

Buridan, John. *Questiones breves super librum de anima.* In *Questiones et decisiones physicales insignium virorum*, edited by George Lockert. Paris, 1518.

______. *Questiones super librum de sensu et sensato.* In *Questiones et decisiones physicales insignium virorum*, ed. George Lockert, Paris, 1518.

Burley, Walter. *De sensibus.* Edited by Herman Shapiro and Frederick Scott. Mitteilungen des Grabmann-Instituts der Universität München, vol. 13. Munich, 1966.

Chalcidius. *See Plato.*

Cohen, Morris R., and Drabkin, I. E., trans. *A Source Book in Greek Science.* Cambridge, Mass., 1958.

Colombo, Realdo. *De re anatomica libri XV.* Venice, 1559.

Damianos. *Schrift über Optik.* Translated by Richard Schöne. Berlin, 1897.

David Armenicus. *See* Pansier, Pierre.

De iuvamentis anhelitus [also *De utilitate respirationis* and *De calore vitali*]. In Galen, *Opera omnia* (Venice, 1541-45), 8 (1545): 365-81. Also in William of Conches, *Dialogus de substantiis physicis.* Strasbourg, 1567. Reprint Frankfurt, 1967. MSS Paris, Bibl. Nationale, Lat. 6865, fols. 118 (117)v-121(120)r; Lat. 11860, fols. 217r-219r; Lat. 15456, fols. 147v-150(151)r; Vatican City, Bibl. Apostolica, Palat. Lat. 1097, fols. 114r-116v.

De Koning, P., ed. and trans. *Trois traités d'anatomie arabes par Muhammed ibn Zakariyya al-Razi, 'Ali ibn al-'Abbas et 'Ali ibn Sina.* Leiden, 1903.

Della Porta, Giovanni Battista *De refractione optices parte libri novem.* Naples, 1593.

______. *Natural Magick.* London, 1658. Reprint, edited by Derek J. Price. New York, 1957.

Descartes, René. *Discourse on Method, Optics, Geometry, and Meteorology.* Translated by Paul J. Olscamp. Indianapolis, 1965.

______. *Oeuvres de Descartes.* Edited by Charles Adam and Paul Tannery. 13 vols. Paris, 1897-1913.

Diels, Hermann, ed. *Doxographi Graeci.* 3d ed. Berlin, 1958.

Diels, Hermann, and Kranz, Walther, eds. *Die Fragmente der Vorsokratiker.* 3 vols. 6th ed. Berlin, 1951-52.

Diogenes Laertius. *Lives of Eminent Philosophers.* Translated by R. D. Hicks. 2 vols. Loeb Classical Library. London, 1925.

Dionysius [Pseudo-]. *Dionysius the Areopagite on the Divine Names and the Mystical Theology.* Translated by C. E. Rolt. London, 1920.

Dominicus de Clavasio. "Les questions de 'perspective' de Dominicus de Clivaxo." Edited by Graziella Federici Vescovini. *Centaurus* 10 (1964): 14-28.

______. *Questiones perspective.* MS Florence, Bibl. Nazionale, Conv. soppr. J.X.19, fols. 44r-55v (ca. 1400).

Durandus of Saint-Pourçain. *In Petri Lombardi Sententias theologicas commentariorum libri IIII.* Venice, 1571. Reprint Ridgewood, N.J., 1964.

Estienne, Charles. *De dissectione partium corporis humani libri tres.* Paris, 1545.

______. *La dissection des parties du corps humain divisée en trois livres.* Paris, 1546. Reprint Paris, 1965.

Euclid. "The Mediaeval Tradition of Euclid's *Optics.*" Edited and translated by Wilfred R. Theisen, O.S.B. Ph.D. dissertation, University of Wisconsin, 1972.

______. *L'Optique et la Catoptrique.* Translated by Paul Ver Eecke. New ed. Paris, 1959.

______. *Opera omnia.* Edited by J. L. Heiberg and H. Menge. 9 vols. Leipzig, 1883-1916. Vol. 7 (1895): *Optica, Opticorum recensio Theonis, Catoptrica, cum scholiis antiquis,* edited by J. L. Heiberg.

______. "The Optics of Euclid." Translated by H. E. Burton. *Journal of the Optical Society of America* 35 (1945): 357-72.

Euclid [Pseudo-]. *See* Björnbo, Axel Anthon, and Vogl, Sebastian.

Fabricius of Aquapendente, Hieronymus (Fabrici of Aquapendente, Girolamo). *Tractatus anatomicus triplex, quorum primus de oculo, visus organo* . . . Venice, 1614.

Farabi, Abū Nasr al-. *Catálogo de las ciencias.* Edited and translated by Angel González Palencia. Madrid, 1932.

______. "A Catalogue of the Sciences." Translated by Marshall Clagett. Mimeographed.

______. *Der Musterstaat von Alfarabi.* Translated by Friedrich Dieterici. Leiden, 1900.

______. *Idées des habitants de la cité vertueuse.* Translated by R. P. Jaussen, Youssef Karam, and J. Chlala. Textes et traductions d'auteurs orientaux, vol. 9. Cairo, 1949.

Fasciculo di Medicina, Venice 1493, The. Edited and translated by Charles Singer. 2 vols. Monumenta medica, no. 2. Florence, 1925.

Filarete, Il (Antonio de Piero Averlino). *Filarete's Treatise on Architecture: Being the Treastise by Antonio di Piero Averlino, Known as Filarete.* Translated by John R. Spencer. 2 vols. New Haven, 1965.

Galen. *De placitis Hippocratis et Platonis.* Translated by Philip De Lacy. In Corpus Graecorum medicorum, forthcoming.

______. *On Anatomical Procedures: The Later Books.* Translated by W. L. H. Duckworth. Cambridge, 1962.

______. *On the Usefulness of the Parts of the Body.* Translated by Margaret T. May. Ithaca, 1968.

______. *Opera omnia.* Edited by C. G. Kühn. 22 vols. Leipzig, 1821-33. Reprint Hildesheim, 1965. Vol. 5 (1823), pp. 181-805: *De placitis Hippocratis et Platonis.* Vol. 7 (1824), pp. 85-272: *De symptomatum causis.*

______. *Opera omnia.* 8 vols. Venice, 1541-45.

______. *See* Pansier, Pierre.

Ghiberti, Lorenzo. *Lorenzo Ghibertis Denkwürdigkeiten (I Commentarii).* Edited by Julius von Schlosser. 2 vols. Berlin, 1912.

Grant, Edward, ed. *A Source Book in Medieval Science.* Cambridge, Mass., 1974.

Grassus, Benvenutus (of Jerusalem). *De oculis eorumque egritudinibus et curis.* Translated by Casey A. Wood. Stanford, 1929.

Gregory of Rimini. *Gregorius Ariminensis super primo et secundo Sententiarum.* Venice, 1522. Reprint Paderborn, 1955.

Grosseteste, Robert. *Die philosophischen Werke des Robert Grosseteste, Bischofs von Lincoln.* Edited by Ludwig Baur. Beiträge zur Geschichte der Philosophie des Mittelalters, vol. 9. Münster, 1912.

______. *On Light.* Translated by Clare C. Riedl. Milwaukee, 1942.

______. "Robert Grosseteste on Refraction Phenomena." Translated by Bruce S. Eastwood. *American Journal of Physics* 38 (1970): 196-99.

Guy de Chauliac. *The Cyrurgie of Guy de Chauliac,* vol. 1. Edited by Margaret S. Ogden. Early English Text Society, no. 265. London, 1971.

____. *The Middle English Translation of Guy de Chauliac's Anatomy, with Guy's Essay on the History of Medicine.* Edited by Björn Wallner. Lund, 1964.

Henry Bate Of Malines. *Speculum divinorum et quorundam naturalium,* vol. 1. Edited by E. Van de Vyver, O.S.B. Louvain/Paris, 1960.

Henry of Langenstein. *Questiones super perspectivam.* In *Preclarissimum mathematicarum opus in quo continentur perspicacissimi mathematici Thome Brauardini arismetica et eiusdem geometria necnon et sapientissimi Pisani carturiensis perspective ... cum acutissimis Ioannis d'Assia* [Henry of Langenstein] *super eadem perspectiva questionibus annexis.* Valencia, 1503. Also in a forthcoming critical edition by H. L. L. Busard and David C. Lindberg. MSS Erfurt, Wissenschaftliche Bibl., Ampl. F. 380, fols. 29r-40v (14th c.); Florence, Bibl. Nazionale, Conv. soppr. J.X.19, fols. 56r-85v (ca. 1400); Paris, Bibl. de l'Arsenal, 522, fols. 66r-88(87)r (*ca.* 1395-1398); Vienna, Österreichische Nationalbibl., 5437, fols. 150r-160v (15th c.).

Henry of Mondeville. *Die Anatomie des Heinrich von Mondeville.* Edited by J. L. Pagel. Berlin, 1889.

Hero of Alexandria. *Opera quae supersunt omnia.* Edited by L. Nix and W. Schmidt. 5 vols. Leipzig, 1899-1914. Vol. 2, fasc. 1 (1900), pp. 301-65: *De speculis,* edited by W. Schmidt. MS Vatican City, Bibl. Apostolica, Ottob. Lat. 1850, fols. 60r-64v (1269).

______. *Pneumatica.* Translated by Bennet Woodcroft. London, 1851.

Herveus Natalis. *Quolibeta.* Venice, 1513. Reprint. Ridgewood, N.J., 1966.

Hieronymo Manfredi. *Anothomia.* In "A Study in Early Renaissance Anatomy, with a New Text: The *Anothomia* of Hieronymo Manfredi (1490)," edited and translated by A. Mildred Westland. In *Studies in the History and Method of Science,* edited by Charles Singer, 1: 79-164. Oxford, 1917.

Hirschberg, Julius; Lippert, J.; and Mittwoch, E., trans. *Die arabischen Augenärzte nach den Quellen bearbeitet.* 2 pts. Leipzig, 1904-05. Pt. 1: 'Ali ibn 'Isa, *Erinnerungsbuch für Augenärzte aus arabischen Handschriften.* Pt. 2: 'Ammar ibn 'Ali al-Mawsili, *Das Buch der Auswahl von den Augenkrankheiten;* Khalifa ibn Abi-l-Mahasin al-Halabi, *Das Buch vom genügenden in der Augenheilkunde*; Salah al-Din ibn Yusuf, *Licht der Augen.*

Hunain ibn Ishaq. *The Book of the Ten Treatises on the Eye Ascribed to Hunain ibn Is-haq (809-877 A.D.).* Edited and translated by Max Meyerhof. Cairo, 1928.

______. "Die aristotelische Lehre vom Licht bei Hunain b. Ishaq." Translated by C. Prüfer and Max Meyerhof. *Der Islam* 2 (1911): 117-28.

______. "Die Augenanatomie des Hunain b. Ishaq." Translated by Max Meyerhof and C. Prüfer. *Sudhoffs Archiv für Geschichte der Medizin und der Naturwissenschaften* 4 (1911): 163-90.

______. "Die Lehre vom Sehen bei Hunain b. Ishaq." Translated by Max Meyerhof and C. Prüfer. *Sudhoffs Archiv für Geschichte der Medizin und der Naturwissenschaften* 6 (1913): 21-33.

______. *Le livre des questions sur l'oeil de Honain ibn Ishaq.* Edited and translated by P. Sbath and Max Meyerhof. Mémoires présentés à l'Institut d'Egypte, vol. 36. Cairo, 1938.

______. *Liber de oculis.* In *Omnia opera Ysaac*, fols. 172r-178r. Lyons, 1515. Also in *Galeni opera omnia latine*, vol. 5, fols. 178r-187r. Venice, 1541. *See also* Pansier, Pierre.

Ibn al-Haytham. *See* Alhazen.

Ibn Masawaih, Yuhanna. *See* Masawaih, Yuhanna ibn.

Ibn Rushd. *See* Averroes.

Ibn Sina. *See* Avicenna.

Ivins, William M., Jr. ed. *On the Rationalization of Sight: With an Examination of Three Renaissance Texts on Perspective.* Metropolitan Museum of Art, paper no. 8. New York, 1938. Reprint New York, 1973.

Jessen, Johannes. *Anatomia Pragensis.* Wittenberg, 1601.

Johannes de Casso. *See* Pansier, Pierre.

Kepler, Johannes. *Dioptrice.* Augsburg, 1611. Reprint, with an Introduction by Michael Hoskin, Cambridge, 1962.

______. *Gesammelte Werke.* Edited by Walther von Dyck and Max Caspar. 18 vols. (in progress). Munich, 1937-. Vol. 2 (1939): *Ad Vitellionem paralipomena, quibus astronomiae pars optica traditur*, edited by Franz Hammer. Vols. 13-18 (1945-59): Briefe 1590-1630.

______. "J. Keplers Behandlung des Sehens." Translated by Ferdinand Plehn. *Zeitschrift fur ophthalmologische Optik mit Einschluss der Instrumentenkunde* 8 (1920): 154-57; 9 (1921): 13-26, 40-54, 73-87, 103-9, 143-52, 177-82.

______. *J. Keplers Grundlagen der geometrischen Optik (im Anschluss an die Optik des Witelo).* Translated by Ferdinand Plehn. Ostwald's Klassiker der exakten Wissenschaften, no. 198. Leipzig, 1922.

______. "Kepler: De modo visionis." Translated by Alistair C. Crombie. In *Mélanges Alexa dre Koyré*, vol. 1, *L'aventure de la science*, pp. 135-72. Paris, 1964.

Khalifa ibn Abi-l-Mahasin al-Halabi. *See* Hirschberg, Julius; Lippert, J.; and Mittwoch, E.

Kindi, Abu Yusuf Ya'qub ibn Ishaq al-. *Propagations of Ray: The Oldest Arabic Manuscript about Optics (Burning-Mirror), from Ya'kub ibn Ishaq al-Kindi.* Translated by Mohamed Yahia Haschmi. Aleppo, 1967.

______. *See* Björnbo, Axel Anthon, and Vogl, Sebastian.

Kirk, G. S., and Raven, J. E., trans. *The Presocratic Philosophers.* Cambridge, 1960.

Lanfranc of Milan. *Lanfrank's "Science of Cirurgie."* Edited by Robert von Fleishhacker. Early English Text Society, no. 102. London, 1894.

Laurens, André Du (Laurentius, Andreas). *A Discourse of the Preservation of the Sight: of Melancholike Diseases; of Rheumes, and of Old Age.* Translated by Richard Surphlet. London, 1599. Reprint, with an introduction by Sanford V. Larkey (Shakespeare Association Facsimiles, no. 15), London, 1938.

Leonardo da Vinci. "Leonardo da Vinci: Of the Eye. An Original New Translation from *Codex D.*" Translated by Nino Ferrero. *American Journal of Ophthalmology,* ser. 3, 35 (1952): 507-21.

______. "Leonardo da Vinci on the Eye: The MS D in the Bibliothèque de l'Institut de France, Paris, Translated into English and Annotated with a Study of Leonardo's Theories of Optics." Translated by Donald S. Strong. Ph.D. dissertation, University of California, Los Angeles, 1967.

______. *The Literary Works of Leonardo da Vinci.* Edited and translated by Jean Paul Richter. 2d ed. 2 vols. London, 1939. Reprint New York, 1970.

______. *The Notebooks of Leonardo da Vinci.* Translated by Edward MacCurdy. 2 vols. New York, 1939. Reprint New York, 1958.

Lucretius Carus, Titus. *De rerum natura.* Translated by W. H. D. Rouse. Rev. ed. Loeb Classical Library. London, 1937.

Maurolico, Francesco. *Photismi de lumine et umbra.* Naples, 1611. MS Lucca, Bibl. Governativa, 2080, fols. 2r-63r (16th c.).

______. *The Photismi de lumine of Maurolycus: A Chapter in Late Medieval Optics.* Translated by Henry Crew. New York, 1940.

Masawaih, Yuhanna ibn (Johannes Mesue). "Die Augenheilkunde des Juhanna b. Masawaih (777-857 n. Chr.)." Translated by C. Prüfer and Max Meyerhof. *Der Islam* 6 (1916): 217-56.

______. *Mesue cum expositione Mondini super canones universales ...* Venice, 1497.

Migne, J.-P., ed. *Patrologiae cursus completus, series latina.* 221 vols. Paris, 1844-64.

Mondino dei Luzzi. *Anatomia.* In *The Fasciculo di medicina, Venice 1493,* edited and translated by Charles Singer. 2 vols. Monumenta Medica, no. 2. Florence, 1925.

______. *Anatomia, riprodotta da un codice Bolognese del secolo XIV e volgarizzata nel secolo XV.* Edited by Lino Sighinolfi. Classici Italiani della Medicina, no. 1. Bologna, 1930.

______. *Anatomies de Mondino dei Luzzi et de Guido de Vigevano.* Edited by Ernest Wickersheimer. Paris, 1926.

Nasir al-Din al-Tusi. "A Statement on Optical Reflection and 'Refraction' Attributed to Nasir ud-Din at-Tusi." Translated by H. J. J. Winter and W. 'Arafat. *Isis* 42 (1951): 138-42.

Nemesius Emesenus. *De natura hominis, Graece et Latine.* Edited by Christian Friedrich Matthaeus. Magdeburg, 1802. Reprint Hildesheim, 1967.

Ockham, William of. *See* William of Ockham.

Oresme, Nicole. "Nicole Oresme and the Marvels of Nature: A Critical Edition of his *Quodlibeta* with English Translation and Commentary." Edited and translated by Bert Hansen. Ph.D. dissertation, Princeton University, 1973.

______. "Nicole Oresme on Light, Color, and the Rainbow: An Edition and Translation with Introduction and Critical Notes, of Part of Book Three of His *Questiones super quatuor libros meteororum.*" Edited and translated by Stephen C. McCluskey. Ph.D. dissertation, University of Wisconsin, 1974.

Pansier, Pierre, ed. *Collectio ophtalmologica veterum auctorum.* 7 fascs. Paris, 1903-33. Fasc. 1 (1903): Arnald of Villanova, *Libellus regiminis de confortatione visus;* Johannes de Casso, *Tractatus de conservatione visus.* Fasc. 2 (1903): Alcoati, *Congregatio sive Liber de oculis.* Fasc. 3 (1903): 'Ali ibn 'Isa, *Epistola de cognitione infirmitatum oculorum sive Memoriale oculariorum.* Fasc. 4 (1904): David Armenicus, *Compilatio in libros de oculorum curationibus Accanamosali.* Fasc. 5 (1907): Zacharias, *Tractatus de passionibus oculorum qui vocatur Sisilacera, id est secreta secretorum.* Fasc. 6 (1908): Anonymous, *Tractatus de egritudinibus oculorum;* Anonymous, *Tractatus de quibusdam dubiis circa dicta oculorum concurrentibus.* Fasc. 7 (1909-33): Hunain ibn Ishaq, *Liber de oculis;* Galen(?), *Littere Galieni ad Corisium de morbis oculorum et eorum curis.*

Pecham, John. *John Pecham and the Science of Optics: Perspectiva communis, edited with an Introduction, English Translation, and Critical Notes.* Edited and translated by David C. Lindberg. Madison, Wis., 1970.

______. *Tractatus de perspectiva.* Edited and translated by David C. Lindberg. Franciscan Institute Publications, Text Series no. 16. St. Bonaventure, N.Y., 1972.

Peter Aureoli. *Commentarium in primum [-quartum] librum Sententiarum.* 2 vols. Rome, 1596-1605.

Plato. *Plato, with an English Translation.* 10 vols. Loeb Classical Library. London, 1914-29.

______. *Plato's Cosmology: The Timaeus of Plato.* Translated, with a running commentary, by Francis M. Cornford. London, 1937.

______. *Plato's Theory of Knowledge: The Theaetetus and the Sophist of Plato.* Translated, with a running commentary, by Francis M. Cornford. London, 1935.

______. *Timaeus a Calcidio translatus commentarioque instructus.* Edited by J. H. Waszink and P. J. Jensen. Plato Latinus, edited by Raymund Klibansky, vol. 4. London/Leiden, 1962.

Platter, Felix. *De corporis humani structura et usu . . . libri III.* Basel, 1583.

Pliny (the Elder). *Natural History.* Translated by H. Rackham. 10 vols. Loeb Classical Library. London, 1938-62.

Plotinus. *The Enneads.*. Translated by Stephen MacKenna. 2d ed. London, 1956.

Porta, Giovanni Battista Della. *See* Della Porta, Giovanni Battista.

Ptolemy, Claudius. *L'Optique de Claude Ptolémée dans la version latine d'après l'arabe de l'émir Eugène de Sicile.* Edited by Albert Lejeune. Université de Louvain, Recueil de travaux d'histoire et de philologie, ser. 4, fasc. 8. Louvain, 1956.

Questio de visu. MS Erfurt, Wissenschaftliche Bibl., Ampl. Q. 369, fol. 4r (14thc.).

Razi, Abu Bakr Muhammad ibn Zakariya al-. *See* De Koning, P.

Reisch, Gregorius. *Margarita philosophica.* Basel, 1517.

Risner, Friedrich. *Opticae libri quatuor.* Kassel, 1606.

______. *Risneri Optica cum annotationibus Willebrordii Snellii.* Edited by J. A. Vollgraff. Ghent, 1918.

Ronchi, Vasco, ed. *Scritti di ottica.* Classici italiani di scienze techniche e arti. Milan, 1968.

Rudoloph of Erfurt. *Questiones de anima.* MSS Cracow, Bibl. Jagiellońska, 635, pp. 237-348 (14th c.); 735, fols. lr-50v (14th c.).

Salah al-Din ibn Yusuf. *See* Hirschberg, Julius; Lippert, J.; and Mittwoch, E.

Seneca, Lucius Annaeus. *Physical Science in the Time of Nero: Being a Translation of the Quaestiones naturales of Seneca.* Translated by John Clarke, notes by Archibald Giekie. London, 1910.

______. *Questions naturelles.* Edited and translated by Paul Oltramare. 2 vols. Paris, 1929.

Snell, Willebrord. *See* Risner, Friedrich.

Solinus, Gaius Julius. *Collectanea rerum memorabilium.* Edited by Th. Mommsen. Berlin, 1895.

______. *The Excellent and Pleasant Worke of Iulius Solinus Polyhistor.* Translated by Arthur Golding. London, 1587. Reprint, with an introduction by George Kish, Gainesville, 1955.

Theophrastus. *Theophrastus and the Greek Physiological Psychology before Aristotle.* Translated by George M. Stratton. London, 1917. Reprint Chicago, 1967.

Tideus. *See* Björnbo, Axel Anthon, and Vogl, Sebastian.

Versoris, Joannes. *Questiones super libros metheororum.* [Cologne], 1493.

______. *Questiones super parva naturalia cum textu Aristotelis.* [Cologne], 1493.

Vesalius, Andreas. *De humani corporis fabrica libri septem.* Basel, 1543. Reprint Brussels, 1964.

Viator, Jean Pélerin. *De artificiali perspectiva.* Toul, 1509. Reprinted in Ivins, William M., Jr., ed., *On the Rationalization of Sight: With an Examination of Three Renaissance Texts on Perspective.* New York, 1938.

Vincent of Beauvais. *Speculum quadruplex sive speculum maius.* 4 vols. Douay, 1624. Reprint Graz, 1964-65.

William of Conches. *Dialogus de substantiis physicis* ... Strasbourg, 1567. Reprint Frankfurt, 1967.

______. *Glosae super Platonem.* Edited by Edouard Jeauneau. Paris, 1965.

______. *Philosophia mundi.* In *Patrologia cursus completus, series latina,* ed. J.-P. Migne, vol. 172, cols. 39-102 (there attributed to Honorius of Autun). Paris, 1895.

William of Ockham. *In quattuor libros Sententiarum*. Lyons, 1495.
Witelo. *Perspectiva*. In *Opticae thesarurus Alhazeni Arabis libri septem . . . Item Vitellonis Thuringopoloni libri X*. Edited by Friedrich Risner. Basel, 1572. Reprint, with an introduction by David C. Lindberg, New York, 1972.
______. "Witelo as a Mathematician: A Study in XIIIth Century Mathematics Including a Critical Edition and English Translation of the Mathematical Book of Witelo's *Perspectiva*." Edited and translated by Sabetai Unguru. Ph.D. dissertation, University of Wisconsin, 1970.
Yuhanna ibn Masawaih. *See* Masawaih, Yuhanna ibn.
Zacharias. *See* Pansier, Pierre.
Zerbi, Gabriele. *Liber anathomie corporis humani et singulorum membrorum illius*. Venice, 1502.

Secondary Sources

Alessio, Franco. "Per uno studio sull'ottica del trecento." *Studi medievali*, ser. 3, 2 (1961): 444-504.
Argan, Giulio Carlo. "The Architecture of Brunelleschi and the Origins of Perspective Theory." *Journal of the Warburg and Courtauld Institutes* 9 (1946): 96-121.
Baeumker, Clemens. *Witelo, ein Philosoph und Naturforscher des XIII. Jahrhunderts*. Beiträge zur Geschichte der Philosophie des Mittelalters, vol. 3, pt. 2. Münster, 1908.
Bailey, Cyril. *The Greek Atomists and Epicurus*. Oxford, 1928. Reprint New York, 1964.
Bauer, Hans. *Die Psychologie Alhazens*. Beiträge zur Geschichte der Philosophie des Mittelalters, vol. 10, pt. 5. Münster, 1911.
Baur, Ludwig. "Der Einfluss des Robert Grosseteste auf die wissenschaftliche Richtung des Roger Bacon." In *Roger Bacon Essays*, edited by A. G. Little, pp. 33-54. Oxford, 1914.
______. *Die Philosophie des Robert Grosseteste, Bischofs von Lincoln*. Beiträge zur Geschichte der Philosophie des Mittelalters, vol. 18, pts. 4-6. Münster. 1917.
Beare, John I. *Greek Theories of Elementary Cognition from Alcmaeon to Aristotle*. Oxford, 1906.
Bednarski, Adam. "Das anatomische Augenbild von Johann Peckham, Erzbischof zu Canterbury im XIII. Jahrhundert." *Sudhoffs Archiv für Geschichte der Medizin und der Naturwissenschaften* 22 (1929): 352-56.
______. "Die anatomischen Augenbilder in den Handschriften des Roger Bacon, Johann Peckham und Witelo." *Sudhoffs Archiv für Geschichte der Medizin and Naturwissenschaften* 24 (1931): 60-78.
Birkenmajer, Aleksander. *Etudes d'histoire des sciences en Pologne*. Studia Copernicana, vol. 4. Wrocław, 1972.
______. "Etudes sur Witelo, I-IV." *Bulletin international de l'Académie polonaise des sciences et des lettres*, Classe d'histoire et de philosophie, année 1918 (Cracow, 1920), pp. 4-6; années 1919-20 (Cracow, 1922-24), pp. 354-60; année 1922 (Cracow, 1925), pp. 6-9.

———. "Witelo e lo studio di Padova." In *Omaggio dell'Accademia Polacca di Scienze e Lettere all'Università di Padova nel settimo centenario della sua fondazione*, pp. 145-68. Cracow, 1922.

Bliemetzrieder, Franz. *Adelhard von Bath:* Munich, 1935.

Block, Irving. "Truth and Error in Aristotle's Theory of Sense Perception." *Philosophical Quarterly* 11 (1961): 1-9.

Bloom, Kathryn. "Lorenzo Ghiberti's Space in Relief: Method and Theory." *Art Bulletin* 51 (1969): 164-69.

Boehner, Philotheus, O.F.M. "*Notitia intuitiva* of Non Existents according to Peter Aureoli, O.F.M. (1322)." *Franciscan Studies* 8 (1948): 388-416.

———. "The Notitia Intuitiva of Non-Existents according to William Ockham: With a Critical Study of the Text of Ockham's Reportatio and a Revised Edition of Rep. II, Q. 14-15." *Traditio* 1 (1943): 223-75.

Böhm, Walter. *Johannes Philoponus Grammatikos von Alexandrien.* Munich, 1967.

Boyer, Carl B. "Aristotelian References to the Law of Reflection." *Isis* 36 (1945-46): 92-95.

———. *The Rainbow: From Myth to Mathematics.* New York, 1959.

Brampton, C. K. "Scotus, Ockham and the Theory of Intuitive Cognition." *Antonianum* 40 (1965): 449-66.

Brion-Guerry, L. *Jean Pélerin Viator: Sa place dans l'histoire de la perspective.* Paris, 1962.

Brizio, Anna Maria. *Razzi incidenti e razzi refressi* (*dal f° 94v ed altri del Codice Arundel*). III Lettura Vinciana, Vinci, Biblioteca Leonardiana, 21 aprile 1963. Florence, 1964.

Brown, Jerome V. "Sensation in Henry of Ghent: A Late Mediaeval Aristotelian-Augustinian Synthesis." *Archiv für Geschichte der Philosophie* 53 (1971): 238-66.

Callus, D. A., ed. *Robert Grosseteste, Scholar and Bishop.* Oxford, 1955.

Casper, Max. *Kepler.* Translated by C. Doris Hellman. London, 1959.

Castigloni, Arturo. "Il trattato dell'ottica di Lorenzo Ghiberti." *Rivista di storia delle scienze mediche e naturali* 12 (1921): 51-68.

Castillo y Quartiellers, Rodolfo del. *La oftalmología en tiempo de los Romanos.* Madrid, 1905.

Caussin de Perceval, G. G. A. "Mémoire sur l'Optique de Ptolémée." *Mémoires de l'Institut Royal de France, Académie des Inscriptions et Belles-Lettres* 6 (1822): 1-39.

Chenu, M.-D., O.P. *Nature, Man, and Society in the Twelfth Century: Essays on New Theological Perspectives in the Latin West.* Translated by Jerome Taylor and Lester K. Little. Chicago, 1968.

Cherniss, Harold. *Aristotle's Criticism of Presocratic Philosophy.* Baltimore, 1935.

———. "Galen and Posidonius' Theory of Vision." *American Journal of Philology* 54 (1933): 154-61.

Clarke, Edwin, and Dewhurst, Kenneth. *An Illustrated History of Brain Function.* Berkeley, 1972.

Crombie, Alistair C. "Early Concepts of the Senses and the Mind." *Scientific American* 210, no. 5 (May, 1964): 108-16.

______. "The Mechanistic Hypothesis and the Scientific Study of Vision: Some Optical Ideas as a Background to the Invention of the Microscope." In *Historical Aspects of Microscopy,* ed. S. Bradbury and G. L'E. Turner, pp. 3-112. Cambridge, 1967.

______. *Robert Grosseteste and the Origins of Experimental Science 1100-1700.* Oxford, 1953.

Dales, Richard C. "Robert Grosseteste's Scientific Works." *Isis* 52 (1961): 381-402.

Day, Sebastian J., O.F.M. *Intuitive Cognition: A Key to the Significance of the Later Scholastics.* Franciscan Institute Publications, Philosophy Series no. 4. St. Bonaventure, N.Y., 1947.

Delambre, J. B. J. "Sur l'Optique de Ptolémée comparée à celle qui porte le nom d'Euclide et à celles d'Alhazen et de Vitellion." In J. B. J. Delambre, *Histoire de l'astronomie ancienne,* vol. 2, pp. 411-32. Paris, 1817.

Dewhurst, Kenneth. *See* Clarke, Edwin, and Dewhurst, Kenneth.

Dictionary of Scientific Biography. 14 vols. (in progress). New York, 1970-.

Duhem, Pierre, *Le système du monde.* 10 vols. Paris, 1913-59.

Easton, Stewart C. *Roger Bacon and His Search for a Universal Science.* Oxford, 1952.

Eastwood, Bruce S. "Averroes' View of the Retina—A Reappraisal." *Journal of the History of Medicine and Allied Sciences* 24 (1969): 77-82.

______. "The Geometrical Optics of Robert Grosseteste." Ph.D. dissertation, University of Wisconsin, 1964.

______. "Grosseteste's 'Quantitative' Law of Refraction: A Chapter in the History of Non-Experimental Science." *Journal of the History of Ideas* 28 (1967): 403-14.

______. "Mediaeval Empiricism: The Case of Grosseteste's Optics." *Speculum* 43 (1968): 306-21.

______. "Metaphysical Derivations of a Law of Refraction: Damianos and Grosseteste." *Archive for History of Exact Sciences* 6 (1970): 224-36.

______. "Robert Grosseteste's Theory of the Rainbow: A Chapter in the History of Non-Experimental Science." *Archives internationales d'histoire des sciences* 19 (1966): 313-32.

______. "Uses of Geometry in Medieval Optics." In *Actes du XIIe Congrès international d'histoire des sciences, Paris 1968,* vol. 3A, pp. 51-55. Paris, 1971.

Edgerton, Samuel Y., Jr. "Alberti's Perspective: A New Discovery and a New Evaluation." *Art Bulletin* 48 (1966): 367-78.

______. "Brunelleschi's First Perspective Picture." *Arte Lombarda: Rivista di storia dell'arte,* 18, nos. 38-39 (1973): 172-95.

______. *The Renaissance Rediscovery of Linear Perspective.* New York, 1975.

Elsässer, W. "Die Funktion des Auges bei Leonardo da Vinci." *Zeitschrift für Mathematik und Physik,* historisch-literarische Abteilung, 45 (1900): 1-6.

Fakhry, Majid. *A History of Islamic Philosophy*. New York, 1970.

Feigenbaum, Aryeh. "Cataract Operation—Its Origin in Antiquity and Its Spread from East to West." In *Actes du VIIe Congrès international d'histoire des sciences, Jérusalem (4–12 août 1953)*, pp. 298–301. Paris, n.d.

Ferraz, Antonio. "Ensayo di analisis gnoseologico de 'Ad Vitellionem paralipomena' de Képler: La function del esquema." *Archives internationales d'histoire des sciences* 17 (1964): 273–89.

Francastel, Pierre. "La perspective de Léonard de Vinci et l'expérience scientifique au XVIe siècle." In *Léonard de Vinci et l'expérience scientifique au XVIe siècle* (Colloques internationaux du Centre national de la recherche scientifique, Paris, 4–7 juillet 1952), pp. 61–72. Paris, 1953.

Fuchs, Oswald, O.F.M. *The Psychology of Habit according to William Ockham*. Franciscan Institute Publications, Philosophy Series no. 8. St. Bonaventure, N.Y., 1952.

Fukala, Vincenz. "Historischer Beitrag zur Augenheilkunde." *Archiv für Augenheilkunde* 42 (1901): 203–14.

Gadol, Joan. *Leon Battista Alberti: Universal Man of the Early Renaissance*. Chicago, 1969.

Garin, Eugenio. "Un trattatello quattrocentesco sulla visione." *Rivista critica di storia della filosofia* 26 (1971): 80–83.

Gauthier, Léon. *Ibn Rochd (Averroès)*. Paris, 1948.

Ghalioungui, Paul. See Nazif, M., and Ghalioungui, Paul.

Gingerich, Owen. "Kepler, Johannes." In *Dictionary of Scientific Biography*, 7: 289–312. New York, 1973.

Gioseffi, Decio. *Perspectiva artificialis: Per la storia della prospettiva, spigolature e appunti*. Trieste, 1957.

Grayson, Cecil. "L. B. Alberti's 'Construzione Legittima.' " *Italian Studies* 19 (1964): 14–27.

Gregorio, S. "Francesco Maurolico e la fisiologia dell'occhio." In *Atti del'VIII° Congresso internazionale di storia della medicina, Roma—dal 22 al 27 settembre 1930*, edited by Pietro Capparoni, pp. 433–35. Pisa, 1931.

Gregory, Tullio. *Anima mundi: La filosofia di Guglielmo di Conches e la scuola di Chartres*. Florence, 1955.

Haas, Arthur Erich. "Antike Lichttheorien." *Archiv für Geschichte der Philosophie* 20 (1907): 345–86.

———. "Die ältesten Beobachtungen auf dem Gebiete der Dioptrik." *Archiv für die Geschichte der Naturwissenschaften und der Technik* 9 (1920): 108–11.

Hahm, David. "Early Hellenistic Theories of Vision and the Perception of Color." In *Perception: Philosophical and Scientific Themes and Variations*, edited by Peter Machamer and Robert G. Turnbull. Columbus, Ohio, forthcoming.

Hammer, Franz. "Nachbericht." In Johannes Kepler, *Gesammelte Werke*, edited by Walther von Dyck and Max Caspar, 2: 393–436. Munich, 1939.

Haskins, Charles H. *The Renaissance of the Twelfth Century*. Cambridge, Mass., 1927.

______. *Studies in the History of Mediaeval Science.* Cambridge, Mass., 1924.

Heiberg, J. L. *Litterargeschichtliche Studien über Euklid.* Leipzig, 1882.

Hirschberg, Julius. "Alkmaion's Verdienst um die Augenkunde." *Albrecht von Graefes Archiv für Opthalmologie* 105 (1921): 129-33.

______. "Arabian Ophthalmology." *Journal of the American Medical Association* 45 (1905): 1127-31.

______. "Die Optik der alten Griechen." *Zeitschrift für Psychologie und Physiologie der Sinnesorgane* 16 (1898): 321-51.

______. "Die Seh-Theorien der griechischen Philosophen in ihren Beziehungen zur Augenheilkunde." *Zeitschrift für Augenheilkunde* 43 (1920): 1-22.

______. "Galen und seine zweite Anatomie des Auges." *Berliner klinische Wochenschrift* 56 (1919): 610-12, 635-38.

______. *Geschicte der Augenheilkunde im Mittelalter und in der Neuzeit.* In *Graefe-Saemisch Handbuch der gesamten Augenheilkunde,* 2d ed., edited by Theodor Saemisch, vol. 13. Leipzig, 1908.

______. "Über die älteste arabische Lehrbuch der Augenheilkunde." *Sitzungsberichte der königlich preussischen Akademie der Wissenschaften* 49 (1903): 1080-94.

Hirschberg, Julius; Lippert, J.; and Mittwoch, E. *Die arabischen Lehrbücher der Augenheilkunde.* Berlin, 1905.

Hoffmans, Hadelin. "La genèse des sensations d'après Roger Bacon." *Revue néo-scholastique de philosophie* 15 (1908): 474-98.

Hoppe, Edmund. *Geschichte der Optik.* Leipzig, 1926. Reprint Wiesbaden, 1967.

Hudeczek, Methodius, O.P. "De lumine et coloribus [according to Albert the Great]." *Angelicum* 21 (1944): 112-38.

Jablonski, Walter. "Die Theorie des Sehens im griechischen Altertume bis auf Aristoteles." *Sudhoffs Archiv für Geschichte der Medizin und der Naturwissenschaften* 23 (1930): 306-31.

Kahn, Charles H. "Sensation and Consciousness in Aristotle's Psychology." *Archiv für Geschichte der Philosophie* 48 (1966): 43-81.

Keele, Kenneth D. "Leonardo da Vinci on Vision." *Proceedings of the Royal Society of Medicine* 48 (1955): 384-90.

______. "Leonardo da Vinci's Physiology of the Senses." In *Leonardo's Legacy: An International Symposium,* edited by C. D. O'Malley, pp. 35-56. Berkeley, 1969.

Kitao, Timothy K. "Prejudice in Perspective: A Study of Vignola's Perspective Treatise." *Art Bulletin* 44 (1962): 173-94.

Klein, Franz-Norbert. *Die Lichtterminologie bei Philon von Alexandrien und in den hermetischen Schriften.* Leiden, 1962.

Koelbing, Huldrych M. "Averroes' Concepts of Ocular Function—Another View." *Journal of the History of Medicine and Allied Sciences* 27 (1972): 207-13.

______. "Felix Platters Stellung in der Medizin seiner Zeit." *Gesnerus* 22 (1965): 59-67.

———. "Kepler und die physiologische Optik: Sein Beitrag und seine Wirkung." In *Internationales Kepler-Symposium, Weil der Stadt 1971*, edited by Fritz Krafft, Karl Meyer, and Bernhard Sticker, pp. 229-45. Hildesheim, 1973.

———. "L'apport suisse à la renaissance de l'ophtalmologie (environ 1550-1630)." *Médicine et hygiène* 22 (1964): 347-49.

———. "Ocular Function, from Aristotle to Kepler." *Sydsvenska Medicinhistoriska Sällskapets Årsskrift*, 1968, pp. 15-30.

———. "Ocular Physiology in the Seventeenth Century and Its Acceptance by the Medical Profession." In *Analecta medico-historica*, 3: *Steno and Brain Research in the Seventeenth Century*, pp. 219-24. Oxford, 1968.

———. *Renaissance der Augenheilkunde 1540-1630*. Bern, 1967.

———. "Zur Sehtheorie im Altertum: Alkmeon und Aristoteles." *Gesnerus* 25 (1968): 5-9.

Krafft, Fritz; Meyer, Karl; and Sticker, Bernhard, eds. *Internationales Kepler-Symposium, Weil der Stadt 1971*. Hildesheim, 1973.

Krautheimer, Richard, and Krautheimer-Hess, Trude. *Lorenzo Ghiberti*. Princeton Monographs in Art and Archaeology, no. 31. Princeton, 1956.

Lackenbacher, Hans. "Beiträge zur antiken Optik." *Wiener Studien* 35 (1913): 34-61.

Leclerc Lucien. *Histoire de la médecine arabe*. 2 vols. Paris, 1876. Reprint New York, n. d.

Lee Edward N. "The sense of an Object: Epicurus on Seeing and Hearing." In *Perception: Philosophical and Scientific Themes and Variations*, edited by Peter Machamer and Robert G. Turnbull. Columbus, Ohio, forthcoming.

Lejeune, Albert. *Euclide et Ptolémée. Deux stades de l'optique géométrique grecque*. Université de Louvain, Recueil de travaux d'histoire et de philologie, ser. 3, fasc. 31. Louvain, 1948.

———. "La dioptre d'Archimède." *Annales de la Société scientifique de Bruxelles*, ser. 1, 61 (1947): 27-47.

———. "Les recherches de Ptolémée sur la vision binoculaire." *Janus* 47 (1958): 79-86.

———. *Recherches sur la catoptrique grecque*. Académie royale de Belgiques, Classe de lettres et des sciences morales et politiques, Mémoires, vol. 2, fasc. 2. Brussels, 1957.

Léonard de Vinci et l'expérience scientifique au XVIe siècle. Colloques internationaux du Centre national de la recherche scientifique, Paris, 4-7 juillet 1952. Paris, 1953.

Lindberg, David C. "Alhazen's Theory of Vision and Its Reception in the West." *Isis* 58 (1967): 321-41.

———. "Alkindi's Critique of Euclid's Theory of Vision." *Isis* 62 (1971): 469-89.

———. "Bacon, Witelo, and Pecham: The Problem of Influence." In *Actes du XIIe Congrès international d'histoire des sciences, Paris 1968*, vol. 3A, pp. 103-7. Paris, 1971.

———. *A Catalogue of Medieval and Renaissance Optical Manuscripts*. Subsidia Mediaevalia, no. 4. Toronto, 1975.

______. "The Cause of Refraction in Medieval Optics." *British Journal for the History of Science* 4 (1968-69): 23-38.

______. "Did Averroes Discover Retinal Sensitivity?" *Bulletin of the History of Medicine* 49 (1975): 273-78.

______. "Introduction" to reprint of *Opticae thesaurus* (Basel, 1572), pp. v-xxxiv. New York, 1972.

______. "The Intromission-Extramission Controversy in Islamic Visual Theory: Alkindi versus Avicenna." In *Perception: Philosophical and Scientific Themes and Variations*, edited by Peter Machamer and Robert G. Turnbull. Columbus, Ohio, forthcoming.

______. "Lines of Influence in Thirteenth-Century Optics: Bacon, Witelo, and Pecham." *Speculum* 46 (1971): 66-83.

______. "New Light on an Old Story" (Essay review of Vasco Ronchi, *The Nature of Light: An Historical Survey,* translated by V. Barocas [Cambrige, Mass., 1970]). *Isis* 62 (1971): 522-24.

______. "Pecham, John." In *Dictionary of Scientific Biography*, 10:473-76. New York, 1974.

______. "The *Perspectiva communis* of John Pecham: Its Influence, Sources, and Content." *Archives internationales d'histoire des sciences* 18 (1965): 37-53.

______. "A Reconsideration of Roger Bacon's Theory of Pinhole Images." *Archive for History of Exact Sciences* 6 (1970): 214-23.

______. "Roger Bacon's Theory of the Rainbow: Progress or Regress?" *Isis* 57 (1966): 235-48.

______. "The Theory of Pinhole Images from Antiquity to the Thirteenth Century." *Archive for History of Exact Sciences* 5 (1968): 154-76.

______. "The Theory of Pinhole Images in the Fourteenth Century." *Archive for History of Exact Sciences* 6 (1970): 299-325.

______. "Witelo." In *Dictionary of Scientific Biography,* forthcoming.

Lindberg, David C., and Steneck, Nicholas H. "The Sense of Vision and the Origins of Modern Science." In *Science, Medicine and Society in the Renaissance: Essays to Honor Walter Pagel*, edited by Allen G. Debus, 1: 29-45. New York, 1972.

Little, A. G. "On Roger Bacon's Life and Works." In *Roger Bacon Essays*, edited by A. G. Little, pp. 1-31. Oxford, 1914.

______. "Roger Bacon's Works." In *Roger Bacon's Essays*, edited by A. G. Little, pp. 375-426. Oxford, 1914.

Lippert, J. *See* Hirschberg, Julius; Lippert, J.; and Mittwoch, E.

Lohne, Johannes A. "Der eigenartige Einfluss Witelos auf die Entwicklung der Dioptrik." *Archive for History of Exact Sciences* 4 (1968): 414-26.

______. "Kepler und Harriot: Ihre Wege zum Brechungsgesetz." In *Internationales Kepler-Symposium, Weil der Stadt 1971*, edited by Fritz Krafft, Karl Meyer, and Bernhard Sticker, pp. 187-214. Hildesheim, 1973.

Long, A. A. "Thinking and Sense-Perception in Empedocles: Mysticism or Materialism?" *Classical Quarterly*, n.s., 16 (1966): 256-76.

Lynch, Lawrence E. "The Doctrine of Divine Ideas and Illumination in Robert Grosseteste, Bishop of Lincoln." *Mediaeval Studies* 3 (1941): 161-73.

Machamer, Peter, and Turnbull, Robert G., eds. *Perception: Philosophical and Scientific Themes and Variations.* Columbus, Ohio, forthcoming.

McKeon, Charles C. *A Study of the Summa philosophiae of the Pseudo-Grosseteste.* New York, 1948.

McMurrich, J. Playfair. *Leonardo da Vinci the Anatomist (1452-1519).* Baltimore, 1930.

Magnus, Hugo. *Die Anatomie des Auges in ihrer geschichtlichen Entwicklung.* Augenärztliche Unterrichtstafeln, edited by Hugo Magnus, vol. 20. Breslau, 1900.

______. *Die Augenheilkunde der Alten.* Breslau, 1901.

Maier, Anneliese. "Das Problem der 'Species sensibiles in medio' und die neue Naturphilosophie des 14. Jahrhunderts." *Freiburger Zeitschrift für Philosophie und Theologie* 10 (1963): 3-32. Reprinted in Anneliese Maier, *Ausgehendes Mittelalter: Gesammelte Aufsätze zur Geistesgeschichte des 14. Jahrhunderts,* 2: 419-51. Rome, 1967.

Maltese, Corrado. "Per Leonardo prospettico." *Raccolta Vinciana* 19 (1962): 303-14.

Meyer, Karl. *See* Krafft, Fritz; Meyer, Karl; and Sticker, Bernhard.

Meyerhof, Max. "Die Optik der Araber." *Zeitschrift für ophthalmologische Optik* 8 (1920): 16-29, 42-54, 86-90.

______. "Eine unbekannte arabische Augenheilkunde des 11. Jahrhunderts n. Chr." *Sudhoffs Archiv für Geschichte der Medizin und Naturwissenschaften* 20 (1928): 63-79.

______. "Eye Diseases in Ancient Egypt." *Ciba Symposia* 1 (1939-40): 305-10.

______. "New Light on Hunain ibn Ishaq and His Period." *Isis* 8 (1926): 685-724.

______. "New Light on the Early Period of Arabic Medical and Ophthalmological Science." *Bulletin of the Ophthalmological Society of Egypt* 19 (1926): 25-37.

______. *See* Prüfer, C., and Meyerhof, Max.

Michaud-Quantin, Pierre. *Etudes sur le vocabulaire philosophique du Moyen Age.* Rome, 1970.

Michel, Paul-Henri. *La pensée de L. B. Alberti (1404-1472).* Paris, 1930.

Millerd, Clara Elizabeth. *On the Interpretation of Empedocles.* Chicago, 1908.

Mittwoch, E. *See* Hirschberg, Julius; Lippert, J.; and Mittwoch, E.

Molland, A. G. "John Dumbleton and the Status of Geometrical Optics." In *Actes du XIIIe Congrès international d'histoire des sciences, Moscou, 18-24 août 1971,* sect. 3-4, pp. 125-30. Moscow, 1974.

Mugler, Charles. *Dictionnaire historique de la terminologie optique des grecs: Douze siècles de dialogues avec lumière.* Paris, 1964.

______. "Sur l'histoire de quelques définitions de la géométrie grecque et les rapports entre la géométrie et l'optique." *L'antiquité classique* 26 (1957): 331-45.

Narducci, Enrico. "Intorno ad una traduzione italiana, fatta nel secolo

decimoquarto, del trattato d'Ottica d'Alhazen, matematico del secolo undecimo, e ad altri lavori di questo scienziato." *Bullettino di bibliografia e di storia delle scienze matematiche e fisiche* 4 (1871): 1-48.
Nazif, M. "Ibn al-Haitham's Work on Optics." In *Ibn al-Haitham: Proceedings of the Celebrations of [the] 1000th Anniversary [of his Birth]*, edited by Hakim Mohammed Said, pp. 285-93. Karachi, n.d. [1970?].
Nazif, M. and Ghalioungui, Paul. "Ibn al Haitham, an 11th Century Physicist." In *Actes du Xe Congrès international d'histoire des sciences, Ithaca, 26 VIII-2 IX 1962*, vol. 1, pp. 569-71. Paris, 1964.
Nebbia, Giorgio. "Ibn al-Haytham nel millesimo anniversario della nascita." *Physis* 9 (1967): 165-214.
O'Brien, D. "The Effect of a Simile: Empedocles' Theories of Seeing and Breathing." *Journal of Hellenic Studies* 90 (1970): 140-79.
O'Neill, Ynez Violé. "William of Conches and the Cerebral Membranes." *Clio Medica* 2 (1967): 13-21.
______. "William of Conches' Descriptions of the Brain." *Clio Medica* 3 (1968): 203-23.
Ovio, Giuseppe. "Cenni d'ottica fisiologica [according to Albert the Great]." *Angelicum* 21 (1944): 302-5.
Panofsky, Erwin. "Das perspektivische Verfahren Leone Battista Albertis." *Kunstchronik*, n.s., vol. 26 (1914-15), cols. 505-16.
______. "Die Perspektive als 'symbolische Form.'" In *Vorträge der Bibliothek Warburg, 1924-1925*, pp. 258-330. Leipzig, 1927.
Pansier, Pierre. "Histoire de l'ophtalmologie," In *Encyclopédie française d'ophtalmologie*, edited by F. Lagrange and E. Valude, 1: 1-86. Paris, 1903.
______. "La pratique de l'ophthalmologie dans le moyen-âge latin." *Janus* 9 (1904): 3-26.
Parent, J.-M. *La doctrine de la création dans l'école de Chartres*. Paris/ Ottawa, 1938.
Parronchi, Alessandro. "La 'costruzione legittima' è uguale alla 'costruzione con punti di distanza.'" *Rinascimento*, ser. 2, 4 (1964): 35-40.
______. *Studi su la dolce prospettiva*. Milan, 1964.
Pedretti, Carlo. "Leonardo on Curvilinear Perspective." *Bibliothèque d'humanisme et renaissance* 25 (1963): 69-87.
Perrod, Giovanni. "La diottrica oculare di Leonardo da Vinci." *Archivio di ottalmologia* 14 (1907): 369-81, 463-97.
Pirenne, M. H. "The Scientific Basis of Leonardo da Vinci's Theory of Perspective." *British Journal for the Philosophy of Science* 3 (1952-53): 169-85.
Polyak, Stephen L. *The Retina*. Chicago, 1941.
______. *The Vertebrate Visual System*. Chicago, 1957.
Priestley, Joseph. *The History and Present State of Discoveries Relating to Vision, Light, and Colours*. London, 1772.
Prüfer, C., and Meyerhof, Max. "Die angebliche Augenheilkunde des Tabit ibn Qurra." *Centralblatt für praktische Augenheilkunde* 35 (1911): 38-41.

Rashed, Roshdi. "Le modèle de la sphère transparente et l'explication de l'arc-en-ciel: Ibn al-Haytham, al-Farisi." *Revue d'histoire des sciences et de leurs applications* 23 (1970): 109-40.

———. "Optique géométrique et doctrine optique chez Ibn Al Haytham." *Archive for History of Exact Sciences* 6 (1970): 271-98.

Reilly, George C. *The Psychology of Saint Albert the Great*. Washington, D.C., 1934.

Rienstra, Miller Howard. "Giovanni Battista Della Porta and Renaissance Science." Ph.D. dissertation, University of Michigan, 1963.

Rintelen, F. "Die Ophthalmologie in Basel zur Zeit des Barocks." *Gesnerus* 14 (1957): 29-39.

Rohmer, J. "L'intentionnalité des sensations de Platon à Ockam." *Revue des sciences religieuses* 25 (1951): 5-39.

Rohr, Moritz von. "Kepler und seine Erklärung des Sehvorganges." In *Johannes Kepler, der kaiserliche Mathematiker*, edited by Karl Stöckl (Bericht des naturwissenschaftlichen Vereins zu Regensburg, vol. 19 [1928-30]): pp. 210-17. Regensburg, 1930.

———. "Keplers Behandlung des beidäugigen Sehens." In *Johannes Kepler, der kaiserliche Mathematiker*, ed. Karl Stöckl (Bericht des naturwissenschaftlichen Vereins zu Regensburg, vol. 19 [1928-30]): pp. 218-24. Regensburg, 1930.

Ronchi, Vasco. *Histoire de la lumière*. Translated by Juliette Taton. Paris, 1956.

———. "Leonardo e l'ottica." In *Leonardo, Saggi e Ricerche*. A cura del Comitato Nazionale per le onoranze a Leonardo da Vinci nel quinto centenario della nascita (1452-1952), pp. 159-85. Rome, 1954.

———. "La prospettiva della rinascita e le sue origini." *Atti della Fondazione Giorgio Ronchi* 24 (1969): 1-21.

———. "L'optique au XVIe siècle." In *La science au seizième siècle* (Colloque international de Royaumont, 1-4 juillet 1957), pp. 47-65. Paris, 1960.

———. "L'optique de Léonard de Vinci." In *Léonard de Vinci et l'expérience scientifique au XVIe siècle* (Colloques internationaux du Centre national de la recherche scientifique, Paris, 4-7 juillet 1952), pp. 115-20. Paris, 1953.

———. *The Nature of Light: An Historical Survey*. Translated by V. Barocas. London, 1970.

———. "Rilievi a proposito dell'ottica di Tito Lucrezio Caro." *Archives internationales d'histoire des sciences* 18 (1965): 161-74.

———. "A Scarcely Known Aspect of the Activity of Leonardo da Vinci in the Field of Optics." *Indian Journal of History of Science* 3 (1968): 80-90.

———. *Storia della luce*. 2d ed. Bologna, 1952.

Sabra, A. I. "The Astronomical Origin of Ibn al-Haytham's Concept of Experiment." In *Actes du XIIe Congrès international d'histoire des sciences, Paris 1968*, vol. 3A, pp. 133-36. Paris, 1971.

———. "The Authorship of the *Liber de crepusculis*, an Eleventh-Century Work on Atmospheric Refraction." *Isis* 58 (1967): 77-85.

______. "Explanation of Optical Reflection and Refraction: Ibn-al-Haytham, Descartes, Newton." In *Actes du Xe Congrès international d'histoire des sciences, Ithaca, 26 VIII-2 IX 1962*, 1: 551-54. Paris, 1964.
______. "Ibn al-Haytham, Abu 'Ali al-Hasan ibn al-Hasan." In *Dictionary of Scientifc Biography*, 6: 189-210. New York, 1972.
______. "Ibn al-Haytham's Criticisms of Ptolemy's *Optics*." *Journal of the History of Philosophy* 4 (1966): 145-49.
______. "The Physical and the Mathematical in Ibn al-Haytham's Theory of Light and Vision." In *Proceedings of the International Conference on the History and Philosophy of Science, Jyväskylä, Finland, 28 June-6 July, 1973*. Forthcoming.
______. "Sensation and Inference in Alhazen's Theory of Visual Perception." In *Perception: Philosophical and Scientific Themes and Variations*, edited by Peter Machamer and Robert G. Turnbull. Columbus, Ohio, forthcoming.
Said, Hakim Mohammed, ed. *Ibn al-Haitham: Proceedings of the Celebrations of* [the] *1000th Anniversary* [of His Birth]. Karachi, n.d. [1970?].
Saint-Pierre, Bernard. "La physique de la vision dans l'antiquité: Contribution à l'établissement des sources anciennes de l'optique médiévale." Ph.D. dissertation, University of Montreal, 1972.
Sambursky, S. "Philoponus' Interpretation of Aristotle's Theory of Light." *Osiris* 13 (1958): 114-26.
______. *The Physical World of Late Antiquity*. London, 1962.
______. *Physics of the Stoics*. London, 1959.
Sarton, George. *Introduction to the History of Science*. 3 vols. Baltimore, 1927-48.
______. "The Tradition of the Optics of Ibn al-Haitham." *Isis* 29 (1938): 403-6.
______. Review of M. Nazif, *Al-Hasan ibn al-Haitham. His Optical Studies and Discoveries* [in Arabic]. *Isis* 34 (1942-43): 217-18.
Sayili, Aydin M. "The Aristotelian Explanation of the Rainbow." *Isis* 30 (1939): 65-83.
Sbath, Paul. "Le livre des questions sur l'oeil de Hunain ben Ishaq, médecin et grand savant Chrétien du IXe siècle (809-887)." *Bulletin de l'Institut d'Egypte* 17 (1934-35): 129-38.
Scalinci, Noè. "La nosocologia e la terapia nell' *Ars probatissima oculorum* di Benvenuto Grasso, Medico-oculista Salernitano del sec. XIII." *Annali di ottalmologia e clinica oculistica*, 64 (1936): 116-39, 188-208.
______. "La oculistica dei Maestri Salernitani." In *Scritti in onore del Prof. P. Capparoni in occasione del 50° anno di laurea*, edited by A. Pazzini, pp. 134-51. Turin, 1941.
Schnaase, Leopold. *Die Optik Alhazens*. Stargard, 1889.
Schneider, Arthur. *Die Psychologie Alberts des Grossen*. Beiträge zur Geschichte der Philosophie des Mittelalters, vol. 4, pts. 5-6. Münster, 1903-6.
Schramm, Matthias. *Ibn al-Haythams Weg zur Physik*. Wiesbaden, 1963.

______. "Zur Entwicklung der physiologischen Optik in der arabischen Literatur." *Sudhoffs Archiv für Geschichte der Medizin und der Naturwissenschaften* 43 (1959): 289-316.

Schüling, Hermann. *Theorien der malerischen Linear-Perspektive vor 1601.* Schriften zur Ästhetik und Kunstwissenschaft, no. 2. Giessen, 1973.

Sharp, D. E. *Franciscan Philosophy at Oxford in the Thirteenth Century.* Oxford, 1930. Reprint New York, 1964.

Shastid, Thomas H. "History of Ophthalmology." In *The American Encyclopedia and Dictionary of Ophthalmology*, edited by Casey A. Wood, 11: 8524-8904. Chicago, 1917.

Siegel, Rudolph E. "Did the Greek Atomists Consider a Non-Corpuscular Visual Transmission? Reconsideration of some Ancient Visual Doctrines." *Archives internationales d'histoire des sciences* 22 (1969): 3-16.

______. *Galen on Sense Perception.* Basel, 1970.

______. "Principles and Contradictions of Galen's Doctrine of Vision." *Sudhoffs Archiv: Zeitschrift für Wissenschaftsgeschichte* 54 (1970): 261-76.

______. "Theories of Vision and Color Perception of Empedocles and Democritus: Some Similarities to the Modern Approach." *Bulletin of the History of Medicine* 33 (1959): 145-59.

Simon, Gerard. "A propos de la théorie de la perception visuelle chez Kepler et Descartes: Réflexions sur le rôle du mécanisme dans la naissance de la science classique." In *Actes du XIIIe Congrès international d'histoire des sciences, Moscou, 18-24 août 1971*, sect. 6, pp. 237-45. Moscow, 1974.

Slakey, Thomas J. "Aristotle on Sense Perception." *Philosophical Review* 70 (1961): 470-84.

Solmi, Edmondo. *Nuovi studi sulla filosofia naturale di Leonardo da Vinci: Il metodo sperimentale, l'astronomia, la teoria della visione.* Mantua, 1905.

Spettman, Hieronymus, O.F.M. *Die Psychologie des Johannes Pecham.* Beiträge zur Geschichte der Philosophie des Mittelalters, vol. 20, pt. 6 Münster, 1919.

Stahl, William H. *Roman Science.* Madison, Wis., 1962.

Steinschneider, Moritz. *Die europäischen Übersetzungen aus dem arabischen bis Mitte des 17. Jahrhunderts.* Graz, 1956.

______. "Zur Oculistik des 'Isa ben Ali (9. Jahrh.) und des sogenannten Canamusali." *Janus* 11 (1906): 399-406.

Steneck, Nicholas H. "Albert the Great on the Classification and Localization of the Internal Senses." *Isis* 65 (1974): 193-211.

______. "The Problem of the Internal Senses in the Fourteenth Century." Ph.D. dissertation, University of Wisconsin, 1970.

______. *Science and Creation: Henry of Langenstein on Genesis.* Notre Dame, Ind., forthcoming.

______. *See* Lindberg, David C., and Steneck, Nicholas H.

Sticker, Bernhard. *See* Krafft, Fritz; Meyer, Karl; and Sticker, Bernhard.

Stöckl, Karl, ed. *Johannes Kepler, der kaiserliche Mathematiker.* Bericht des Naturwissenschaftlichen Vereins zu Regensburg, vol. 19 (1928-30). Regensburg, 1930.

Straker, Stephen M. "Kepler's Optics: A Study in the Development of Seventeenth-Century Natural Philosophy." Ph.D. dissertation, Indiana University, 1970.

Sudhoff, Karl. "Augenanatomiebilder im 15. und 16. Jahrhundert." In *Studien zur Geschichte der Medizin*, edited by Karl Sudhoff, 1: 19-26. Leipzig, 1907.

______. "Augendurchschnittsbilder aus Abendland und Morgenland." *Sudhoffs Archiv für Geschichte der Medizin und der Naturwissenschaften* 8 (1915): 1-21.

Taylor, A. E. *A Commentary on Plato's Timaeus*. Oxford, 1928.

Tea, Eva. "Witelo, prospettico del secolo XIII." *L'arte* 30 (1927): 3-30.

Temkin, Owsei. "On Galen's Pneumatology." *Gesnerus* 8 (1951-52): 180-89.

Ten Doesschate, Gezienus. *De derde Commentaar van Lorenzo Ghiberti in Verband met de middeleeuwsche Optiek*. Utrecht, n.d. [1940?].

______. *Perspective: Fundamentals, Controversials, History*. Nieuwkoop, 1964.

Thorndike, Lynn. *A History of Magic and Experimental Science*. 8 vols. New York, 1923-56.

Trouessart, J. *Recherches sur quelques phénomènes de la vision, précédées d'un essai historique et critique des théories de la vision, depuis l'origine de la science jusqu'à nos jours*. Brest, 1854.

Turbayne, Colin M. "Grosseteste and an Ancient Optical Principle." *Isis* 50 (1959): 467-72.

______. *The Myth of Metaphor*. Rev. ed. Columbia, S.C., 1971.

Turnbull, Robert. "The Role of the 'Special Sensibles' in the Perception Theories of Plato and Aristotle." In *Perception: Philosophical and Scientific Themes and Variations*, edited by Peter Machamer and Robert G. Turnbull. Columbus, Ohio, forthcoming.

Unguru, Sabetai. "Witelo and Thirteenth-Century Mathematics: An Assessment of His Contributions." *Isis* 63 (1972): 496-508.

van Hoorn, Willem. *As Images Unwind: Ancient and Modern Theories of Visual Perception*. Amsterdam, 1972.

Venturi, G. B. *Commentari sopra la storia e la teoria dell'ottica*. Bologna, 1814.

Verbeke, G. *L'évolution de la doctrine du pneuma du Stoicisme à S. Augustin*. Paris, 1945.

Verdenius, W. J. "Empedocles' Doctrine of Sight." In *Studia varia Carolo Guilielmo Vollgraff a discipulis oblata*, pp. 155-64. Amsterdam, 1948.

Vernet, A. "Un remaniement de la *Philosophia* de Guillaume de Conches." *Scriptorium* 1 (1946-47): 243-59.

Vescovini, Graziella Federici. "Contributo per la storia della fortuna di Alhazen in Italia: Il volgarizzamento del MS. Vat. 4595 e il 'Commentario terzo' del Ghiberti." *Rinascimento*, ser. 2, 5 (1965): 17-49.

______. "L'inserimento della 'perspectiva' tra le arti del quadrivio." In *Arts libéraux et philosophie au moyen âge* (*Actes du quatrième Congrès international de philosophie médiévale, Université de Montréal, Canada, 27 août-2 septembre 1967*), pp. 969-74. Montreal/Paris, 1969.

______. *Studi sulla prospettiva medievale*. Università di Torino, Pubblicazioni della facoltà di lettere e filosofia, vol. 16, fasc. 1. Turin, 1965.

Virieux-Reymond, Antoinette. "Quelques réflexions à propos d'un texte de Démocrite concernant la théorie de la vision." In *Actes du XIIe Congrès international d'histoire des sciences, Paris 1968*, vol. 3A, pp. 145-49. Paris, 1971.

Vogl, Sebastian. "Roger Bacons Lehre von der sinnlichen Spezies und vom Sehvorgange." In *Roger Bacon Essays*, edited by A. G. Little, pp. 205-27. Oxford, 1914.

von Fritz, Kurt. "Democritus' Theory of Vision." In *Science, Medicine, and History: Essays on the Evolution of Scientific Thought and Medical Practice, Written in Honour of Charles Singer*, edited by E. A. Underwood, 1: 83-99. London, 1953.

von Simson, Otto. *The Gothic Cathedral: Origins of Gothic Architecture and the Medieval Concept of Order*. Rev. ed. New York, 1962.

Wallace, William A., O.P. *Causality and Scientific Explanation*, vol. 1: *Medieval and Early Classical Science*. Ann Arbor, 1972.

______. *The Scientific Methodology of Theodoric of Freiberg*. Studia Friburgensia, n.s., no. 26. Fribourg, 1959.

Weissenborn, H. "Zur Optik des Eukleides." *Philologus* 45 (1885): 54-62.

White, John. *The Birth and Rebirth of Pictorial Space*. 2d. ed. London, 1967.

______. "Developments in Renaissance Perspective, I." *Journal of the Warburg and Courtauld Institutes* 12 (1949): 58-79.

Wiedemann, Eilhard. "Aus al Kindis Optik." *Sitzungsberichte der physikalisch-medizinischen Societät in Erlangen* 39 (1907): 245-8.

______. "Eine Zeichnung des Auges bei dem Bearbeiter der Optik von Ibn al Haitam, Kamal al Din al Farisi, und Merkverse über den Bau des Auges." *Zentralblatt für praktische Augenheilkunde* 34 (1910): 204-8.

______. "Ibn al Haitam, ein arabischer Gelehrter." In *Festschrift, J. Rosenthal zur Vollendung seines siebzigsten Lebensjahres gewidmet*, pt. 1, pp. 147-78. Leipzig, 1906.

______. "Ibn Sina's Anschauung vom Sehvorgang." *Archiv für die Geschichte der Naturwissenschaften und der Technik* 4 (1913): 239-41.

______. "Roger Bacon und seine Verdienste um die Optik." In *Roger Bacon Essays*, edited by A. G. Little, pp. 185-203. Oxford, 1914.

______. "Sull'ottica degli arabi." Translated by A. Sparagna. *Bullettino di bibliografia e di storia delle scienze matematiche e fisiche* 14 (1881): 219-25.

______. "Über die Camera obscura bei Ibn al Haitam." *Sitzungsberichte der psysikalisch-medizinischen Societät in Erlangen* 46 (1914): 155-69.

______. "Über die Erfindung der Camera obscura." *Verhandlungen der deutschen physikalischen Gesellschaft* 12 (1910): 177-82.

______. "Zu Ibn al Haitams Optik." *Archiv für die Geschichte der Naturwissenschaften und der Technik,* 3 (1910-11): 1-53.

______. "Zur Geschichte der Lehre vom Sehen." *Annalen der Physik und Chemie*, n.s., 39 (1890): 470-74.

Wilde, Emil. *Geschichte der Optik.* 2 vols. Berlin, 1838-43. Reprint Wiesbaden, 1968.
Winter, H. J. J. "The Arabic Optical MSS. in the British Isles (Mathematical and Physical Optics)." *Centaurus* 5 (1956): 73-88.
______. "The Optical Researches of Ibn al-Haitham." *Centaurus* 3 (1954): 190-210.
Winter, Martin. *Über Avicennas Opus egregium de anima* (*Liber sextus naturalium*). Wissenschaftliche Beilage zu dem Jahresbericht des K. Theresien-Gymnasiums in München für das Schuljahr 1902/1903. Munich, 1903.
Wittkower, Rudolf. "Brunelleschi and 'Proportion in Perspective.'" *Journal of the Warburg and Courtauld Institutes* 16 (1953): 275-91.
Wolfson, Harry Austryn. "The Internal Senses in Latin, Arabic and Hebrew Philosophic Texts." *Harvard Theological Review* 28 (1935): 69-133.
Würschmidt, Joseph. "Roger Bacons Art des wissenschaftlichen Arbeitens, dargestellt nach seiner Schrift *De speculis*." In *Roger Bacon Essays*, edited by A. G. Little, pp. 229-39. Oxford, 1914.
______. "Zur Theorie der Camera obscura bei Ibn al Haitam." *Sitzungsberichte der physikalisch-medizinischen Societät in Erlangen* 46 (1914): 151-54.

INDEX

The principal treatment of a subject is indicated by *italic* numerals.

CPSIA information can be obtained
at www.ICGtesting.com
Printed in the USA
LVHW01s2206210118
563333LV00003B/16/P

9 780226 482354